Genetic Information

Acquisition, Access, and Control

Genetic Information
Acquisition, Access, and Control

Edited by

Alison K. Thompson

University of Toronto
Toronto, Ontario, Canada

and

Ruth F. Chadwick

University of Central Lancashire
Preston, United Kingdom

Kluwer Academic / Plenum Publishers
New York, Boston, Dordrecht, London, Moscow

Proceedings of an International Conference on Genetic Information: Acquisition, Access, and Control, held December 5 – 6, 1997, in Preston, United Kingdom

ISBN 0-306-46052-1

233 Spring Street, New York, N.Y. 10013

10 9 8 7 6 5 4 3 2 1

A C.I.P. record for this book is available from the Library of Congress.

Printed in the United States of America

PREFACE

It is difficult to think of an example of an advancement in the biological sciences that has had an impact on society similar to that of the new genetics. Recent developments in biotechnology have occasioned much discussion among academics, professionals, and lay people alike. In particular, many questions and concerns have arisen over the acquisition, access, and control of genetic information. There are several reasons why the new genetics has commanded such widespread attention, and why it is now the subject of considerable debate. Special reference is given in this volume to the implications of genetic information for five different subject areas: eugenics, the insurance industry, the commercialisation of genetic testing, strategies for raising public awareness, and the value of theoretical ethical and sociological frameworks in the debate. This diverse collection of papers attempts to address and critically discuss issues surrounding the control of, and access to, genetic information from ethical, medical, legal, and theoretical points of view.

The first and shortest section of the book attempts to address concerns over the eugenic potential of new biotechnologies. It also provides a historical context for the debate, for controversy over the subject of eugenics predates the current debate over genetic information by a considerable length of time. Indeed, by the time the first patent was issued for Chakrabarty's strain of oil eating bacteria in the early 1970s, the term *eugenics* had already acquired strong pejorative connotations. One of the fundamental questions looked at in this section is whether or not the new genetics, when coupled with current biotechnological capabilities, do in fact pose a eugenic threat, and if so, whether the eugenic capability of these developments is intrinsically sinister.

One very vocal group that subscribes to the view that genetics is dangerous and inherently disciminatory in its eugenic implications is the disability rights movement. Indeed, at the time of the conference, several representatives of this group expressed concern that the paper given by Helga Kuhse would not present a balanced account of the arguments surrounding the issue of whether or not genetic screening discriminates against the disabled. In fact Kuhse's paper does go to great lengths to explicate the arguments and concerns of those who feel genetic screening does pose a threat to those who live with disabilities, and there was some lively debate during the conference, including a radio debate, on this very issue.

Access to genetic information is one issue that has particular relevance for the insurance industry. There has been a significant level of anxiety over developments in genetic testing and screening that may allow the industry to discriminate against potential and actual policy holders on the basis of their genetic make-up alone. It is feared that the insurance industry will come to require genetic testing and screening as a part of the normal underwriting process, and subsequently use this genetic information to justify the denial of access to and the raising of premiums on insurance policies. Although the question of whether the health insurance industry should have access to policy holders' genetic information is of considerably more gravity in countries, like the United States, which do not have universal healthcare systems, human genome analysis has similar implications for the life insurance industry as well. It is feared that discrimination against those deemed to have "risky" genotypes will have severe consequences not just for the industry itself, but also for individuals, their families and for society as a whole. The papers in this section of the book provide a range of contexts for exploring the empirical and theoretical sides to these issues, including perspectives from several different countries.

Discussion of whether or not it is ethically, legally, or socially acceptable to use genetic information for the purposes of insurance underwriting has led some to consider the implications of what access to, say, a gene data bank would mean for the industry. We can deal with such questions, however, within the context of the debate over the commercialization of genetic information. The following section of the book, therefore, deals specifically with questions having to do with the nature of genetic information, (i.e. is it different from any other information that we may have commercialized previously?), and with issues arising from the fact that human genome analysis is indubitably being carried out in a commercial environment. A variety of questions pertaining to the relative advantages and disadvantages of the commercialization of genetic information are dealt with in this section. Included are papers not only on the more obvious topic of gene patenting, but also on questions regarding commercialization at the point of service to the client, as well as concerns over our conceptions of normalcy and human disease.

It is often argued that much of the anxiety over the new genetics would be diminished if the public were better informed. Indeed, the implications of commercialization at the point of service to the client would be altogether different for the well informed client, as opposed to the uninformed client. Questions such as how best to raise and gauge public awareness are currently being addressed in many nations, but explorations of the social factors influencing attitudes toward public education, and levels of public awareness need to be carried out. Thus, in the fourth section of the book, we look at the findings from many practical studies of the levels of public awareness as they pertain to specific genetic disorders. In addition, a report from a citizens' jury is given, and issues related to public awareness in the context of ante- and neonatal screening programmes are explored.

In the final section of the volume, the focus shifts from the more practical issues related to genetic information to the theoretical. There is a tendency to talk about scientific information in general, and genetic information in particular, in an amoral, neutral way. Consideration of the need for, and utility of, using different theoretical frameworks to explore and locate the ethical issues surrounding the acquisition, access, and control of genetic information forms the substance of this section. Theoretical frameworks improve our understanding of the moral and social foundations of bioethical debates. Indeed, theory is essential to the process of developing a bioethical discourse, as well as being a valuable aid to individuals who wish to make autonomous, informed decisions. Central to the thrust of many of the papers in this section is the principle of justice; in some cases as it relates to the new genetics in developing countries. Thus, the final section is meant to be the theoretical complement to what comes before, and ranges over a wide variety of issues. It is hoped that this will serve to expand the scope of debate surrounding genetic information and its uses, and that it will provide a unique collection of perspectives on the issues from an international and multidisciplinary perspective.

The papers in *Genetic Information: Acquisition, Access, and Control* are based on an international conference of the same name, held in December, 1997 at the University of Central Lancashire, which was organized by the Centre for Professional Ethics on behalf of the International Association of Bioethics and Euroscreen II.

We should like to thank all the contributors to this volume, as well as the Board of Directors of the International Association of Bioethics and the members of Euroscreen II for agreeing to a conference on this topic. Thanks are also due to the members of the Scientific Programme Committee: Mairi Levitt, Tony McGleenen, Alastair Campbell, Susan Sherwin, and Domenico Coviello. In addition, we would like to thank Plenum Publishing for their patience in awaiting this copy.

Alison Thompson
Ruth Chadwick
University of Central Lancashire
Preston, UK
1998

CONTENTS

Eugenics

1. Can We Learn from Eugenics? 1
   Daniel Wikler

2. Preventing Genetic Impairments: Does It Discriminate against People with Disabilities? 17
   Helga Kuhse

Genetics and Insurance

3. Private Parties, Public Duties? The Shifting Role of Insurance Companies in the Genetics Era 31
   Trudo Lemmens

4. Coercion, Control, and Consequence in Genetic Testing: Views on Insurance among Tested Individuals and the General Public 41
   Robert N. Mayer, Ken R. Smith, Cathleen D. Zick, and Jeffrey R. Botkin

5. Genetic Testing and Adverse Selection in the Market for Life Insurance: Preliminary Findings for the BRCA1 Gene Mutation 57
   Ken R. Smith, Cathleen D. Zick, Robert N. Mayer, and Jeffrey R. Botkin

6. Genetic Engineering and German Health Insurances 71
   Thomas Uhlemann

7. Selling Souls: Ethical Theory and the Commercialisation of Genetic Information 79
   Martin Johnston

The Commercialization of Genetic Information

8. The Ethics of 21st Century Bioinformatics: Ethical Implications of the Vanishing Distinction between Biological Information and Other Information 85
   Dick Holdsworth

9. There Is Nothing Special about Genetic Information 97
Søren Holm

10. High Speed Genetic Testing Technology and the Computerization of the Medical Record 105
Curtis R. Naser

11. Ethical Questions in the Pursuit of Genetic Information: Geneticization and BRCA1 121
Susan Sherwin and Christy Simpson

12. The Ethics of Gene Patenting 129
Ruth Macklin

13. Paradigms of Author/Creator Property Rights in Intellectual Property Law: Ethical Implications for the Acquisition, Access, and Control of Genetic Information 139
Robin Mackenzie

14. Regulating the Commercialization of Human Genetics: Can We Address the Big Concerns? 149
Timothy Caulfield

15. Ethical Impacts of Human Health-Related Biotechnology in Brazil 163
Marilia Bernardes Marques

Public Awareness

16. Adolescents and Carrier Testing: Attitudes and Ethical Presuppositions 175
Kris Dierickx

17. Down's Syndrome Screening: How Do They Know? 183
Ros Bramwell and Susan Wade

18. Public Perspectives of the New Genetics: The Citizens' Jury Experiment 191
Rachel Iredale and Marcus Longley

19. Genetics and Journalism: A View from the United States 201
Joan Stephenson

Theoretical Concerns

20. Genetic Information and "Genetic Identity" 207
Richard E. Ashcroft

21. Genetic Information and Knowing when You Will Die 219
Margaret P. Battin

22. Influences of Genetic Testing on a Person's Freedom 235
Elisabeth Hildt

23. Genetic Knowledge: The Contribution of Sociologies 243
Simon Dyson

24. Germ-Line Gene Therapy: Is the Existing UK Norm Ethically Valid? 255
Sheila Dziobon

25. Negotiating the Dilemmas of Prenatal Testing for Genetic Disorders: What Is the Virtuous Person to Do? 267
Ann Gallagher

26. Genetic Information: Questions and Worries from an African Background 275
Godfrey B. Tangwa

27. Genetic Knowledge in a Just Society 283
Nils Holtug

28. Biotechnology, Genetic Information, and Community: From Individual Rights to Social Duties? 297
Sirkku Kristiina Hellsten

29. Linear Destiny and Geometric Fate 309
Emilio Mordini

30. A New Framework for the Use of Genetic Information 321
Jacquelyn Ann K. Kegley

Index 331

1

CAN WE LEARN FROM EUGENICS?

Daniel Wikler

University of Wisconsin

1. EUGENICS YESTERDAY AND TODAY

The word "eugenics" may be unfamiliar to most people today, but for a period of about sixty-five years, roughly 1880 to 1945, both that term and the theories of human "improvement" which it denotes were in wide currency. Francis Galton, a cousin of Darwin, invented the term and launched a movement to improve the human race, or at least to halt its perceived decline, through selective breeding. His ideas spread quickly, and by the 1920s eugenics movements existed all over the world. Eugenics, a movement for social betterment clothed in the mantle of modern science, claimed the allegiance of most genetic scientists and drew supporters from political right, left, and center. Unfortunately for that movement, and indeed for much of mankind, eugenics was embraced by Hitler and his Nazi followers, tarnishing its name forever. After the fall of the Third Reich, eugenic ideas quickly lost their cachet, becoming virtually taboo in the United States and Europe, where the term "eugenic" is now used primarily as an epithet.

We should not forget eugenics. We can learn much by studying the history of the movement and by engaging in careful moral analysis and assessment of its doctrines. Eugenics has many lessons. It provides a valuable case study of the way in which the prestige of science can be used to disguise the moral premises and motives for a social movement, and how class, racial, and other biases can exert powerful and damaging influence over such a movement while remaining virtually invisible to its advocates. And it is another illustration of the sad thesis that good (or, at least, high-minded) intentions can lead to evil consequences.

Eugenics also offers a perspective on the practices of our own era, the second moment in history in which the prospect for using the science of heredity to reshape society presents itself. Indeed, critics of certain practices in clinical genetics, and of some contemplated uses for the genetic technology of the future, maintain that these are eugenics in disguise. We must be able to evaluate this claim so that we can avoid the errors and wrongs of the past as we frame public policies for genetics in the future.

In our reconsideration of the movement, we should avoid an unthinking rejection of every eugenic thought or value. The fact that eugenicists were in favor of a particular

Genetic Information, edited by Thompson and Chadwick.
Kluwer Academic / Plenum Publishing, New York, 1999.

measure or goal is not in itself sufficient reason to oppose it. We need a good analysis of which eugenic aims were wrong-headed, and why. We might judge that some of the questions to which eugenicists proposed answers ought not be ignored, and indeed that they are now given too little attention, in part because of their eugenic associations.

This paper provides both a brief history of the eugenics movement and a moral analysis of some of its tenets. I begin by recounting briefly the rise and fall of this complex international movement. I do not in any way wish to revise the very bad reputation which the eugenics movement currently suffers, and where old-style eugenics is advocated today, as in Singapore (Chee and Chee 1984) and China (Beardsley 1997; Dong-Sheng 1981; Wertz 1996), conventional criticisms of these ideas still apply. When we turn to contemporary practices in clinical genetics whose status as "eugenic" is in dispute, however, the arguments must be more subtle. I argue that the label "eugenic" does in some instances apply, but that when this is the case the "eugenic" effect or intent of the practices ought not to always engender alarm or opposition. Though we rightly reject all of the programs practiced or proposed by the eugenics movement in its heyday, I will argue that this retrospective evaluation does not point unequivocally to a rejection of any and all eugenics for the future.

2. EUGENICS PAST

2.1. The Rise of Eugenics

Though the literature of eugenics extends back to Plato, the modern movement took its cue from biology: first, Darwin's theory of natural selection, with a boost later from mendelian genetics. Galton understood that the theory of natural selection had important implications for understanding the development of the human species, and sought to investigate the possibility that talents and virtues of character and personality were inherited along with other traits, offering their bearers advantages in natural selection. Galton coined the term "eugenics" in 1883, defining it as the "science of improving stock--not only by judicious mating, but whatever tends to give the more suitable races or strains of blood a better chance of prevailing over the less suitable than they otherwise would have had." His research, enhanced by statistical methods developed as he needed them, convinced him that society's stock of talent could be greatly enlarged if members of favored families were to increase their rate of childbearing ("positive eugenics"). The balance should be further improved, he believed, by discouraging from reproducing those who had less to offer ("negative eugenics").

Galton's influence was nearly immediate. Darwin declared himself persuaded by his cousin's eugenic arguments, and Galton attracted a number of distinguished disciples. In Germany, the Racial Hygiene society was formed in Berlin by 1905 (Weindling 1989); the English Eugenics Education Society was founded in 1907, with Galton elected honorary president the next year (Kevles 1985, 59). In the United Kingdom and the United States, the movement drew on the middle and upper middle classes, including many professionals and academics (Rafter 1988; MacKenzie 1981; Kevles 1985; Mazumdar 1992). During the decades 1890–1920, eugenic ideas were advanced also in numerous non-English-speaking countries as diverse as Norway, Brazil, and the Soviet Union. Both a research program and a popular movement, eugenics was taught at leading universities, and received attention in standard biology textbooks.

The popular eugenics movements, meanwhile, succeeded in rapidly introducing eugenic ideas into public discourse. Accounts of generations of misfits in such "white trash" family lines as the "Jukes" and the "Kallikaks" were widely publicized, warning that an unwise reproductive act could wreak havoc for generations (Rafter 1988). Following British successes at health exhibitions before the turn of the century, American eugenic organizations took a particular interest in maintaining exhibits and events at state fairs and public expositions. "Fitter Families" competitions were mounted at state fairs, with governors and senators handing out awards (ibid).

The content of the eugenic programs varied considerably. Eugenicists tended to agree that the human race was in decline, but they differed over both cause and remedy. The French and Brazilian eugenics movements were at least as concerned about neonatal care as with heredity, and their hereditarian thinking was Lamarckian--that is, they believed that parents passed on to their children characteristics acquired during their lifetimes (Schneider 1990; Stepan 1991). Most eugenicists elsewhere accepted Galton's view, buttressed by the "germ plasm" hypothesis of August Weismann, that selection rather than environment determined heredity. Eugenicists tended to draw from this account the implication that medical care frustrated evolution by permitting the unfit to survive and reproduce (though Darwin and a number of others who held this view nonetheless continued to support humanitarian measures).

Eugenicists differed also in their practical proposals and legislative aims. While action on behalf of positive eugenics was limited to such mild measures as family allowances, some eugenicists (particularly in the United States and, later, Germany and Scandinavia) did not hesitate to call for coercive measures, either sexual segregation or, later, involuntary sterilization, to prevent those imagined to have undesirable genes from propagating their kind.

In Germany, eugenics became an integral element of medical thinking, which envisioned a three-way division of health care involving medical care for the individual, public heath for the community, and eugenics for the race (Weiss 1990; Proctor 1988). Eugenics, for some, was an extension of a tradition of a social orientation in German medicine that had produced Rudolf Virchow and other pioneers of public health.

Historians have generally followed Daniel Kevles's (1986) classification of eugenicists, at least in England and the United States, as either "mainline" or "reform." In the United States and Britain, mainline eugenics was largely (but not exclusively) conservative in political orientation. Galton was but the first of a long line of eugenicists who believed that those who achieved (at least in fields such as science and literature, where social position was insufficient for advancement) were distinguished from others in their possession of great natural, inherited talent. Indeed, the mainline eugenicists tended to believe that a person's station in life reflected his or her capabilities and could thus be used as an indication of the genes likely to be passed down to subsequent generations. To the extent that eugenics is remembered at all, what is recalled tends to be the "mainline" movement, with its conservative politics and its tendencies toward class bias, racism, and xenophobia — all of which foreshadowed the Nazi's embrace of eugenic doctrines.

In actuality, however, there once were eugenicists all over the political spectrum. The "reform" contingent, often socialists, and including many of the leading figures in the science of human genetics, accepted eugenic goals, but were unsparingly critical of the mainline eugenicists' research, biases, and proposals. Hermann Muller, an American geneticist who later won a Nobel prize for demonstrating the effect of radiation on chromosomes, insisted that natural talent could not be assessed in a society such as the United

States, which did not offer equal opportunities for advancement to its citizens; only under socialism could the fit be identified as such, and then encouraged to multiply.

The labels "mainline" and "reform" do not do justice to the great variety of viewpoints and goals associated with the eugenics movements. Indeed, as Diane Paul has observed, one sign of the ubiquity of eugenic thinking was the attempt by parties on all sides of particular social disputes to further their cause by demonstrating that their recommendations would have the strongest eugenic effect. Eugenics, seen as an avenue for the application of science to social problems, was attractive to some of the architects of the modern welfare state, such as the Progressives in the United States and the Scandinavian Social Democratic parties (Broberg and Roll-Hanson 1996).

Indeed, much of the opposition to eugenics during that era, at least in Europe, came from the right. The eugenicists' legislative successes in Germany and Scandinavia were not matched in such countries as Poland and Czechoslovakia, even though measures had been proposed there, largely because of the conservative influence in these countries of the Catholic Church (Roll-Hansen 1988). The Church opposed eugenics in principle (and they were virtually the only institution to do so), but this was of a piece with their opposition to abortion and contraception: then, as now, the Church was opposed to limitations on fertility, and their opponents were often on the left.

To be sure, early eugenicists were also opponents of birth control, since they believed that its use by the upper classes exacerbated the degeneration of the gene pool. But not all eugenicists took this position. The eugenic banner was seized also by feminists who argued that control over fertility, along with emancipation generally, permitted women to improve the race through sexual selection.

2.2. The Nazi Debacle

Eugenics in Germany, while distinctive in having a medical leadership, had been marked by much the same divergences of opinion as the movements in other countries. Though numerous prominent eugenicists were racist and anti-Semitic, others were avowedly anti-racist (and some were Jews), and a number stood on the political left (Weindling 1989). The Nazis imposed a uniformity of viewpoint, securing the allegiance of the many eugenicists who rallied to its cause for a thoroughly racist, nationalist eugenic program that recognized no limits in the pursuit of "racial hygiene."

Eugenics was central to the entire Nazi enterprise, joined with romantic nativist and racist myths of the pure-bred Nordic. The emphasis on "blood" called for a purifying of the nation's gene pool so that Germans could regain the nobility and greatness of their genetically pure forbears (Burleigh and Wipperman 1991).

As Robert Proctor (1988) and other historians have shown, the subsequent programs of sterilization, "euthanasia" of the unfit (a program that took the lives of tens of thousands of "Aryans," mostly young children), and eventually the Holocaust itself were part of the unfolding of this central idea. The sterilization and "euthanasia" programs, which did not initially target Jews and other minorities, were an exercise in negative eugenics designed to improve the native German stock from its degenerated condition. Legislation barring sexual relations between Jews and "Aryans," and ultimately the Holocaust, were intended to prevent further adulteration of the "pure" German nation with inferior genes. Jews and others who contributed evil genes were the disease afflicting the German nation, which Hitler, the physician, would cure.

These measures were complemented by a range of other genetic interventions, ranging from an elaborate system of Genetic Courts passing judgment on the genetic fitness of those thought to harbor defective genes, to marriage advice clinics, to the *Lebensborn* breeding program for SS men and other racially motivated initiatives in positive eugenics (Weindling 1989). The academic fields of anthropology, biology, and medicine were reformulated in racial and eugenic terms, and the profession of medicine in Germany was compromised by its participation in government programs of identification, sterilization, and murder of those deemed unfit (Aly and Prosch 1994, Weindling 1989, Burleigh 1994, Wikler and Barondess 1994).

2.3. Decline and Fall

In its first years, Nazi eugenic programs and propaganda won the acclaim of eugenic leaders in the United States. The Nazis flattered their counterparts overseas by pointing to legislation in California and elsewhere not only as precedents but also as models, and the authors of these statutes toured Germany and filed favorable reports upon their return (Kuhl 1994). After the Holocaust and the defeat of the Germans, however, eugenicists in most other countries were quick to distance themselves from German eugenics; since the Germans had presented themselves as the most consistent and purposeful of eugenicists, the movement itself fell into general disrepute. American eugenics organizations experienced amnesia over their prewar affinity with their German counterparts, spoke out against racism, and urged Americans to consider eugenics as a source of national strength. Nevertheless, the eugenics societies soon lost their followers; the American society's journal was renamed *Journal of Social Biology*, and what had in prewar years been a virtual consensus in favor of eugenics among genetic scientists disappeared within a decade. The movements' offices were shut down, and the Rockefellers and other funding sources turned their attention to related but more reputable concerns, such as world population control, the prevention of birth defects--and to genetics and molecular biology (Paul 1991).

There is some controversy over the explanation of the sudden disappearance of eugenics from our national consciousness. The account given in the first histories of the eugenics movement was that eugenics was abandoned as the science of genetics progressed, leaving genetic scientists increasingly dubious of the central factual claims of the movement. A revisionist tradition points to the strikingly rapid repudiation of eugenics by reputable geneticists in the mid-1940s, a period marked not by any sudden increase in scientific knowledge but by the scientist's strong interest in distancing themselves from the Nazis.

These accounts have different implications for the future of genetic policy. If eugenics succumbed to the advancement of science, perhaps the lid on its coffin is nailed as tightly shut as it needs to be. If, however, the retreat from eugenics was simply one of fashion, the movement has not been repudiated on the basis of fact or even principle, and we might unthinkingly (or, worse, consciously) return to eugenics when and if fashion changes again. Finally, if clinical genetics is simply eugenics under a different name, we must achieve a clear understanding of the morality of both.

3. IS EUGENIC DOCTRINE INHERENTLY EVIL?

The history of the eugenics movement is marked by a sorry record of pseudoscience, prejudice and bias, and, in its Nazi version, even mass murder. We can learn from eugenics that at least one movement dedicated to the betterment of humankind through genetic improvement led to terrible wrongs. But must this goal point us in the direction of evil? In the remainder of this essay, my question is whether there was, and is, a moral misjudgment, an inherent wrong, at the heart of eugenic doctrine; and, if so, in what it consists. The attempt to answer this question presents an opportunity to assess the choices open to us in the coming decades of progress in genetics. If we are to avoid the errors and sins of the eugenics movement, we will need an account of what these were. And the same holds true if we are to avoid the converse danger of refraining from justifiable remedies and interventions because we mistakenly believe them to share eugenics' taint.

This inquiry is an uneasy hybrid of history and moral philosophy. Since our goal is to discern where the shadow of eugenics falls, the analysis has begun with a (brief) record of what eugenicists actually believed. But to comment on the applicability of their beliefs, goals, and values for the future, we must abstract from their historical context, trying instead to find themes which might apply to our own time and yet which can reasonably be attributed to the eugenicists of a century ago.

3.1. Easy Targets

So much that the eugenicists believed, said, and did has been repudiated that one need not look far to find their "errors." The eugenicists' scientific claims and pretensions are a case in point. Indeed, present-day warnings of a return to eugenics often amount to cautions over untenable claims in behavioral genetics, in particular the heritability of personality traits, and both genetic essentialism and determinism. Though debate continues on such claims—new discoveries of "the gene for" diverse behavioral characteristics appear frequently, and almost as frequently are later withdrawn—the bulk of the eugenicists' claims of the genetic basis of personality are now believed erroneous.[1]

Similarly, there are few defenders of the violations of reproductive rights, and rights of bodily integrity, involved in eugenic involuntary sterilization programs—to say nothing of eugenic euthanasia, as practiced on small numbers of infants in the United States and on a mass scale by the Nazis (Pernick 1996). Diane Paul (1995) has pointed to the development of strong guarantees of reproductive autonomy as a key difference between our own era and that of the eugenicists, one which, it would seem, would preclude the kind of artificial selection which the eugenicists had proposed.

Much the same can be said of the class biases and racism which so marked the mainstream eugenics movements in the US and UK (to the extent that one historian defined eugenics as a war on the lower classes (Mazumdar 1992). While these biases certainly persist, anyone in the United States or Britain who openly advocated a eugenic program that explicitly endorsed such attitudes would be quickly reprimanded.

Each of these attributes of eugenics—genetic determinism, disregard for individual rights, and racial and class bias—is so closely linked to the reputation of eugenics that warnings of a return of eugenics are often simply accusations of one of these fallacies and wrongs. If these were all that eugenics amounted to, the analogy of eugenics to prohibition, an historical curiosity of no particular importance for our time, would be sustained, and the present paper could end at this point.[2] Put differently, we might argue that if we try to imagine a eugenics movement from which we remove the class and racial biases, the faith that personality traits were fixed by heredity, and the conviction that the freedom of the individual to decide whether and with whom to procreate must be overridden in the name of genetic improvement, then it is not the eugenics movement we are imagining. For these attributes defined the movement.

This kind of analysis, however, comes at the cost of rejecting the definition of eugenics given by Galton, who coined the term and initiated the movement. Galton's several definitions over the years varied, but they were variations on a simple theme: using our understanding of the laws of heredity to improve the stock of humankind. In itself, this notion is not necessarily committed to genetic determinism, violations of reproductive liberty, class bias or racism. And though common in the eugenics movements of 1883 or 1933, these beliefs and attitudes also affected other social movements and programs, and indeed the discourse of the educated classes generally. If we may carry Galton's core notion to the present, presumably more enlightened day, what sort of program would it entail? And will we find any hint of a eugenic original sin, a wrong present even in Galton's original conception?

3.2. Five Candidate Wrongs

I will survey five wrongs, or putative wrongs, which might be or have been alleged to be inherent in the core eugenic doctrine of improving the stock of humankind by application of the science of human heredity. Most, I believe, are not good candidates: either they are not inherent, or they are not necessarily wrongs. But in the end, caution toward eugenics is still advised.

3.2.1. Replacement. The first candidate wrong faults the core doctrine of eugenics on the grounds that it seeks "better" (or "fortunate") people rather than people who are made "better" (or "fortunate"). This complaint faults eugenics for posing as a doctrine of benevolence. While "human betterment" is the name of eugenics' game, according to this view, it actually betters no humans. No person's diseases are cured, and no individual's intelligence is raised, by eugenic interventions even when (and if) they are successful in their own terms. Instead, the programs cause the world to be populated by individuals who have these advantages from their beginnings. In essence, eugenics favors healthy people over unhealthy people, and smart ones over stupid ones. That may be acceptable as a basis for choosing friends, or even employees, according to this complaint, but it is not a particularly noble social aspiration. Eugenics, in this view, does not involve any hopes for our fellow human beings, but rather a preference for the sort of fellow human beings we could have.

Does this charge identify an inherent wrong in eugenics? I think not. It does locate something inherent in the doctrine—selection is what eugenics was about[3]—but the complaint does not succeed in showing that it is really a wrong. A host of unquestionably benevolent social programs, from natural conservation to a high savings rate, also replace people rather than help people . As Derek Parfit (1984) has taught a generation of moral philosophers, this is simply (and trivially) due to the fact that interventions with large-scale effects inevitably affect the circumstances of human reproduction, such as the moments at which people engage in sexual intercourse. This in turn determines who will be born, for when it comes to identity and fertilization, timing is everything. Each of us is the unique product of the union of a particular sperm and a particular egg; the product of a different pair would be someone else. Macroeconomic interventions, along with most other large-scale measures, result in different sperm being united, in sexual reproduction, with different eggs, and thus change the cast of characters which will populate that part of the globe in the next generation. Yet good social policies are not a bit objectionable for that reason. It is true that these policies, unlike eugenic programs, do not *aim* to determine who will be conceived and born. Nevertheless, the effect is largely the same, and it is noteworthy that they are no less laudable for that.

In any case, this complaint against eugenics proves too much. It would find fault with a woman's decision to marry one suitor rather than another because the first would be the superior parent; or another parent's decision to delay having a child until he or she was financially and emotionally ready to be a good provider and parent; and with parents who discouraged the maternal urges of an unmarried teenage daughter on similar grounds. Yet surely these choices are perfectly defensible. If the "replacement" complaint against eugenics applies to these as well, we must reject its claim to have found a serious moral flaw in that doctrine.

3.2.2. Value Pluralism. Wilhelm Johannsen, the Danish geneticist, asked in 1917, who was to set the criteria for ideal man: "But what is the ideal? Who shall be responsible for the decision? The complexity of society makes it impossible that one single human type should be the best. We need all different types of humanity (Hansen 1996)." It is not uncommon to find the eugenicists blamed for promoting a particular conception of human perfection, failing to appreciate the essential plurality of values and ideals of human excellence. Like others, they assumed that the ideal would be similar to themselves, or at least to those whom they most admired. Mainline eugenicists in the UK and US, largely members of the upper-middle professional classes, hoped for a society in which each person would attain their level of virtue, and despised those who failed to display the proper bourgeois values. Nazi racial hygienists, many of whom considered themselves to be of "the Nordic type", valued the Nordic type. Hermann Muller, the socialist geneticist and eugenicist, extolled a wide range of models, including Lenin, Gandhi, and Sun Yat-Sen. Surely an heterogeneous group; but all of these were, like Muller himself, exceptionally brilliant men.

As the question attributed to Johannsen, a scientist and reluctant eugenicist, demonstrates, the difficulty of defining human perfection was not entirely lost on the eugenicists, but the strident rhetoric of much of the mainline eugenics literature brooked no opposition and admitted to no doubt over what constituted a "healthy" and virtuous style of life.

We might suppose, therefore, that what is wrong with eugenics is a denial of the plurality of ideals of a valuable human life. Eugenics, according to this complaint, must inevitably impose a particular vision of human perfection. Those who urge eugenics show a limitation of moral understanding and fail to realize that theirs is but one of a multiplicity of such visions, shaped differently by diverse cultural traditions and circumstances and by moral reason. This limitation in understanding is potentially harmful to people of the sort the eugenicists hope not to reproduce, since it denies to them the self-respect which accompanies the aspiration to raise children in one's own image, should this be the desire of those parents.

Is this the wrong, or a wrong, inherent in eugenics? Understood as a claim about the historical eugenics movement, as opposed to the pure Galtonian ideal, I believe that the complaint is partly right but mostly wrong. If directed to the ideal itself, as it might be realized in the future, it is again mostly wrong.

The complaint is right about the historical movement in that the mainline eugenicists made no secret of their ferocious, and in some cases, murderous, disdain for the very kinds of people whose fertility they wish to curb. Davenport (1922) celebrated the death of a child born to a prostitute:

> I recall the impassioned appeal of a sociologist for assistance in stopping the frightful mortality among the children of prostitutes. But the daughters of prostitutes have hardly one chance in two of being able to react otherwise than their mothers. Why *must* we start an expensive campaign to keep alive those who, were they intelligent enough, might well curse us for having intervened on their behalf? Is not death nature's great blessing to the race?

Oliver W. Holmes, America's celebrated Supreme Court judge, wanted no more of the sort represented by the petitioner Carrie Buck, the third of "three generations of imbeciles" who had propagated "enough" (Holmes 1927), and in refusing her petition to remain fertile, opened the floodgates of sterilization in American institutions (Reilly 1991).

Today, we rightly abhor these sentiments, and, of course, the even more repugnant judgments about human "types" which animated the Nazis. Nevertheless, the failure to respect the plurality of values was not the central problem even of mainline eugenics. The traits which the eugenicists believed heritable and worthy of cultivation were ones which are valued by people with widely varying ideals of personal development, plan of life, and family structure. Though some eugenicists did believe there to be particular genes for drunkenness, "shiftlessness", and the like, in the main the eugenicists focused on a very short list of traits about which there is little controversy. Intelligence dominated the list, or was the only item on it; self-control and a few other very general virtues were sometimes added. For many eugenicists, a long list of objectionable phenotypic traits, ranging from sloth to immorality, were the result of the lack of the genes thought to be necessary for these cardinal virtues. There is little real dispute over the value of these all-purpose talents, even among those who reject the class snobbery of the mainline eugenicists. Whatever one's favored pursuit or style of living, intelligence and self-control helps one make the most of it. When we consider a future eugenics program, based on Galton's core idea, we can easily envision one that would focus exclusively on these all-purpose advantages. Value pluralism need not be an issue.

It remains true that the mainline eugenicists were anything but tolerant of personal and social ideals which differed from their own. They favored breeding humans with an eye to intelligence and self-control because they thought that these traits were necessary if a person were to lead a "proper" kind of life, i.e, one like their own. Claims of this kind, e.g. that the poor are too stupid to understand the difference between right and wrong, or to exercise the restraint necessary for the nuclear family, resurface today in such works as Herrnstein and Murray's (1994) book, *The Bell Curve*. But the transmissible characters targeted by the intervention remains one on which there is agreement regardless of differing ideals of human perfection.

Value pluralism *could* become an issue in eugenics, even if it is not inherent in the core idea. Deaf parents who wish to abort fetuses which do not test positive for inherited deafness, and dwarf parents who want only a child with the gene for achondroplasia, hold unconventional values, and their freedom to act on them is at issue in the ethics of clinical genetics. The European Parliamentary panel on genetic engineering, headed by a Green representative to the German Bundestag, held that genetic screening requires us to decide what is "normal and abnormal, acceptable and unacceptable, viable and non-viable forms of the genetic make-up of individual human beings before and after birth"(quoted in Kevles 1992). If we ever acquire an ability to influence personality and character through genetic choice or manipulation—to choose, for example between aggressive and gentle dispositions—this debate will be of crucial importance.

Everyone supports the goal of health, and though we do not share precisely the same concepts of health (and of disability), diversity of opinion is limited to a few disputed instances. When genetic interventions are aimed at enhancing the genome of the healthy individual, however, the scope of potential disagreement is nearly unlimited. Some of us may live long enough to see the genetic advances which will occasion such debates. Nevertheless, eugenic programs could avoid the problem of value pluralism simply by limiting its focus to those human characters on whose desirability there is universal or widespread agreement.

3.2.3. Statism. In a recent address, James Watson (1997) reviewed the odious history and possible future of eugenics and concluded that the most important safeguard is to eliminate any role for the state. He provided a strong case. The great wrongs visited on vulnerable people in the name of eugenics—institutionalization, sexual segregation, sterilization, and, in Germany, murder on a mass scale—could not have occurred without the agency of the state. In England, where the state's role was minimal, eugenics may have been offensive but it did not violate individual rights.

Many would take issue with Watson's contention that the state is the chief enemy. What Troy Duster (1990) has called "backdoor eugenics" threatens to visit harm on the genetically disfavored through the cumulative effect of many private decisions on the part of employers, insurers, and prospective parents. As Robert Wachbroit (1987) has observed, government and society might conceivably switch roles, with the former intervening in private choice in order to preserve the liberties and well-being of those whose genes threaten disease or disability. In such a scenario, denying a role to the state might hasten eugenic evils rather than protecting against them. If the "backdoor" concern is justified, we ought not conclude that the wrongs of eugenics can be avoided as long as the state forswears any eugenic intent.

In any case, a strong state role is not essential for a eugenic program. True, it may be difficult to win compliance with eugenic prescriptions without the long arm of the law. That is why Galton, imagining a fully voluntary regime, mused that eugenics might have to be instated as a civil religion in order to induce members of society to make the sacrifices required. Eugenics never attained this status, whether in the UK or elsewhere (not even in contemporary Singapore, where the head of state has been an enthusiast). As Paul (1995) notes, the British eugenics movement was no less "eugenic" for being a citizen's movement relying on voluntary measures, and from this fact it follows that statism is not a source of wrongs inherent in the core of the eugenic program.

3.2.4. "Collectivism". An alternative analysis locates the wrong inherent in eugenics in its concern for the genetic well-being of the group rather than that of the individual. In this view, concern for the individual is benign. Indeed, genetic intervention might be mandatory, from the moral point of view, in certain cases. Parents who knowingly bring into being children who suffer agonizing and deadly defects might be accused of "wrongful life." The fateful turn toward eugenics occurs, in this view, when we widen our interest from the individual child to the group, hoping not that our own sons and daughters will be healthy but that the population's gene pool will be improved. This is a "collectivist" vision in the sense that the object of our concern is the group as such, while our concern with the individuals who constitute the group is primarily in the contribution they might make toward the well-being of the collective.

Narrowly defined, "collectivism" doctrines are those according to which interests inhere in the collective entity or group in addition to the group's members. Much of eugenic writing, whether "mainline" or "reform", was collectivist in this sense. More loosely, we might understand the label "collectivist" for eugenic doctrines or policies which locate interests only in individuals, but which condone trading of the well-being of some for that of others. Social Democratic eugenicists in Scandinavia, for example, were often candid in noting the burden imposed by eugenic sterilization upon those sterilized, but justified the practice in terms of the reduced burden of dependents (Broberg and Roll-Hansen 1996).

According to this understanding of where the wrong in eugenics lies, a bright line can be drawn here, one that both distinguishes medical genetics from eugenics and locates the wrong inherent in the latter. If we draw the line here, we reject the notion that parents who seek "the perfect baby" are themselves engaging in eugenics. This understanding of eugenics provides a green light to medical genetics, which can be permitted to continue its rapid development without the worry that it is revisiting the errors of the past.

But what, precisely, is the wrong which this view attributes to eugenics? Consider these three statements:

1a. I favor a genetic intervention because I want my child to have the "best" (healthiest, etc.) genes.
1b. We favor genetic interventions (on behalf of each of us) because we want our children to have the "best" (healthiest, etc.) genes.
1c. I favor genetic interventions (for each person in our group) because I want our children to have the "best" (healthiest, etc.) genes.

If 1a is morally acceptable, surely it doesn't become wrong when voiced by several people (in the form of 1b). And how can I be faulted by endorsing that group's hope (1c)? 1b and 1c are merely the aggregate of many instances of 1a. One might expect to hear 1c uttered by, say, a health official, or a legislator who sponsors a measure which would provide genetic services to large numbers of people. Concern for the welfare of large numbers of people is part of their job description.

Consider, by contrast:

2. The sum total of benefits involved in a program of genetic interventions will be greater than the costs.

Here we seem to come closer to a "collectivist" view, for 2 does not claim that the benefits for each individual might outweigh the costs. It leaves open the possibility that some may lose while others benefit, promising only that the magnitude of the latter will be greater.

However, this appearance may, I believe, be misleading. Statements such as 2 are often made by way of justifying the use of public funds. The point of the intervention in such cases is not to save public money, for the professed (and, we may assume, the actual) goal is to ensure that as many children as possible are born with genes which make their lives go well. Given the endless competition which exists for public funds, however laudable their purpose, it always helps if one can argue that the net social cost is zero or better. This calculation has been a trump card in debates over health care allocation when played by advocates for perinatal medical care, and it might apply equally well for a program aiming to provide better genes (see also Paul t.v.).

Consider, finally:

3. A program of genetic intervention will limit the number of people who are a burden to others.

Have we, with this step, crossed the line to the "collectivist" position? And if so, does this claim partake of eugenics's original sin? The answer, I believe, is not as straightforward as it might appear.

We might begin by noting that if this claim does in fact put us on the wrong side of that imaginary moral line, we may have stepped over it a bit earlier. I just argued that a cost-benefit calculation, such as 2, need not be motivated by a wish to save society some money. But of course it could be. This might be the real goal in a particular instance even when the advocates of a program offer it merely as a justification for the use of public funds. In either case, 2 would be in the same moral company as 3.

But is 3 "collectivist," where we understand that term as betraying concern for individuals only insofar as they add or detract from the well-being of the group, and is it morally repugnant for that reason? We should note, first, that 3 is not necessarily "collectivist" in the narrow sense defined above, according to which the beneficiary is a collective entity, be it the Reich, the Revolution, or The Race, for which no sacrifice of individual well-being can be too great. Nazi eugenics, of course, was a collectivism of this other sort, obsessed with the glory of the reified Volk. But that is no part of the original Galtonian eugenics, at least at its core.

More to the point, the core notion of eugenics does not necessarily ask for sacrifice of any sort. Programs which isolated or sterilized tens of thousands of people, and of course those which resorted to murder, imposed the greatest of sacrifices, but Galton's original proposal did not call for these measures. English eugenetics, for all their concerns over the excess fertility of the unfit, generally proposed voluntary curbs on reproduction (Kevles 1985).

In any case, the sacrifice which a eugenic program might ask of prospective parents is likely to be much less onerous as technology develops. In Galton's day, eugenics was mainly concerned with who mated with whom and how many children resulted. For the "unfit", childlessness (even if voluntary) was the price of eugenic correctness. Today, a eugenic principle might call for prospective parents to screen pregnancies so that the children they bring to term have the greatest feasible genetic advantages. Tomorrow, these same parents might be encouraged to avail themselves of genetic interventions to cure and to enhance. Excepting perhaps the fetuses which are aborted as a result of such a program, no one would be asked to make sacrifices. Because parents almost always seek advantages for their children—health above all—there is a congruence between a eugenicist's concern for the public and a parent's concern for his or her child. Where there is not, a voluntary program would leave the decision to the parent. The potential child whose conception or birth is avoided by this intervention does not count in the moral calculation which "collectivism" insists we make. Common sense must concur.

Nevertheless, this kind of eugenic program might claim some actual, living victims. As disability-rights advocates have insisted, it is difficult to argue for public programs on the basis of claims like 3 without suggesting, in the same act of speech, that the existence of people who are dependent on others is a fact to be regretted; and this sends the message that these lives are not, in some sense, valuable. I will not take the trouble to argue that this sentiment is reprehensible and that the opposite message ought to be the cornerstone of public policy, both in genetics and elsewhere. Every person is valuable, and not only for any contribution which he or she might make to others. The rhetoric of mainline eugenics in the UK and the US, with its denunciations of "human filth" and "human rubbish," are justification enough for the abysmal reputation of these movements, even apart from the programs of mass sterilization and murder which followed in their wake.

Perhaps we have found, therefore, some hint of an "original sin" of eugenics. In the series 1a-1b-1c-2–3, it occurs somewhere between 2 and 3, when we begin to calculate the value of genetic improvement not in terms of the well-being of the individual whose genes are less likely to cause that individual to suffer, and more likely to enhance that individual's well-being, but for the effect which the existence of that individual might have on the well-being of others. But this is not quite "collectivism", and I would urge that this wrong, if it is that, be given a different rubric: unfairness.

3.2.5. Fairness. In the United States and England (though not in Germany), the fields of eugenics and public health involved different people, expert professions, journals, and aims. But the two movements shared many assumptions and attitudes. As Charlotte Muller noted in her insightful review, differences in health status across racial and income lines tended to be explained by public health scientists in terms of heredity. Burdens imposed by eugenics were justified by the analogy to public health, as Justice Holmes did in comparing sterilization to vaccination. Martin Pernick (1997) has noted extensive overlap even in the jargon of the two fields, each of which resorted to "isolation" and "sterilization" of the individuals who were thought to pose threats to the well-being of the public. Eugenics was often described in medical terms (Kamrat-Lang 1995), e.g. as an effort to prevent the spread of (genetic) disease from generation to generation. Hitler was lauded as the great doctor of the German nation, rescuing the Aryan gene pool from the genetic disease introduced by Jewish infestation (Proctor 1988).

Public health had one more characteristic in common with eugenics: it created and struggled with many of the same moral problems. A persisting theme in the ethics of public health is the greater effectiveness often achievable if the interests of some are sacrificed to the interests of others. Despite the great protection Americans enjoy in the inviolability of the person, public health requirements sometimes have priority, as Holmes's reference to vaccination policy shows. How to balance these benefits and burdens is a question of distributive justice which public health programs will always have to face.

Despite the fact that genetic technology will permit some eugenic goals to be achieved without burdening prospective parents, a public policy of providing "better" genes to future generations is bound to impose social costs. Even a fully voluntary, medically-oriented program—what is called "clinical genetics" today and which strenuously avoids any association with the eugenics of old—must answer to advocates for the disabled who claim that the well-being of the disabled is put at risk when genetic screening programs try to ensure that none with their disabilities will be conceived or born. Bioethicists have warned of decreasing tolerance of differences, once we acquire the power to choose "the best", and this intolerance might impose social sanctions on those who declined to make such choices.

More concretely, it is not unreasonable to fear that if it once again becomes respectable to advocate eugenics, the wrongs of the past will return in full force. I have argued that eugenics, considered as a set of principles, need not assume genetic determinism, nor advocate or condone racism or class bias. In actual practice, however, what guarantee can there be that a eugenic program would not be guided by these still-prevalent beliefs and attitudes? Similarly, we could imagine, in principle, a eugenic program which avoids coercive measures, particularly sterilization, but what assurance do we have that these measures might not eventually be viewed as justified, if public policy seeks to provide "better genes" and the benefits they might bring to society?

The ethics of eugenics and the ethics of public health, therefore, are closely related. Neither, unfortunately, has received the same attention as the ethics of personal health care. What standard of justice should be used in guiding any new eugenics? The first pages of Rawls' *A Theory of Justice* include this famous passage:

> ...Each person possesses an inviolability founded on justice that even the welfare of society as a whole cannot override. For this reason justice denies that the loss of freedom for some is made right by a greater good shared by others. It does not allow that the sacrifices imposed on a few are outweighed by the larger sum of advantages enjoyed by many. Therefore in a just society the liberties of equal citizenship are taken as settled; the rights secured by justice are not subject to political bargaining or to the calculus of social interests. (Rawls 1971, 3–4).

This is not a bad starting position: public policy in genetics, whether or not it is termed eugenics, ought not to infringe on personal liberty. But this does not necessarily call on us to avoid any risk of burdening some individuals for the sake of the genetic well-being of future generations. I am not personally persuaded, for example, that the threat of stigmatizing the disabled requires us to abandon the effort to ensure that future generations are free of avoidable disability. But this kind of concern points us to a valid question of justice and also to an irony.

The point about justice is that genetic benefits provided in services engaged by a particular pair of parents may have adverse effects on others, and we are bound to reflect on the fairness of the resulting distribution of benefits and burdens. The irony is that this very admonition pulls us toward, and possibly over, the bright line which bounds that which we identified as a possible wrong inherent in the core notion of eugenics. This line is crossed when the goal of our genetic intervention is not only the well-being of the individual, but also the effect on others of bringing this person into the world. If we are required by distributive justice to consider the effects upon all members of the community when we contemplate genetic interventions, this moral imperative is in effect telling us to consider not only the benefit of a contemplated intervention for a particular individual, but also for others. If there is a wrong inherent in the core Galtonian eugenic project, it surely has to do with this very move — one which takes us beyond the "medical" or "clinical" focus on the patient at hand (or, in the case of procreation, on the child to be) to the society as a whole. In the latter, wider view, the patient recedes from the foreground and a moral judgment is made on the basis of a calculation which takes into account the claims of many.

4. CONCLUSION

What follows from this series of observations? Is there a wrong inherent in the core Galtonian eugenic program? And what guidance might this give us in deploying the resources of the new genetics?

One respectable position which I have not taken up directly in the above is that the core notion of eugenics may be benign, because it is trivial. No one objects in principle, according to this view, to using what we know of the science of heredity to improve the chances of future generations for achieving greater well-being. What rouses passionate debate are the means to be used; or the problem of value pluralism; or one of the other "easy targets" which are discussed in section 2 above.

Perhaps so. Much of the controversy over China's Law on Maternal and Infant Health Care has indeed focused on its apparent threat of coercion, rather than its goal of a generation of healthy Chinese children (Nature 1994). But clearly some who express concern about a return of eugenics in the West are worried by the move from "medical" concern for the individual to "eugenic" ambitions for improving the gene pool in general, even if coercion is not proposed as a means to this end. Perhaps the worry is roused by the fact that this move was accompanied, early in this century, by great wrongs, harms justified by the greater good, and the fear that once the "collective" goal is established, the demand for sacrifices by individuals will not be long in coming. In some cases, however, the complaint against "eugenics" seems to be lodged against those who profess concern for the genes of humankind apart from the genes of one individual patient, whether or not the broader concern be advanced by coercion or other harms.

I do not see that much hangs on the resolution of this question. In either case, we can draw the important (if obvious) lesson that progress in genetics must pay attention to these questions of distributive justice. This very general, yet morally crucial, requirement ought to guide us now, as we decide which programs of genetic testing and screening to undertake, and also in the future, as we contemplate the possibility of refashioning the human genome to engineer a new, perhaps improved version of *homo sapiens*. Done justly, the genetic well being of "the group" is a proper object of concern. The question of moral importance is not whether this constitutes eugenics; it is whether it can be done fairly and justly. It wasn't, the last time it was tried.

ACKNOWLEDGMENTS

Much of this essay is drawn from my paper, "Eugenic Values", forthcoming in *Science in Context*, Autumn 1998; and a forthcoming volume, tentatively entitled *In the Shadow of Eugenics*, co-authored by Allen Buchanan, Dan Brock, and Norman Daniels. Particular thanks is due to Diane Paul for advice, instruction, and guidance on historical sources.

ENDNOTES

1. The contention over behavioral genetics is vigorous and acrimonious, but the rhetoric sometimes obscures what is really being debated. James Watson (1997) has asserted that "the scientific left" fights every claim of behavioral genetics because these threaten the left's dogma concerning the socioeconomic explanations of all social problems, including social problems which can be traced to human behavior. Given the eugenic Nazi past, however, it is understandable that claims such as that of a violence-disposing gene (in a Dutch family) would be scrutinized closely, and even with hostility, as would fears that this research could be used to revive the mainstream eugenicists' exculpation of social institutions in explaining social problems.
2. Or we might extend the analysis a bit by asking whether we are really so certain that these beliefs and practices deserve complete repudiation (If the reader, for example, supports laws barring marriage between sister and brother, is any part of this support based on concern for the genetic results for offspring? Is this not a violation, on eugenic grounds, of the fundamental personal right to choose whom to marry?).
3. In the future, a eugenic intervention might take the form of alterations in the genome of an embryo, which would not (on this account) affect the identity of the individual involved. Hence a future eugenic program need not involve selection in this sense.

REFERENCES

Aly, Gotz; Chroust, Peter; and Pross, Christian (1994), *Cleansing the Fatherland: Nazi Medicine and Racial Hygiene*, Baltimore: Johns Hopkins University Press.

Beardsley, Tim (1997), 'China Syndrome: Chinas eugenics law makes trouble for science and business', *Scientific American*, March.

Broberg, Gunnar, and Roll-Hansen, Nils (1996), *Eugenics and the Welfare State: Sterilization Policy in Denmark, Sweden, Norway, and Finland*, East Lansing: Michigan State University Press.

Burleigh, Michael, and Wippermann, Wolfgang (1991), *The Racial State: Germany 1933–1945*, Cambridge: Cambridge University Press.

Chan, Chee Khoon, and Chee, Heng Leng (1984), 'Singapore 1984: Breeding for Big Brother', in Chee, Heng Leng and Chan, Chee Khoon (eds.), *Designer Genes: I.Q., Ideology, and Biology*, Kuala Lumpur: Institute for Social Analysis, pp. 4–13.

Davenport, Charles (1922), 'The Eugenics Programme and Progress in its Achievement', In Aldrich, Moron A., et al (eds.), *Eugenics: Twelve University Lectures*, New York: Dodd, Mead and Co.
Dong-Sheng. Sun (1981), 'Popularizing the Knowledge of Eugenics and Advocating Optimal Births Vigorously', *Renkou Yanjiu* (Population Research)(Beijing), No. 4, pp. 37–41.
Duster, Troy (1990), *Backdoor to Eugenics*, New York: Routledge
Galton, Francis (1883), *Inquires into Human Faculty and Its Development*. London: Macmillan.
Herrnstein, Richard J., and Murray, Charles (1994), *The Bell Curve*, New York: Free Press.
Holmes, Oliver Wendell (1927), Opinion in *Buck v. Bell*, 274 U.S. 205, 207
Kamrat-Lang, Debora (1995), 'Healing Society: Medical Language in American Eugenics', *Science in Context*, Vol. 8. No. 1, pp. 175–196.
Kevles, Daniel J. (1985), *In the Name of Eugenics*. Berkeley: University of California Press
Kuhl, Stefan (1994), *The Nazi Connection: Eugenics, American Racism, and National Socialism*, Oxford: Oxford University Press
MacKenzie, Donald (1976), 'Eugenics in Britain', *Social Studies of Science* 6, pp. 499–532.
Mazumdar, Pauline M. H. (1992), *Eugenics, Human Genetics and Human Failings: The Eugenics Society, its Sources and its Critics in Britain*. London: Routledge
Nature (1994), 'China's Misconception of Eugenics (editorial)', *Nature*, Vol. 6. January, pp. 1–2.
Parfit, Derek (1984), *Reasons and Persons*, Oxford: Oxford University Press
Paul, Diane B. (1991), 'The Rockefeller Foundation and the Origins of Behavior Genetics.' In *The Expansion of American Biology*, Benson, K., et al.(eds.), New Brunswick: Rutgers University Press (1995), *Controlling Human Heredity*, New York: Academic Press
Pernick, Martin (1996), *The Black Stork*, Oxford: Oxford University Press
Proctor, Robert (1988), *Racial Hygiene: Medicine Under the Nazis*. Cambridge: Harvard University Press
Rafter, Nicole H. (1988), *White Trash: The Eugenic Family Studies 1877–1919*, Boston: Northeastern University Press
Rawls, John (1971), *A Theory of Justice*, Cambridge: Harvard University Press
Reilly, Philip R. (1991), *The Surgical Solution: A History of Involuntary Sterilization in the United State*, Baltimore: Johns Hopkins University Press
Roll-Hansen, Nils (1989), 'Eugenic Movement in Scandinavia', *Brit. J. Hist. Sci*, Vol. 22, pp. 335–346.
Schneider, William (1990), 'The Eugenics Movement in France, 1890–1940', In Adams, Mark (ed.), *The Well-Born Science*, Oxford: Oxford University Press
Stepan, Nancy L. (1991), *The Hour of Eugenics: Race, Gender and Nation in Latin America*, Ithaca: Cornell University Press
Wachbroit, Robert (1987), 'What is Wrong with Eugenics?', *QQ : Report from the Institute for Philosophy and Public Policy*, Vol. 7. No. 2–3, pp. 6–8.
Watson, James (1997), Genes and Politics: Presidential Address. *Annual Report 1996: Cold Spring Harbor Laboratory*, Cold Spring Harbor (New York): Cold Spring Harbor Laboratory
Weindling, Paul (1989), *Health, Race and German Politics between National Unification and Nazism 1870–1945*, Cambridge: Cambridge University Press
Weiss, Sheila (1987), 'The Race Hygiene Movement in Germany', *Osiris*, Vol. 3, pp. 193–236.
Wertz, Dorothy (1996), 'Eugenics: Alive and Well in China', *The Gene Letter*, Vol. 1. No. 1, July
Wikler, Daniel and Barondess, Jeremiah (1993), 'Bioethics and anti-Bioethics in Germany: What Must We Remember?', *Kennedy Institute of Ethics Journal*, Vol. 3. No. 1, pp.39–55.

2

PREVENTING GENETIC IMPAIRMENTS

Does It Discriminate against People with Disabilities?

Helga Kuhse

Centre for Human Bioethics
Monash University, Clayton/Vic., Australia

1. INTRODUCTION

The eugenic philosophies and policies of the 19th and early 20th century cast a dark shadow over contemporary genetics. In Europe, Great Britain and the United States, and in particular in Nazi Germany in the 1930's and early 1940's, eugenic practices were widespread. These practices included programs directed at the mass elimination of certain populations and groups, as well as the non-voluntary sterilization of the mentally disabled and criminally insane. In the United States, for example, the constitutionality of sterilization laws was upheld in the infamous 1927 US Supreme Court decision *Buck v. Bell*, when justice Oliver Wendell Holmes, jr. defended a compulsory sterilization order for Carrie Buck with the words: "Three generations of imbeciles are enough" (1927, pp. 1000–1002).

Whereas proponents of the "old genetics" often held racist and elitist views, proponents of the "new genetics" are—as Allen Buchanan observes—"explicitly—almost self-consciously—*universalistic*: genetic science is to serve human beings generally, rather than any particular people or nationality or race" (Buchanan, 1996, p. 18).[1] Moreover, whereas the old genetics was associated with coercive government schemes, the new genetics is open to individual choice. It does not, for example, forcibly prevent prospective parents from bringing genetically impaired children into the world, it merely gives them the opportunity to do so.

On the face of it, the avowed goal of the new genetics—"to deliver today's children and future generations from genetic diseases" (Rowe, 1994)—would seem unobjectionable. Not everyone is convinced, however, that this is so. Rather, there is a cluster of distinct dissenting voices—voices found in the critical disability rights literature—which charge that the very endeavour that seeks to prevent genetic disability and disease is morally flawed because it discriminates against, and devalues the lives of, disabled people. The source of this discrimination lies not, or not only, in the fact that (at least for now) the only feasible way of eliminating genetic impairment and disease is by eliminating genetically disabled individuals, or by preventing them from coming into existence, but in the very idea that genetic impairment requires a medical response. Disability, it is claimed, is not so much a biological given as a social construct. It ought to be prevented not by ridding the world of disabled individuals, but by creating enabling social conditions.[2]

It is because of this connection between the new genetics and the prevention of the existence of disabled individuals that disabilities rights advocates have charged that genetic testing and screening programs are far from universalistic. The new genetics, like the old eugenics, is, they say, based on the morally objectionable belief that some human lives are more valuable than others. As Christopher Newell, an Australian disabilities rights advocate, puts it: in determining "what sort of people there should be", genetic testing and screening programs are based on the proposition that certain "forms of life" should not exist. This is inherently discriminatory and unjust; it devalues the lives of disabled people because it denies them the very right to exist (Newell, 1997, p. 1).

These are serious charges. But are they sound? Before taking a closer look at these arguments, it is important to provide some context.

2. THE BACKGROUND

Some 8–10 percent of all infants are born with some impairment, and between 2 and 4 percent are born with a serious disability. About half of all serious disabilities have a genetic basis. Genetic disabilities can take the form of hereditary conditions, such as Tay Sachs disease, Duchene muscular dystrophy, Lesch-Nyhan syndrome and Huntington's chorea, and of congenital conditions, such as Down's syndrome. They may involve a single gene, or a number of genes, as is likely to be the case in spina bifida. Some conditions can be corrected or ameliorated by diet, medication or surgery. Others, however, will destroy or seriously hamper physical and/or mental functioning and lead to premature death. In short, then, there are many and varied genetic conditions that can have a very different effect on the degree of disability a person will experience, and on the quality and quantity of an affected individual's life.

The prevention of genetic disabilities has long been regarded as a desirable goal and has been the primary motivation behind genetic screening and counselling programs, and behind prenatal testing. While prior knowledge of a fetal disability can give couples the opportunity to prepare for the birth of a disabled child, and time to adjust to her disability, there is little doubt that the major reason for couples undergoing genetic screening and prenatal testing is to prevent the birth of a disabled infant. A 1988 study from New England University, for example, found that 86% of Australian and British respondents, and 80% of Americans, regarded fetal disability as an acceptable ground for lawful abortions (Kelley and Bean, 1988, p. 7).

Initially, the detection of fetal disabilities was limited to chromosomal and chemical analyses, based on such factors as maternal age and detailed family histories. More recently, genetic probes have become available that permit the detection of some genetic disabilities in the developing fetus, while single-cell diagnostic biopsy techniques can be applied in the context of *in vitro* fertilization procedures. Couples at risk can make use of the latter technique, and opt for so-called "*in vitro* abortions" of affected embryos, in preference to later prenatal diagnosis and *in vivo* abortion.

This means that today's prospective parents, who are known to be at risk, have (other things being equal) at their disposal a number of ways by which they can attempt to prevent the birth of a genetically handicapped infant:

1. they can decide not to conceive a child;
2. they can make use of donor gametes or embryos;
3. they can undergo IVF, and opt for the *in vitro* abortion of affected embryos;
4. they can conceive a child in the normal way, and opt for abortion if the fetus is found to be disabled;
5. they may also, in future, be able to have recourse to germ-line or somatic gene therapy, although this is not at present a realistic possibility.

It is against this backdrop of different modes for the prevention of a large and very diverse number of genetically based disabilities that the question of discrimination must be examined.

I shall begin by looking at the argument that disability is a social construct, and that it is primarily society rather than biology that handicaps people.

3. DISABILITY AS A SOCIAL CONSTRUCT

Those who see disability as in some sense socially constructed, do not regard it as a mere physiological or biological phenomenon; rather they see the degree to which a person is disabled, or whether she is disabled at all, as depending on a number of socially determined conditions—for example, on the way in which societies have organized themselves in terms of various modes of social interaction and cooperation. Society, it is claimed, *constructs* disabilities and thereby creates, or contributes to, the unfair treatment and oppression of disabled people.[3] It is because of the social nature of disability that we should stop concentrating on how the existence of disabled people can be prevented; rather, we should focus on how best to eliminate disabling social conditions (Newell, 1997).

There are various ways in which the term "disability as a social construct" can be understood. The following two stories will serve to distinguish between some plausible and some implausible understandings. The first story features Nunez the protagonist of one of H.G. Wells's stories, which has recently been retold by Robert A. Crouch in his article on cochlear implants for prelingually deaf children (Crouch, 1997). Nunez, finds himself the only person with sight in a community of people who have been blind for fifteen generations. "In the Country of the Blind", Nunez thought, "the One-Eyed Man is King" (Wells, 1911).

> Surrounded by persons he considers disabled, Nunez sets out to convince the inhabitants of the country of the blind that they are missing out on a great deal because of their blindness. Despite his best efforts, however, the blind are not persuaded by his rhetoric, and Nunez, exasperated by their lack of understanding, shouts: "You don't understand ... You are blind, and I can see." Broken, Nunez gives up his attempts to convince the blind of his superiority and in an interesting role reversal *himself* becomes the subject of an attempt to be assimilated into the community of the blind. Convinced that all of Nunez's talk about such obvious nonsense as "sight" and "blindness" is due to the effect of Nunez's prominent eyes on his brain function, the community doctor proclaims: "And I think I may say with reasonable certainty that, in order to cure him completely, all that we need do is a simple and easy surgical operation—namely, to remove these irritant bodies"—his eyes. To which a blind elder replies: "Thank Heaven for science!" (Crouch, 1997, p. 14).

Wells' story can be read in various ways. One way would be to take it as a salutory reminder that the lives of people with disabilities need not be miserable, and that the lack of certain capacities or functions need not be experienced as a loss.

Another way would be to take it as denying the very idea of a socially or standpoint-independent notion of disability. I think we should reject this general idea. Even if the absence of a significant function or capacity (or an increased susceptibility to disease and other limiting conditions) is not felt as a loss and is compatible with a subjectively satisfying life, we can still sensibly speak of a disability as a mental or physical condition that we have good rational reasons to avoid.[4] Disabilities are best seen as limiting factors—factors that will, to varying degrees, set limits on an individual's options and opportunities. These limits may consist in reduced experiential opportunities (as would be the case in blindness, for example) and in reduced opportunities for social interaction, but they may also be set by serious ill-health and the burdens imposed by medical and surgical treatments, and the knowledge of a significantly shortened life.[5]

If we accept this, we must reject accounts that deny the very idea of socially independent functional limitations and impairments. A more plausible way of understanding the claim that disability is a social construct is expressed in the second story I want to recount. In this story, the author, Vic Finkelstein, asks us to imagine a world—a fully functioning large village—where wheelchairs are the norm. A small number of able-bodied people find themselves in the village, by no choice of their own. In this wheelchair world, Finkelstein suggests, these able-bodied people would soon become disabled. They would lack wheelchair mobility, suffer from backproblems and head-injuries (as a consequence of finding themselves in an environment characterised by low ceilings and doorways designed for wheelchair use rather than for upright walking). While provided with free helmets to protect their heads, and with backbraces that bend them to standard height, they would, due to various social constraints and prejudices, be unable to find employment and charitable institutions would need to raise funds to support the "able-bodied disabled".

One day, Finkelstein continues his story, the able-bodied realize that their disability is *social* in nature. They suggest to the wheelchair-users that the door and ceiling heights be changed and "even argued that perhaps, just perhaps, their disabilities could be overcome (and disappear!) with changes in society" (Finkelstein, 1981, pp. 34–36).

In this ironic sketch, Finkelstein draws attention to the fact that particular social arrangements and modes of interaction may cater for the needs and interests of one group of people, while neglecting those of another. In his upside-down world, able-bodied people are disabled because they find themselves in a society that has devised infra-structures and modes of cooperation that meet the needs of wheelchair-users, rather than their own. And—the crux of the story—just as the disabilities of the "able-bodied disabled" can be overcome by social change, so, disability rights advocates argue, can many of the disabilities of the disabled.

Allison Davies, a British disability rights advocate, who describes herself as having severe spina bifida, articulates this view when she writes:

> ... if I lived in a society where being in a wheelchair was no more remarkable than wearing glasses and if the community was completely accepting and accessible, my disability would be an inconvenience and not much more than that. It is Society which handicaps me, far more seriously and completely than the fact that I have spina bifida (Davis, 1989, p. 19).

This perspective is a valuable one. It alerts us to the fact that there is no necessary connection between an individual having a functional impairment and that individual being socially disabled or handicapped. Rather, whether and to what extent functional impairment will manifest itself in disablity will frequently depend on institutional arrangements and social accommodation and acceptance.[6] Recognizing that disability can be "socially constructed" rather than objectively given, will open the door to an increased sensitivity to the possibility of a social rather than purely medical response (Leist, 1997, p. 20).

At the same time, however, claims about the relational character of disability[7] may also obscure two important points. The first one is this: Some of the most profound problems experienced by individuals with certain impairments are difficult, if not impossible, to ameliorate by social interventions. A clear case would be infants born with such conditions as Tay-Sachs disease, Lesch-Nyhan syndrome or anencephaly (even the most enthusiastic social program could not ensure that these infants' disabilities related to physical and mental impairment would be solved), but there are many other conditions as well where social interventions can only partially overcome the impairment. This means that the distinction between impairment and disability must, contrary to what some disability rights advocates suggest or imply[8], not be collapsed (French, 1996).

Secondly, claims about the relational character of disabilities often fail to draw attention to the diversity and range of disabilities—both physical and mental—that can limit or thwart social interaction. People affected by different disabilities require very different social responses. The needs of a person with hearing impairment, for example, are quite different from the needs of a person affected by visual impairments, or by motor impairment, or intellectual impairment.

Do present social arrangements unjustly discriminate against people with disabilities? Statistical evidence, and the personal experiences of people with disabilities suggest a positive answer. On many, if not most, indicators of participation in mainstream social life—such as "employment, income levels, suitable housing and access to public transport, buildings, information (newspapers, radio and television) and leisure facilities"—disabled people as a group fare badly (Finkelstein, 1996, p. 11).

On the face of it, this would seem discriminatory and unjust. But can we draw this conclusion? There is little doubt that there are many instances where people with disabilities are treated unfairly—where relatively simple aids, provision of access to buildings and the like—ought to be provided to remove or ameliorate social disadvantages. But this approach will not, and cannot, remove all disadvantages and achieve equal opportunity for all. To achieve that end—if it were possible at all—would require a much more radical approach.

The revised German Constitution, operative since November 1994, proclaims that "Nobody must be disadvantaged because of his or her disability" (*Grundgesetz der Bundesrepublik Deutschland*, Artikle 3, Paragraph 3). But what would social life be like if we were to try to ensure that nobody would be disadvantaged, on account of her or his disability? This is not merely a question of finance or economics, of providing some people—to follow Vic Finkelstein's example—with safety helmets or of changing door and ceiling heights, rather, it is ultimately also about such fundamental issues as social goals and values, and about the appropriateness or otherwise of replacing our present emphasis on intellectual and physical excellence and competition by different goals and models of cooperation.

One radical mode of eliminating disadvantage is ridiculed in one of Kurt Vonnegut's story, "Harrison Bergeron" (Vonnegut, 1968, pp. 7–13)[9], set in the year 2081, when "everybody was finally equal":

> They weren't only equal before God and the law. They were equal every which way. Nobody was smarter than anybody else. Nobody was better looking than anybody else. Nobody was

> stronger or quicker than anybody else. All this equality was due to the 211th, 212th, and 213th Amendments to the Constitution, and to the unceasing vigilance of agents of the United States Handicapper General (Vonnegut, 1968, p. 7).

To ensure that the intellectually well-endowed would not have an "unfair advantage on account of their brains", they were required by law to wear little mental handicap radios in their ears, and the frequent transmission of sharp noises, gun blasts and sirens made certain that any thoughts the wearer might have would "flee in panic". The physically strong and able-bodied had to wear weights around their necks. Those endowed with pleasant features were required to wear hiddeous masks, while sash weights and bags of birdshot ensured that ballerinas displayed no free and graceful movements and were "no better than anybody else would have been" (Vonnegut, 1968, pp. 8–10).

If disabling or handicapping those who have certain abilities is one way of trying to remove social disadvantages in those who lack them, it is not a plausible one. The reason is not only that we value excellence in art, in intellectual endeavours, and in many of the other pursuits that give special meaning and value to human life, but also that it would involve a severe curtailing of options and opportunities for many. A much more plausible way of trying to remove or ameliorate disadvantages and inequalities is by enabling—through education and training programs, through the provision of access and various aids—those who have disabilities. This is, of course, the contemporary approach; and while there is, as I noted above, little doubt that more ought to be done to remove what constitutes unjust discrimination against people with disabilities, this is not to endorse the view—arguably entailed by the claim that disability is but a social construct—that social institutions and modes of cooperation must be such that all people with disabilities have equal opportunity to participate in social life.

In her reflections on the view that disability is but a "socially imposed restriction" that could and should be overcome by different social arrangements, one visually impaired writer in the field, Sally French, sceptically asks:

> How helpful or practical, then, are ideas of social and physical adjustment when we look beyond the simple examples of adapted buildings, braille production and large-print books? Even if it were possible to transform the world to eliminate the disabilities of a small minority of people, would there not be a danger of disabling the rest of the population, including many of those with similar impairments? (French, 1996, p. 21).

In other words, if we established a world that admirably meets the needs not only of wheelchair users, as the fictional world created by Vic Finkelstein did, but also those of the visually impaired or blind, those of the deaf, and those of the mildly as well as the profoundly intellectually impaired, we would be creating other groups of people who would inevitably find themselves disabled. As has been convincingly argued by Dan Wikler (1983) and Allen Buchanan (1996), some form of disablement cannot, ultimately, be avoided. The reason is that choosing the "the dominant cooperative scheme" (Buchanan, 1996)—also entails choosing who will and who will not be disabled.

Do existing cooperative schemes unjustly discriminate against disabled people? The answer is not clear. What is being asked here is not simply whether the provision of supportive services, such as education and training and physical aids such as ramps for wheelchairs and large print books are adequate or not, but rather whether present social institutions and arrangements (the rules and frameworks that determine who can and who cannot participate in the "social game") are unjust insofar as they exclude some people with disabilities from participation. The question is particularly perplexing because it presupposes not only that we are able to appeal to a satisfactory theory of justice, but also—and more perplexingly—to a theory of value that will justify the choice of the cooperative scheme itself. And such a theory is, to the best of my knowledge, still outstanding.

Fortunately none is required in the present context. The question before us is not what our proper social response to already existing people with disabilities ought to be; rather it is whether the *prevention* of genetic impairment, through prenatal testing and screening, unjustly discriminates against disabled people. Discriminating unjustly against disabled people once they exist may be one thing; preventing the existence of genetic disability in those that will exist may be quite another.[10]

4. PREVENTION OF GENETIC DISABILITIES WILL WORSEN THE PLIGHT OF DISABLED PEOPLE

Another prominent argument against attempts to prevent genetic disabilities rests on the claim that such endeavours are likely to worsen the plight of disabled people (Houghton, 1994). Allen Buchanan has called this argument "the loss of support argument"(Buchanan, 1996, pp. 21–24).

This argument—a kind of "slippery slope argument"—is based on the empirical claim that genetically handicapped people will be worse off if there is a reduction in their numbers, and on the moral claim that attempts to reduce the existence of genetically based handicaps are wrong because they constitute unjust discrimination.

But is there a necessary relationship between the number of disabled people who need social support, and the level of support that is likely to be available? It seems the answer is no. While it may well be the case that larger numbers of people who share common interests can exert increased political pressure on governments and service providers, they will very often also be faced by very real fiscal constraints. The fact, for example, that many Western nations have a rapidly increasing aged population does not mean that more and better social services will be available to all aged people. On the contrary, experience from Australia suggests the reverse. There are long waiting lists for hip replacement operations and similar age-related services, and newly introduced government policies seek to recoup nursing home costs from an increasing population of the aged through the levelling of up-front fees and bonds.

In this connection it is also important to note that prenatal diagnosis for some conditions, such as Down's syndrome and spina bifida, has been available for some time and that abortion laws, in countries such as Australia and the United Kingdom, allow abortions on the ground of fetal abnormality. While the availability of prenatal diagnosis has resulted in fewer children being born with these conditions, there is no evidence that the reduction in the number of people living with the relevant conditions has resulted in a loss of support. Rather, parallel with the employment of prenatal diagnosis, countries such as Australia have witnessed an expanding recognition of the interests and rights of disabled people and, more generally, an increasing tolerance towards, and respect for, various minorities.

These anecdotal examples are not, of course, conclusive, but they suffice to show that the relationship between the number of people who have particular interests and needs and the level of support that is available to cater for those needs is merely a contingent one. Many factors, other than numbers of interest holders, would also seem to play a role.

But let us assume that the "loss of support" argument is correct: that there would, in fact, be some loss of support. What would follow from this? It would *not* follow that our attempts to prevent genetic handicap are discriminatory and unjust. As Allen Buchanan correctly notes, "the problem with the 'lack of support' argument is that it considers only the interests of those who will have disabilities after a reduction in the incidence of the disabilities occurs. It entirely overlooks the interests that others have in not having disabilities" (Buchanan, 1996, p. 22).

Allen Buchanan develops the argument in the following way: in some cases, it may be possible to correct the incidence of genetic disabilities by correcting defects in the embryo, and by developing chemicals that mimic or counteract the products of faulty genes. In these cases, genetic science would be employed for a clearly therapeutic end: to make the lives of individuals better by preventing genetically based diseases.[11] It is, however, important to note that genetic science is only one way in which the incidence of disabilities can be reduced. There are many other non-genetic ways as well. Buchanan refers to treating babies' eyes at birth to prevent blindness from contact with gonococcus bacteria. Another example would be the prevention of mental retardation from phenylketonuria by way of a special diet during infancy. Given that individuals have a legitimate interest in avoiding genetic and other damages that would occur if we did not intervene, we have a moral obligation to prevent such damages when we can do so safely and efficiently. This means that the "loss of support" argument ought to be rejected because it takes only *some* of the legitimate interests into account—the interests of those who will have a genetic disability after the reduction in the disability occurs. Far from establishing that genetic interventions are discriminatory and unjust, disability rights advocates who advance the "loss of support argument" may, ironically, be making the same mistake they attribute to proponents of the "new genetics": "advocating practices that are unjust because they are discriminatory and exclusionary" (Buchanan, 1996, pp. 22–24).

It seems to me that Allen Buchanan is correct: the "loss of support" argument cannot legitimately be marshalled against all interventions that prevent the existence of genetic handicap. But should the charge that the new genetics is discriminatory and exclusionary perhaps be understood in a somewhat different way? As being directed not against *all* interventions that remove or ameliorate impairments, but only against those that prevent genetic disability by preventing the existence of disabled individuals?[12] It is true, disability rights advocates sometimes write as if they were opposed to all interventions that reduce the incidence of genetic disability, but this interpretation is inconsistent with another view widely supported in the literature: that the alleviation of disability is, other things being equal, a good thing. As one disability rights advocate puts it:

> [V]ery many disabled people would welcome physical interventions which *guarantee* elimination of an impairment. This is surely demonstrated by the continuing attraction of rehabilitation programmes to return function; support for research into modifying multiple sclerosis, epilepsy or spinal injury, etc.; the frequency of corrective surgery (such as the removal of cataracts) and use of equipment to approximate normal behaviour (such as hearing and walking aids). Even disability organisations sceptical about experiments to make disabled people "normal" do not campaign against the prospect of eliminating impairment (Finkelstein, 1996, p. 10).

What many disability rights advocates are centrally concerned about when they allege discrimination is thus not a reduction in the incidence of disability *per se*; rather, they are concerned about the fact that such disabilities are prevented by having recourse to abortion. Prenatal testing and screening procedures, it is said, do not have a therapeutic intent; rather, in most cases the detection of a genetic abnormality in a zygote, embryo or fetus, will, as Christopher Newell puts it, be premised on the norm: "that such an entity will be flushed or terminated. Hence the detection is for the reason of destruction rather than for the sake of therapy—except that according to the ideology such destruction becomes therapeutic" (Newell, 1997, p. 4).

There are a number of ways in which the claim that it is the *destruction* of genetically abnormal embryos that involves discrimination against disabled people can be understood. I shall focus on two such interpretations. The first one holds that the destruction of an embryo or foetus discriminates against those who, if not aborted, could have lived satisfying lives; the second one holds that prenatal diagnosis discriminates against people living with disabilities. Both views are articulated by a young Australian journalist, Helen Houghton, when she writes:

> I was born 29 years ago with spina bifida myelomeningocle. As a consequence of this congenital condition, I have side effects such as hydrocephalus; a paralysed soft palate; paraplegia—I wear callipers and crutches to walk, or use a wheelchair; [I have] double incontinence of bowel and bladder; and more recently I have also experienced slight difficulty in grasping objects.
>
> If I had been born in the last couple of years and not 29 years ago, my parents would have been able to gain far more knowledge before my birth (spina bifida is determined at around 3 weeks after conception and can be detected in prenatal tests)... Realistically, I have to accept that my parents may have taken the decision to terminate their pregnancy. This is a pretty frightening thought from where I sit today, and has caused me to question issues such as the nature of discrimination and disability...
>
> If prenatal diagnosis is used as a tool to eradicate as many disabilities as possible in society, then it does discriminate against people with disabilities: [against] those who are already living with disabilities and [against] those potential babies who may have lived very successfully with disabilities (Houghton, 1994, p. 98).

I shall begin by taking a look at Helen Houghton's last claim—that prenatal testing discriminates against "those potential babies who may have lived very successfully with disabilities".

5. PARENTAL DISCRIMINATION?

Reflecting on her own life, Helen Houghton writes:

> What has it been like for a person born with severe disability? Has it been worthwhile, despite the pain (27 operations), discrimination and negative community attitudes during much of my 29 years? I believe it has been worthwhile for me, but that is not to say that it has been an easy life...Prenatal diagnosis and the termination of pregnancy on the basis of an identifiable disability does discriminate, because there has to be a decision about which disabilities to terminate, and a decision has to be made about what quality of life actually means...Doctors may have expertise on the medical side of disability, but they can't really know what a person's quality of life will amount to... People with disabilities should be given the chance to succeed or fail on their own terms (Houghton, 1994, pp. 94–101).

Helen Houghton's account makes it clear that she is glad to be alive and that, more generally, a life even with severe disabilities can be worth living. And while there may well be some disabilities that are incompatible with a subjectively worthwhile life—so-called "wrongful life" cases—the vast majority of genetic impairments that are or will be subject to genetic testing and screening will not be for such conditions. Rather, they involve conditions that are compatible with a life worth living.

This suggests that the claim that prenatal genetic testing and screening involves unjust discrimination against those disabled babies who might have lived worthwhile lives has some plausibility. The reason lies in the fact that parents who make use of prenatal diagnosis and abortion will generally choose to have another later child which does not have the impairment in question. In other words, the abortion of the disabled child involves an implicit choice between two possible children—one disabled, the other not. Choosing the "normal" child over the disabled child, who would have had a life worth living, it might be said, is discriminatory and unjust because it rests on the erroneous assumption often held by able-bodied people that the lives of disabled people are less worthwhile than the lives of "normal" people.

While the idea of judging the value of people's lives and of "replacing" one possible person by another, on account of that judgment, is intuitively deeply repugnant to many people, judgments rather similar to those involved in aborting an impaired fetus are also involved in other reproductive decisions aimed at avoiding the birth of a disabled child. They are rather straightforwardly involved in making use of donor gametes and embryos, in at least some forms of gene therapy, in the decision to remain childless[13], and can be involved in the timing of one's pregnancy.

A well-known argument by Derek Parfit will illustrate the last point. Suppose, Parfit says, that a woman is planning to stop taking contraceptive pills so that she can have a child. Before doing so she is told by her doctor that, because of a temporary medical condition, any child she conceives now will be disabled. If the woman will wait three months, the doctor says, the condition will pass and any child she conceives then would be normal.

Many of us would, I think, take the view that in these circumstances the woman ought to postpone her plans to become pregnant. We would think this even if the disability the child would have would be compatible with a worthwhile life. We would think this even if the disability the child would have is not so terrible as to make the child's life one of unredeemed misery. Suppose that the child would be unable to walk without callipers, but would have no other disabilities. A life with limited mobility with no other disabilities can still be a life very definitely worth living. Yet there seems to be wide agreement that even if an impairment were of this relatively mild kind, the woman ought to wait three months before she conceives.

Behind this response may lie the thought that if the woman were to decide against waiting the three months, she would be needlessly causing her child to have a disability; but the point to notice about Parfit's example is that if the woman waits before becoming pregnant, the child she will then conceive is a *different* child from the one she would have had if she had not postponed becoming pregnant.

This means that the woman could not be said to have harmed the child if she had refused to postpone her plans to become pregnant. The child conceived without delay would still have a life which would be worth living; and if the woman would not have become pregnant at that time, this particular child would not have existed at all. So if we want to say that the woman has acted wrongly we cannot do this by claiming that she has been harming "her child"—because this claim cannot be true of any particular child which the woman might have. If she does not wait, there will be no child who is harmed by her decision.

It is significant that spelling this out does not lead people to modify their judgment that the woman should wait before becoming pregnant. Since life with this kind of disability would still be worthwhile, our judgment cannot be based on the interests of the child who will be born only if the woman does not wait. So if we still judge that the woman should wait, our judgment must be influenced by the fact that we are taking into account the future prospects of "the next child" who will be born only if the woman does not wait.

Does the taking into account of alternative children's prospects necessarily involve the judgment that the lives of persons with disability are less valuable than the lives of non-disabled persons, or that "certain forms of life" should not exist? (Newell, 1997, p. 1). I think not. A parent's decision to avoid the birth of a disabled child and to have a non-disabled one instead, would more typically be motivated by the desire not to handicap one's child, thereby by curtailing his or her future options and opportunities. It might also be motivated by the desire to avoid unduly straining one's marriage, or burdening one's family, given that the raising of a disabled child will often require considerable emotional, physical and financial resources. Given this, one may consistently hold the belief that it is better, all things considered, that one avoid the birth of a significantly impaired child, while also holding that the lives of all *persons*—disabled or not—have the same worth. The two beliefs are rendered consistent by the view that possible people—fetuses and the unconceived—are not persons and that we do not have a moral duty to bring possible people into existence, even if their lives would be subjectively worthwhile. On this view, the bringing-into-existence of people who would be able to lead satisfying lives is seen as morally neutral, that is, there would be no moral requirement to bring any child, whether disabled or not, into the world, and there would equally be no moral grounds that speak against parents bringing such a child into existence.[13] Parents who prevent the existence of a possible disabled child in favour of a non-disabled one would, on this view, be discriminating between having a disabled and a non-disabled child, but their doing so would not be unfair or unjust. Notions of fairness and justice apply to persons, not to merely possible persons.

If this is correct, then Helen Houghton's charge that reproductive choices aimed at preventing the birth of disabled children unjustly discriminate against those disabled people who might have existed and lived worthwhile lives cannot stand. But what about Helen Houghton's second objection—the claim that prenatal testing and related technologies unjustly discriminate against people already living with disabilities? One way of understanding this objection would be to focus on the possibly negative effects a reduction in the number of disabled people would have on disabled people—on those already living with disabilities and also on people who might in future be born with disabilities. I have already suggested that we should dismiss this objection on the grounds that it fails to give adequate consideration to the interests others have in *not* having disabilities. There is, however, a second and more subtle strand to this objection as well. It relates to the implicit message genetic technologies aimed at the prevention of handicaps are said to be sending to disabled people. While philosophers, like myself, may be able to defend on conceptual grounds the view that the goal of the new genetics is universalistic and does not unjustly discriminate against people with disabilities, many people with disabilities are interpreting the message differently. They see the new genetics not so much aimed at the elimination *disability* as at the elimination of disabled *people.* This is felt most strongly in the context of prenatal diagnosis that is followed by abortion. Here, it is said, we quite clearly have a an "intervention premised upon destruction [with] various forms of life being regarded as better off dead" (Newelll, 1997, p. 1).

I have argued that this view is *not* entailed by decisions to prevent the existence of impairment in the children one is going to have, even if this involves abortion. Many people with disabilities will undoubtedly disagree. One reason may be that not everyone will share with me the view that abortion is, from the future child's point of view, on a par with non-conception; another reason is that irrespective of one's views on abortion, the fact remains that the new genetics is, contrary what is often said, not value-neutral. It is, if my arguments are correct, universalist and does not unjustly discriminate against disabled people. It is also non-coercive insofar as the choice to bring a disabled child into the world rests with the parents, not with the state, as it sometimes did with the "old genetics". But it is not value neutral. It clearly suggests that it is better, other things being equal, that people not have impairments and offers various means by which parents can avoid the birth of a disabled infant. While the goal of preventing disabilities is undoubtedly good—and no different from that inherent in traditional medicine—the means are different. Traditional medicine seeks to prevent disability by preventing disabling diseases, the new genetics seeks to prevent disability by preventing the existence of disabled people. In seeking the reduction of genetic disabilities in this way, the new genetics is in fact saying that the world should not contain so many people with disabilities. Despite all the philosophical protestations to the contrary, many disabled people are therefore likely to remain steadfast in their belief that the thrust of the new genetics implies that *they* should not exist (Holland, 1998, p. 1; Buchanan, 1996, p. 32; Kaplan, 1993, p. 610).

CONCLUSION

The disability rights literature leaves little doubt that many people with disabilities feel deeply threatened by the new genetics and by the possibility that the employment of these technologies will further reinforce existing negative attitudes to disability. No doubt born out of these fears and concerns, disability rights advocates urge that rather than focus our energies and resources on the discriminatory and oppressive technologies associated with prenatal diagnosis, genetic testing and screening, we should be placing increased emphasis on the removal of socially constructed disabilities, as well as on the removal of negative social attitudes towards people with disabilities. Very often it is these attitudes—the stigma attached to disability—that imposes greater limits on the lives of people with disabilities than their mental or physical impairments.

This would be sound advice, where it not for the fact that much would be lost if the new genetics were abandoned. Rather than abandon our efforts to prevent disability, we ought to be working towards two goals. Continue our endeavours to prevent genetically based handicap, to help individuals live their lives free of serious impairments; and seek to remove social stigma and unjust discrimination against people living with disabilities. The second goal has more far-reaching dimensions than is often recognized. Despite the increasing capabilities of the new genetics there will always be people who have various impairments. This means that taking the question of justice seriously would ultimately also mean addressing the vexed issue of what kinds of dominant social frameworks and structures ought to facilitate and regulate the living together of disabled and non-disabled people.

NOTES

1. My paper owes much to Allen Buchanan's excellent paper 'Choosing who will be disabled: genetic intervention and the morality of inclusion'; particularly to his reflections on the social construction of disability.
2. The disabilities literature is vast and there is no single disabilities rights position. The cluster of related views I am referring to is increasingly articulated in the critical disabilities literature. A sample is found in the references to this paper.

3. For some of the differing positions within "social constructivism" see, for example, Collin, 1997; Goffman, 1975; Söeder, 1989; Flucher, 1989. The view that the social construction of disability constitutes a form of oppression is widely held in the critical disabilities studies literature. See, for example, Abberley, 1987, Newell, 1997.
4. The distinction between "impairment", "disability" and "handicap" is fraught with problems. I have seen no entirely satisfactory account as to how the distinctions ought to be drawn, and will not attempt to defend one here. Hence, I am not claiming consistent use of these terms—although I will generally use the term "impairment" to refer to loss or malfunction at a merely physiological, anatomical or psychological level. For some recent discussions of this issue, see, for example: World Health Organization's *International Classification of Impairments, Disabilities and Handicaps* (1993); Edwards (1997); Nordenfelt (1997).
5. For somewhat similar formulations, see e.g., Brock (1995); Buchanan (1996); Davis (1997); Harris (1993).
6. According to Nora Ellen Groce's historical account, hereditary deafness did not, given certain social arrangements, constitute a disability on the Martha's Vinyard Island (Groce, 1985). See also Wikler (1983) and Buchanan (1996).
7. The term "relational character of disability" is Allen Buchanan's. See Buchanan, 1996, pp. 38–39.
8. See, for example, Hoyningen-Suess (1987). [I owe this reference to Anton Leist (1997, p. 19).
9. Ruth Macklin drew my attention to this story, when I was reading this paper at the University of Central Lancashire.
10. Of course, the problem of answering here the question of the kind of dominant cooperative scheme we should adopt can be avoided only if we accept that genetic disability is not merely a social construct, but has an objective basis in physiological impairment—and that impairments are undesirable because they limit socially independent options and opportunities. If we accept this, then we should also accept that it would, other things being equal, be better to prevent or cure impairments (thereby giving individuals the opportunity to participate equally in complex and rich cooperative schemes) rather than devise less complex and less rewarding cooperative schemes that will allow individuals with a wide range of disabilities equal social participation. We have long accepted this conclusion with regard to the prevention and treatment of disease and disability in general, and convincing reasons would have to be found why we should depart from it when the prevention of genetic disability is at issue.
11. Particularly with regard to germ-line gene therapy interesting questions of identity arise. Is this in fact a form of therapy, or is one individual destroyed to create another? For some of these discussions, see, for example, Zohar, 1991; Kahn, 1991; and Elliott, 1993.
12. This could, depending on the conceptual answers we give, also include genetic engineering. See Note (11) above.
13. While the decision to remain childless does not involve choosing a non-disabled child over a disabled one, it nonetheless involves the decision to remain childless *because* the child one could have would be disabled. In other words, one would have a child were it not for the fact that there was a substantive risk, or near certainty, that the child would be disabled.
14. For a defence of this kind of view, see, for example, Brock, 1995; Tooley, 1983.

REFERENCES

Abberley, P. (1987), 'The Concept of Oppression and the Development of a Social Theory of Disability', *Disability, Handicap and Society*, Vol. 2, pp. 5–20.

Brock, Dan (1995), 'The Non-Identity Problem and Genetic Harms—The Case of Wrongful Handicaps', *Bioethics*, Vol. 9, 1995, pp. 269–275.

Buck v. Bell (1927) [Supreme Court] Reports 274, pp. 1000–1002.

Buchanan, Allen (1996), 'Choosing who will be disabled: genetic intervention and the morality of inclusion', *Social Philosophy and Policy*, Vol. 13, pp. 18–46.

Collin, F. (1997), *Social Reality*, London, Routledge.

Crouch, Ronald A. (1997), 'Letting the deaf be deaf—Reconsidering the use of cochlear implants in prelingually deaf children', *Hastings Center Report*, Vol. 27, No. 4, pp. 14–21.

Davis, A. (1989), *From Where I Sit*, London: Triangle, 1989.

Davis, D. S. (1997), 'Genetic Dilemmas and The Child's Right to an Open Future', *Hastings Center Report*, Vol. 27, March/April, pp. 7–15.

Edwards, S.D. (1997), 'Dismantling the Disability/Handicap Distinction, *Journal of Medicine and Philosophy*, Vol. 22, pp. 589–606.

Elliott, R. (1993), 'Identity and the Ethics of Gene Therapy', *Bioethics*, Vol. 7, pp. 29–40.

Finkelstein, Vic (1981), 'To deny or not to deny disability', in Brechin, A., Liddiard, P. and Swain, J. (eds.), *Handicap in a Social World*, Sevenoaks: Hoddder & Stoughton", pp. 34–36.
Finkelstein, Vic (1996), 'The commonality of disability', in Swain, J., Finkelstein, V., French, S., and Oliver, M. (eds.), *Disabling barriers—enabling environments*, London: Sage Publications, 1996.
Flucher, G. (1989), *Disabling Policies?*, London, Falmer Press, 1989.
French, Sally (1996), 'Disability, Impairment or something in between?', in Swain, J., Finkelstein, V., French S., and Oliver, M. (eds.), *Disabling Barriers—Enabling Environments*, London: Sage Publications, pp. 17–25.
Goffman, E. (1975), *Stigma: Notes on the Management of Spoilt Identity*, New Jersey, Mosby & Co.
Groce, Nora Ellen (1985), *Everyone Here Spoke Sign Language*, Cambridge, Mass.: Harvard University Press.
Grundgesetz der Bundesrepublic Deutschland, Bonn, Bundeszentrale für politische Bildung.
Harris, John (1993), 'Is Gene Therapy a Form of Eugenics?', *Bioethics*, Vol. 7, pp. 178–180.
Holland, A. (1997), 'Genetically Based Handicap', unpublished paper.
Houghton, H. (1994), 'Does Prenatal Diagnosis Discriminate Against the Disabled?', *Ethical Issues in Prenatal Diagnosis and the Termination of Pregnancy*, McKie, J. (ed.), Melbourne: Centre for Human Bioethics, 97–102.
Hoyningen-Suess, U. (1987), 'Behinderung auf den Begriff bringen: Moeglichkeiten und Grenzen', *Vierteljahresschrift fuer Heilpädagogik und ihre Nachbargebiete*, Vol. 56, pp. 459–464.
Kahn, J.P. (1991), 'Genetic Harm: Bitten by the Body that Keeps you?', *Bioethics*, Vol. 5, pp. 289–308.
Kaplan, Deborah (1993), 'Prenatal Screening and Its Impact on Person's with Disabilities', *Clinical Obstetrics and Gynecology*, Vol. 36, pp. 605–612.
Kelley, J. and Bean, C. (1988), *Australian Attitudes*, Sydney: Allen and Unwin.
Leist, A. (1997) 'Dimension einer Ethik der Behinderungspaedagogik' in Bonfranchi, R. (ed.), *Zwischen allen Stühlen–Die Kontroverse zu Ethik und Behinderung*, Erlangen: Harald Fischer Verlag.
Newell, C. (1997), "Critical Reflections on Disability, Difference and Genetic Testing", unpublished paper presented to the "Genetic Testing, Ethics and Public Health" Seminar of the Public Health Association of Australia NSW Branch at Royal North Shore Hospital, Sydney, 15 August, 1997.
Nordenfelt, L. (1997), 'The Importance of a Disability/Handicap Distinction', *Journal of Medicine and Philosophy*, Vol. 22, pp. 607–646.
Rowe, P. (1994), 'Jeans for Genes: An Event for the Children's Medical Research Institute', a circular letter to Australian Politicians dated 8th June 1994, by the Lorimer Dods Professor and Director of the Children's Medical Research Institute, New South Wales, as quoted by Newell, 1997.
Söeder, M. (1989), 'Disability as a social construct: the labelling approach revisited', *European Journal of Special Needs Education*, Vol. 4, pp. 117–129.
Tooley, M. (1983), *Abortion and Infanticide*, Oxford: Clarendon Press.
Vonnegut, Jr., K. (1968), 'Harrison Bergeron' in Vonnegut, Jr., K., *Welcome to the Monkey House—A Collection of Short Stories*, USA: Delacorte Press, pp. 7–13.
Wells, H.G. (1911), *The Country of the Blind and Other Stories*, London: T. Nelson & Sons, pp. 535–68.
Wikler, D. (1983), 'Paternalism and the Mildly Retarded', in Sartorius, R.E. (ed.), *Paternalism and the Mildly Retarded*, Minneapolis: University of Minnesota Press, pp. 83–94.
World Health Organization (1993), *International Classification of Impairments, Disabilities and Handicaps*, Geneva: WHO.
Zohar, N. (1991), 'Prospects for "Genetic Therapy"—Can a Person Benefit from Being Altered?', *Bioethics*, Vol. 5, pp. 275–288.

3

PRIVATE PARTIES, PUBLIC DUTIES?

The Shifting Role of Insurance Companies in the Genetics Era

Trudo Lemmens

Clarke Institute of Psychiatry
University of Toronto Joint Centre for Bioethics
88 College Street, Toronto, Ontario, M5G 1L4, Canada

What is the nature of insurance and how should it be distributed? Do private insurers have public duties in this respect, and if so: to what extent? This paper addresses these questions in the context of the debate on the use of genetic information by insurance companies in countries with a universal health care system. I will develop an argument based on Michael Walzer's theory of justice. The attention to the cultural determinants of justice in his approach are particularly relevant in light of what I qualify as the "shifting role of insurance companies in the genetic era." It should become clear that I do not argue that the role of insurance is necessarily changing *as a result of* the developments in genetics. Rather, genetic developments have to be placed in the context of systematic reductions in health care and welfare budgets in many states which claim to have a generous system in that respect. The increasing private involvement in health care and, in general, the essential role that several forms of insurance start to play as essential means to provide protection against misfortune, might offer sufficient grounds to argue that insurance companies are now playing a different role in society. Developments in genetic testing simply add another dimension to this societal change. I will argue that genetic developments could enable an increased individual differentiation on the basis of risk-status. As a result, issues of equitable access to health care and social services will be highlighted.

Until now, genetic testing has not been systematically used by insurers, but that could change soon. The development of a genetic biochip that would allow one-time testing for a multitude of genetic factors, brings genetic testing closer to all of us. It is becoming a standard diagnostic tool. Few of us will remain untouched by it. In light of this, some argue that the "threat" of genetics will disappear, as all of us will be susceptible to some risk. Others on the contrary, warn of the development of systematic discrimination of the "asymptomatically ill" and of the creation of a new social class, a "genetic proletariat," a class of people who are excluded from employment and can no longer obtain insurance (Dreyfuss and Nelkin, 1992, p.318; Billings *et al.*, 1992).

Genetic Information, edited by Thompson and Chadwick.
Kluwer Academic / Plenum Publishing, New York, 1999.

It is in this alarmist context that we should place the writings of Paul Billings, Joseph Alper and others, who have analysed over the last decade occurrences of genetic discrimination (Alper *et al.*, 1994; Billings *et al.*, 1992; Geller *et al.,*1996; Natowicz, Alper and Alper,1992). These studies have indicated how socio-economic actors such as insurers and employers have already excluded people on the basis of their genetic constitution. However, they do not confirm that genetic tests have been systematically used and that *many* people have been excluded on the basis of genetic test results. The most valuable aspect of these studies is that they are an indicator of the fact that genetic information is often misunderstood and misinterpreted and then wrongly applied by individual insurers or employers and also show that insurers and employers might have particular interests in using this information. This lack of knowledge sometimes seems extreme: A person with a family history of Huntington informed me that he received a letter from an insurer telling him that he could not offer life insurance at this stage. The insurer encouraged him, though, to apply again "should his health improve."

"Informed and learned insurers" will argue that these are isolated cases and that insurers have every interest in selling policies and in offering reasonable insurance contracts. Lowden, a Canadian insurance expert states as follows: "We need to sell insurance to stay in business. We do not seek to deny applicants but to assess risk and to place business" (1996, p.436). According to him, insurers have a major interest in determining the level of risk-increase realistically; and to take into consideration any reduction of risk by the development of preventive measures. In the context of breast cancer, this could mean that insurers would offer women who test positive for BRCA 1&2 insurance contracts that are perhaps more expensive than for other women of the same age group, but roughly equal to premiums for men of the same age. It also means that if preventive strategies become available—and so far these remain very controversial in the context of breast cancer—these will reduce the risk and thus could bring down the insurance premium.

Does this mean that genetics will not change anything fundamental, as some insurers suggest, and that it will be "business as usual"? It is difficult to predict what will happen precisely. It is credible, that insurers will use information rationally, that applicants will be able to use genetic information often to their advantage and that premiums will not substantially increase. But I also believe that genetics challenges the business, particularly in the context of changes in the health care system, such as those that are taking place in Canada as well in some European countries.

Genetics will challenge insurance because of a combination of factors, many of which are not unique. Most of these factors, e.g. the family and ethnic relevance of the information, the fact that genetic mutations most often indicate a risk-factor and no actual disease, the lack of control over genetic predisposition, and so on, have been discussed extensively elsewhere (Sandberg, 1995; Lemmens and Bahamin, 1996). It is not relevant for my argument here to discuss why these aspects are not unique to genetics. The most significant difference is that genetics will enable insurers to establish more detailed individual risk assessments than ever before. Insurance pools will be reduced and more detailed underwriting through genetic tests will become cheaper and thereby financially interesting. It is likely to become an ordinary underwriting tool.

According to those who hold a more pessimistic vision of the use of genetic information by third parties, these developments will stimulate discriminatory practices. In reaction to this, some argue that human rights laws, providing a special clause for the protection of people with a genetic susceptibility, would solve the problem. Several European countries already prohibit the use of genetic testing for insurance purposes (Lemmens and Bahamin, 1996) and similar legislation—be it mainly in the context of health insurance—exists within the US (Pear, 1997; Davis and Mitrius, 1996; Rothenberg 1995).

But there are problems with this approach. Susan Wolf, for example, claims that this policy suffers from several shortcomings (Wolf, 1995). Her most important criticism deals with the fact that selective protective legislation for genetic susceptibility does not deal with the underlying problem of "geneticism", which she describes as the current "eagerness to draw genetic conclusions, the search for supposedly deviant genes" (1995, p.347). She reacts against this tendency to create genetic categories and dividing people on the basis of genetic factors. Drawing from the literature on race and gender, she argues that transforming a genetic mutation, carrier status, or a general increased susceptibility into a deviation that deserves protection, contributes to the establishment of a standard of "normality." It creates a difference between the genetically "us" and "them."

Importantly, she also indicates that anti-discrimination provisions cannot be effective. Indeed, prohibiting genetic testing for insurance purposes -unless carefully framed- does not deprive insurers of the possibility to use genetic information from family histories, information that insurers always obtained. It is further worth mentioning that it will simply become impossible to distinguish "genetic information" from other information in medical files. Genetic testing is being integrated in clinical care, and it is inconceivable how one can keep it separate from other medically relevant information. Keeping important diagnostic information outside medical files might also put people at serious risk and thus be unacceptable for medical reasons.

Moreover, it seems arbitrary and unjust to prohibit the use of genetic data while allowing insurers to increase premiums or to exclude people on the basis of other health data. When a prohibition on the use of genetic testing in insurance was discussed in the Belgian Parliament, a senator expressed the view that it made an inequitable distinction between genetic and other diseases. Indeed, why would people be protected only as soon as, and to the extent that their susceptibility to disease is declared as having a genetic origin or component? Protection would then depend on the level of research into specific diseases. As soon as a genetic component would be found, protection would be granted. This seems arbitrary and could have surprising results. What if dependence on nicotine or alcohol really is genetically determined? Could someone addicted to smoking argue that it is a genetic disease and challenge the traditional higher insurance premiums for smokers?

This indicates that it might be impossible to deal with the social problems resulting from the application of genetic technology with patchwork legislation. A general approach is necessary, and one that takes away the financial and social pressure on individuals who are affected by disease or by an increased risk for disease. The problem is that many people rely on insurance as this web of protection that will catch them when they have been deprived of all further support. The reduction in universal health care and welfare provisions are a major force that pushes them in this direction. With the increasing differentiation that genetics will bring us, this web of protection will become increasingly versatile, and increasingly different from person to person. It is not a web that promotes equity or equal protection against the negative results of disease or risk for disease.

The question I want to ask here is: should this push us to implement rules in insurance that might prove ineffective and inequitable to those who suffer from non-genetic diseases and that do not deal with the underlying problem of the more general societal impact of genetic testing? Wolf and others who argue for a wider approach seem to touch upon the idea that inequity on a broader social level cannot be dealt with adequately by trying to impose a partial equity in the domain of insurance.

In order to clarify this issue, I would like to develop an argument based on Michael Walzer's theory of justice along the following lines: 1. insurance policies, as all goods shared among people have a specific value only in a given social, cultural and economic context. They also stand in a relation to other goods and their value has to be measured in relation to these goods. 2. In most European countries and in Canada, insurance contracts are traditionally seen as goods sold on the marketplace, identifiable as commercial items and bought for private desires of certainty. 3. To the extent that these contracts are commodities, their transaction should fulfill specific marketplace criteria of fairness. Actuarial fairness then, i.e. giving people insurance on the basis of their different risk-status, might be an appropriate criterium of fairness for these insurance policies in that particular context. 4. As with the value of any other good, the value of insurance may change over time. 5. In the case of insurance, changes in social policies related to health and welfare have changed the role and value of insurance policies. 6. To the extent that our society decides that insurance companies should have a public redistributive role, the nature of insurance as a good changes and the rules of equity of the marketplace may no longer apply.

I end with an open question which should be further developed: can a private insurance system ever be submitted to stringent regulations and can we expect that the private sector redistributes appropriately goods that belong in the sphere of health and welfare? Is there a way to ensure the appropriate distribution of social goods by private parties?

In his influential "Spheres of Justice", Walzer develops a theory of justice based on an idea of complex equality (1983). Equality, he argues, is not achieved by an equal distribution of all goods, an idea that he qualifies as "simple equality". A just society is not obtained through offering people an equal number of things. That, he argues, has shown to be ineffective. Equality can only be approached, but probably never completely realized, by freeing society from domination. A just society is one in which domination has been eliminated or seriously limited. We have a situation of domination, Walzer argues, when the possession of one good (e.g. money, or the status of nobility) allows its possessors to control and command all other goods (e.g. political power). This situation is avoided when we limit the power that goes with the possession of specific goods. In order to do so, he suggests a theory in which the existence of different spheres of justice is central. The theory is based on a differentiation of several types of goods, carrying different meanings in a specific cultural and social context. He further argues that because different goods have different meanings and different functions in a given society, different principles of justice apply to them.

I cannot enter into much detail here, but it is important to mention some of the ideas about goods that Walzer enumerates in his book. First, distributive justice is concerned with social goods. Goods obtain meaning because they are shared, because they have a particular function in a given society. Goods are always *social* in nature. They are always the object of exchange, both physically and symbolically. Second, people have a particular relation to goods. How goods are perceived and exchanged in a particular culture determines who these people are and how they interact. Third, goods have a particular meaning dependent on the culture in which they circulate. "There is no single set of primary or basic goods conceivable across all moral and material worlds." (1983, p.8) Walzer argues, e.g. that even necessary goods such as food carry different meanings. Food can be a basic life-provision, but it can also obtain a religious status. A cow can be a mere "milk producer" or a "big chunk of meat" or it can have divine-like status, dependent on the culture in which it lives. Fourth, it is the meaning of goods that determines how they are to be distributed. Some goods are outside any commercial distributive sphere in one culture, and clear objects of commercial transaction in other cultures. Distributive criteria and arrangements are intrinsic not to the "good-in-itself" but to the social good, to the good as it obtains its meaning in a particular culture. Fifth, even though Walzer will recognize that some goods have "characteristic normative structures, reiterated across the lines of time and space" (1983, p.9), the status and meaning of goods may change. "Social meanings are historical; and so distributions, and just and unjust distributions, change over time." Sixth, based on the fact that the meanings of goods are distinct, Walzer argues that their distributions must be autonomous and respond to different criteria of justice.

Thus, every set of goods constitutes a distributive sphere within which only certain forms of exchange are acceptable. Justice is obtained when we respect the boundaries between these different spheres. Or, as Walzer states it succinctly "good fences make just societies" (1983, p.319). For example, the possession of money should not give more voting rights and political office should not give the right to confiscate money and property without justification. Within these different spheres of justice, different distributive principles apply. Walzer distinguishes three principles: free exchange, desert and need. The rule of Desert applies in the domain of public recognition, the sphere of awards and honours. This domain is not of much importance for our discussion here. The principle of free exchange applies to goods that we can qualify as Money and Commodities. The rules of the Marketplace are based on free exchange. Walzer thus recognizes that our liberal economic system has its reason of existence and follows specific distributive criteria that are appropriate for goods that we can qualify as commodities.

However, the same rules of free exchange should not apply, Walzer argues, in other domains. They should not offer control outside its sphere, or we fall into a situation of dominance, which is the real threat to equality. People should receive an honorary degree or the Noble Prize because they deserve it, not because they possess money or have political power. More importantly for our discussion, money should not dominate the sphere of Security and Welfare.

Walzer spends much time discussing this last sphere. He argues that any society recognizes a sphere of security and welfare, in which the distributive criterion is need: to give to everyone according to need. Provisions for the communal good can be found in every culture, they are universal in existence. But their shape is particular. Different societies focus on the communal provision of different things. Walzer gives the example of the Athens democracy. It had a sophisticated system of public provisions, comparable to our system. It organized publicly more traditional goods such as an army, courts, road repair and temples. But it also considered very particular things such as public baths and drama festivals as essential elements of public and religious life. There was, however, no public education and little social security, contrary to many current democracies. Walzer uses this and other examples to show that as societies develop over time and differ across territories, so do the meanings of the goods. In other words: what different societies include in the sphere of Security and Welfare might very well differ from country to country, but also from time to time. Some goods have gained social importance far beyond the scope they previously had. Walzer sketches an interesting example of temporal change with a brief history of the development of medicine. Medicine, for example, did not have the social role in a medieval society that it now plays. When eternal salvation was a core value, as was the case in the Middle Ages, the construction of churches and temples was more important than the search for a long and healthy life and medicine did not need to be socially provided. But medical care developed over time, in interrelation with a shift in our value system. Society invested in medicine's development, and it has become an essential part of the Sphere of Security and Welfare. Walzer stresses its importance in the following way: "deprivation [of medical care] is a double loss—to one's health and to one's social standing. Doctors and hospitals have become such massively important features of contemporary life that to be cut off from the help they provide is not only dangerous but also degrading" (1983, p.89). But the precise content of the category of medical care, and how it has to be provided, again, differs from place to place. The US, for example, clearly has not chosen the same road to guarantee access to medical care and social welfare as European countries and Canada have.

But their choice, as any choice, is subjected to adjustments and as in any democracy, subject to debate. It is through democratic decision making that a given community decides on the content of its social contract. The perception of what should be offered may change over time, and the development of genetics may be one reason to adjust the concrete, historic implementation of the US health care system. That, in fact, has been suggested by several US health care experts. They have argued that the increasing possibility of differentiation between people individually on the basis of genetic risk factors, risks to augment social inequality and to exclude many people from health care coverage (National Human Genome Research Institute, 1997; 'Genetic Testing and Insurance,' 1995; NIH-DOE Working Group, 1993; Anderson, 1992). It is ironic that we simultaneously experience in Canada and Europe a tendency towards at least a partial privatization of health care.

The problem in determining what belongs to a specific sphere, thus also in determining how it should be distributed, is not only that the social meaning of goods changes. Walzer points out that the spheres of justice are not hermetically closed. First, some goods that we value as social goods will be sold on the market place. Society can choose to use the market, in which free exchange is the key distributive factor, to distribute some parts of goods to which all should have access. But there is a clear limit to what the market should provide. In Walzer's words: "Needed goods are not commodities. Or, more precisely, they can be bought and sold *only insofar as they are available above and beyond whatever level of provision is fixed by democratic decision making (and only insofar as the buying and selling doesn't distort distributions below that level)*" (1983, p 90; my emphasis). This needs further reflection. It could be an argument in favour of universal health care, but one could also use it to claim a minimum of financial support by the government.

Second, even in the marketplace, there are some blocked exchanges, transactions on which the criterion of free exchange is not applicable. Some things are clearly not for sale, and in others, sale is limited out of a principle of redistribution. Limited blocks within the market are for Walzer essential, in that they should correct power imbalance. Exchanges out of desperation should be barred. Exchanges in the market should be free bargains, between more or less equals. This means that the inherent inequity in society should be tempered through state intervention. Politics comes in to reassure an equilibrium of the market. Walzer points out that by providing basic welfare, the state promotes equality of bargaining power. By doing so, it avoids that people bargain without any resources and possibility of survival without accepting a job. Minimum wage is another example Walzer gives to show how governments intervene in the market to promote basic standards from which one can negotiate according to market criteria.

Where does insurance fit into this scheme? How do we apply this to insurance? When do we have to interfere with insurance? If insurance is a mere commodity, a choice of people who choose through that channel some form of risk-reduction, it falls entirely within the sphere of the market. However, does it always remain completely within that sphere? What if it is the only way of guaranteeing that your family has a decent life in case you die? Or the only way to let your children study in the future? Or the only way to pay your funeral? Or an essential condition to buy a house? Should politics impose limits on it?

Anti-discrimination statutes, for example those in Canada, reflect the idea that insurance belongs to the sphere of commercial transactions and that the distributive principle of free exchange applies to this sphere. While the legislation creates some limits to what insurers can do, the essential power of insurance companies to distinguish according to health and risk to health is maintained. The limits on insurance practices can be expressed as some blocked transactions, based on a common understanding of what type of distinctions would be inappropriate, even in the domain of free exchange. Anti-discrimination laws try to provide a general framework of protection to enable individuals to participate as full citizens in activities that are valued in our society, to build out a valuable -or good-life (Lemmens, 1997, pp.66–67). The Canadian Human Rights Act, for example, wants "to give effect to the principle that every individual should have an equal opportunity with other individuals to make for himself or herself the life that he or she is able and wishes to have, consistent with his or her duties and obligations as a member of society, without being hindered in or prevented from doing so by discriminatory practices." (Section 2). By excluding most insurance practices from the scope of equality protection, there seems to be a recognition that -in part at least- insurance is not within the sphere of goods that should be provided to members of the community in accordance with needs, but rather should be left in the sphere of free exchange. It seems to belong to the domain of commerce, rather than security and welfare and is not seen as essential for the construction of a valuable life in the community.

Could this change? Throughout the limited discussion of Walzer's theory, we have seen that three elements are of key interest in the context of insurance: (1) boundaries between different spheres of justice are vulnerable to shifts in social meanings. What is a social good now, might have circulated in the sphere of money and commodities before (e.g. health care); (2) Within the sphere of money and commodities, there are some blocked exchanges. Governments do interfere to guarantee an equilibrium and to obtain a contractual equilibrium; and (3) Some of the goods belonging to the sphere of Security and Welfare can be submitted to market criteria (and thus to different rules of distributive justice) but only to the extent that they are sufficiently provided to the public on the basis of the criterion of need. It seems interesting to apply these ideas in the context of the shifting role of insurance companies in the genetics era. In the context of this paper, I can only touch upon some issues. While I believe Walzer's theory reveals some very relevant and useful elements for our discussion, different conclusions can be drawn from it. I only want to offer here some observations that should be the subject of further reflection.

Walzer stresses the importance of communal provisions in the sphere of security and welfare. He clearly supports the inclusion of any good in this sphere, as soon as it is somehow perceived to play an essential role in society, filling a need for individual well-being in a given societal context. He also stresses that in order to have equity in the sphere of money and commodities, you need to somehow create an equality of bargaining power. That means that people should not be desperate. This seems to call for some strong basic welfare provisions. In other words: private insurance schemes should not be the only means to secure the future of family members, the only means to pay a decent funeral service, the only means to obtain some form of financial security. The same is true for health care. Additional health insurance and drug insurance, functioning on the basis of a different distributive criterion, should not become the only means to having decent health care.

That being said, a question remains. Could it be that private insurance, as an additional privately organized safeguard against bad luck, has become a social need? That may be so, in particular if the safeguard of social security has been weakened. To the extent that we can reconcile this with the rules of insurance, limitations (or blocked exchanges) might be appropriate. This has already been done. We can think here of those countries in which a system of basic insurance has been introduced. The Dutch insurance industry, for example, agreed not to request the results of genetic tests for contracts under D.Fl. 200,000 (Lemmens and Bahamin, 1996). Similarly, the Association of British Insurers issued a Code of Practice on Genetic Testing, prescribing that genetic test results need not be disclosed for life insurance contracts under £100,000 which are directly linked to a mortgage for the purchase of a house (Association of British Insurers, 1997). (In the literature, the link between a mortgage to buy a house and life insurance was cited as an example of how life insurance can become a social good (Nuffield Council on Bioethics, 1993, p.67; Sandberg, 1995, p.1554). This indicates how the private industry and government can work together to respect the distributive criteria of "providing according to need." In my opinion, the government should keep final responsibility in imposing and controlling these limitations on the market mechanisms. This is not yet the case in the countries I just mentioned, in which restrictions are self-imposed by the industry.

Finally, even if insurance is entirely within the sphere of money and commodities, it does not mean that there are no limits to what insurers can do. But these limitations cannot be such that they affect entirely "free exchange" and they have to be rational and coherent. The use of genetics as a separate entity might therefore not be acceptable, since it seems difficult to justify differential treatment depending on whether a health condition is genetic or not.

Overall, it seems that Walzer comes up with a strong defence of the sphere of security and welfare, giving a place to market mechanisms in distributing money and commodities. But the real value of a society lies in the appropriate development of a protective sphere. This is something we should not forget when we make changes to health care systems that are currently based on, or rather aim at universality. The inequality that could be created if genetics is used in a society in which welfare and health care are partially privatized highlights the importance of his view. As Walzer states: The social contract "is an agreement to redistribute the resources of the members in accordance with some shared understanding of their needs, subject to ongoing political determination in detail. The contract is a moral bond. It connects the strong and the weak, the lucky and the unlucky, the rich and the poor, creating a union that transcends all differences of interest, drawing its strength from history, culture, religion, language, and so on. Arguments about communal provisions are, at the deepest level, interpretations of that union. The closer and more inclusive it is, the wider the recognition of needs, the greater the number of social goods that are drawn into the sphere of security and welfare" (1983, pp.82–83).

Developments of genetic testing urge us to reconsider a development towards privatizing health care and welfare. Embracing private health insurance schemes and applauding benefactors without firm governmental support for those who are genetically *and* otherwise disadvantaged or without any restriction on the market will not help to connect the strong and the weak. On the eve of a new, genetics millennium, it seems appropriate to call for a different moral bond.

ACKNOWLEDGMENT

Research for this paper and participation at this conference was supported by a grant from the Canadian Breast Cancer Research Initiative.

This paper is dedicated to the memory of Benjamin Freedman.

REFERENCES

Alper, J.S. *et al.* (1994), 'Genetic Discrimination and Screening for Hemochromatosis', *Journal of Public Health Policy*, vol. 15, pp.345–358.

Anderson, M.C. (1992), 'Genetic Testing in Insurance Underwriting: A Blessing or a Curse? An Examination of the Tension between Economics and Equity in Using Genetic Testing in Risk Classification.' *Creighton Law Review*, vol.25, pp.1499–1535.

Association of British Insurers (1997), *Genetic Testing: ABI Code of Practice*, London.

Billings, P. *et al.* (1992), 'Discrimination as a consequence of genetic testing', *American Journal of Human Genetics*, vol.50, pp.476–482.

Davis, H.R. and Mitrius, J.V. (1996), 'Note: Recent Legislation on Genetics and Insurance' *Jurimetrics*, vol.37, pp.69–82

Dreyfuss, R.C. and Nelkin, D. (1992), 'The Jurisprudence of Genetics', *Vanderbilt Law Review*, vol.45, pp.313–348.

Geller, L.N. (1996), 'Individual, Family, and Societal Dimensions of Genetic Discrimination: A Case Study Analysis' *Science and Engineering Ethics*, vol.2, pp.71–88.

'Genetic Testing and Insurance. The Ad Hoc Committee on Genetic Testing/Insurance Issues' (1995) *American Journal of Human Genetics*, vol.56, pp.327–331.

Lemmens, T. (1997) ' "What About Your Genes?" Ethical, Legal and Policy Dimensions of Genetics in the Workplace', *Politics and the Life Sciences*, vol.16, pp.57–75.

Lemmens, T. and Bahamin, P. (1996) *Genetics in Life, Disability and Additional Health Insurance in Canada: a Comparative Legal and Ethical Analysis (Report Submitted to the Medical, Ethical, Legal and Social Issues Advisory Committee of the Canadian Genome and Analysis and Technology Programme)* Montreal: Centre de Recherche en Droit Public.

Lowden, R. (1996), 'Certificates of Confidentiality' (letter to the editor) *American Journal of Human Genetics*, vol. 58, pp.436–437.

National Human Genome Research Institute (1997), *Health Insurance in the Age of Genetics*, Department of Health and Human Services (http://www.nhgri.nih.gov/NEWS/Insurance/)

Natowicz, M.R., Alper J.K. and Alper J.S. (1992), 'Genetic Discrimination and the Law' *American Journal of Human Genetics*, vol.50, pp.465–475.

NIH-DOE Working Group on Ethical, Legal and Social Implications of Human Genome Research (1993) *Genetic Information and Health Insurance,* Washington, D.C.: Human Genome Project, HIH Pub. No. 93–3686.

Nuffield Council on Bioethics (1993) *Genetic Screening: Ethical Issues*, London.

Pear, R. (1997), 'States Pass Laws to Regulate Use of Genetic Testing' *New York Times*, October 18, pp.A1&A8.

Rothenberg, K.H. (1995), 'Genetic Information and Health Insurance: State Legislative Approaches', *Journal of Law, Medicine and Ethics*, vol. 23, pp.312–319.

Sandberg, P. (1995), 'Genetic Information and Life Insurance: A Proposal for an Ethical European Policy', *Social Sciences and Medicine*, vol.40, pp.1549–1559.

Wolf, S. (1995), 'Beyond Genetic Discrimination: Towards the Broader Harm of Geneticism', *Journal of Law, Medicine and Ethics*, vol.23, pp.345–353.

4

COERCION, CONTROL, AND CONSEQUENCE IN GENETIC TESTING

Views on Insurance among Tested Individuals and the General Public

Robert N. Mayer,[1] Ken R. Smith,[1] Cathleen D. Zick,[1] and Jeffrey R. Botkin[2]

[1]Department of Family and Consumer Studies
[2]Department of Pediatrics
University of Utah, Salt Lake City, Utah 84112-0080

1. INTRODUCTION

Recent discoveries about the genetic basis of disease and well-being have raised important public policy issues involving insurance markets. These concerns involve coercion, control, and consequence. In the insurance context, *coercion* refers to the possibility that individuals will be forced to undergo genetic tests as a precondition for obtaining insurance. When testing occurs, *control* refers to questions about who will have access to genetic test results and under what conditions. Finally, *consequence* takes into account the various ways in which genetic test information can be used; paramount here is a concern about possible discrimination in access to insurance.

In the United States, a variety of state and federal laws have been either passed or considered in an effort to address concerns about coercion, control, and consequence. While lawmakers have attempted to protect the public from potential problems associated with genetic testing, public views in this area remain largely unknown. This paper presents results from a study that includes several measures of public opinion regarding the *life* insurance issues raised by genetic testing. Specifically, respondents are asked questions involving the issues of coercion, control, and consequence: (a) whether insurance companies should be allowed to require that the people they insure obtain genetic testing; (b) whether, if individuals chose to have a genetic test, their life insurers should have access to the results; and (c) whether insurance rates should be lowered or raised depending on the results of a genetic test for breast cancer. Members of two distinct groups of women provide responses: members of the general public and women in high-risk cancer families who have undergone genetic testing for a BRCA1 gene mutation. (This mutation accounts for a relatively small percentage of all breast cancer cases, but the presence of the gene mutation indicates a 50–85% lifetime chance of breast or ovarian cancer.) The responses of these women are examined in relation to various demographic, economic, and health status variables.

2. U.S. LEGISLATIVE BACKGROUND

The importance of coercion, control, and consequence in the insurance context is evidenced by the recent flurry of state legislation regarding genetic testing. As of mid-1996, at least sixteen states had such laws (Andrews, 1997; Holmes, 1996/97). The statutes vary in the extent to which they address health disability, and/or life insurance and whether they consider genetic testing in general or tests for particular diseases (e.g., sickle-cell anemia). Nevertheless, they all address one or more issues of coercion, control, and/or consequence.

Ten state laws have addressed some aspect of the nexus between genetic testing and life insurance. Five of these laws address genetic testing generally, while the remaining five focus on a narrow set of diseases. *Coercion* has been addressed in several state laws that bar insurers from requiring genetic tests. In May, 1997, Arizona's governor signed a new law that, in addition to other provisions, precludes insurers from requiring genetic testing as a condition of coverage (Schmidt, 1997). A 1996 Minnesota law has a similar prohibition, while a 1996 New jersey law allows life insurers to require genetic tests but only with notification and consent.

Other state laws address *control* issues regarding who may have access to genetic test results and under what conditions. A 1994 Florida law, for instance, describes genetic results as the "exclusive property of the person tested." In October, 1995, California's governor signed a law that recognizes a right to privacy regarding genetic testing and forbids the disclosure of test results. A Colorado law addressing health, disability, and long-term care insurance describes genetic testing information as "confidential and privileged." This information can only be released after specific written consent.

The potential *consequences* of genetic testing are addressed in state laws more often than either coercion or control. Virtually all state laws ban as unfair the use of genetic information in health insurance decisions, and several state statutes prohibit life insurers from denying coverage or charging excessive rates based on genetic test results. Yet, several state laws (e.g., Arizona and Montana) allow the use of genetic test results in life insurance decisions as long as there is a sound actuarial basis for the practice.

Discrimination has been the primary focus of recent federal legislation involving genetic testing. The Health Insurance Portability Act of 1996 prevents insurance companies from charging higher premiums or denying coverage to people because of their genetic history, although the restriction applies only to people enrolled in group health insurance plans, and it doesn't prevent insurance companies from raising rates for entire groups. Current legislation sponsored by Louise Slaughter in the House of Representatives and Olympia Snowe in the Senate could address potential discrimination by life insurers as well. In July, 1997, President Clinton expressed support for this legislation when he stated, "It's wrong when someone avoids taking a test that could save a life just because they're afraid the genetic information will be used against them ("Clinton Fighting Bias...," 1997).

3. LITERATURE REVIEW

3.1. Aggregate Public Opinion

State and federal laws, both passed and pending, have addressed issues of coercion, control, and consequence in relation to genetic testing and life insurance. While discrimination is presumably a "hot button" issue for the general public, issues of coercion and control may be salient as well. What evidence has been accumulated with respect to public views of these issues?

The short history of genetic testing precludes a large number of public opinion studies. When people are asked questions about genetic testing, it may be the first time they have ever heard of it or given it serious thought. An April, 1995 poll conducted by Louis Harris and Associates (*Index to International Public Opinion: 1995–96*, 1997) found that 27% of a national sample of adults had heard or read "quite a lot" about "new genetic tests that predict the possibility that a person will develop certain genetically-influenced diseases or conditions, such as heart disease, cancer and Alzheimer's." Another 42% said they had heard "some" about these tests, but it is possible that this percentage is inflated by the desire of respondents to appear knowledgeable about current social issues. Unfortunately, this survey did not assess the extent to which self-reported knowledge measures actual knowledge.

Awareness of new technological developments always spreads within a population at an uneven pace, leaving some people without knowledge for a long period of time. In the area of genetic testing, the problem of public unawareness is compounded because many people wish to remain ignorant, if not about genetic testing in general, at least about their personal risk of diseases amenable to genetic testing. In a Time/CAN poll (Elmer-DeWitt, 1994), the sample was evenly split between people who said they would take the opportunity to have a genetic test that could "tell them what diseases they were likely to suffer later in life" and those who would turn down this opportunity. Generally, inclusion in a survey of people who are unaware of an issue biases upward estimates of public concern (Herrmann, Sterngold, & Warland, in press; Sterngold, Warland, & Herrmann, 1994), so this problem is especially relevant to public opinion regarding genetic testing.

No survey to date has asked a sample of the general public about mandatory genetic testing as a precondition for life insurance coverage. A question in a 1995 Harris Poll question (*Index to International Public Opinion: 1995–96*, 1997) raises the possibility of mandatory testing at birth:

> Proposals have been made for state department of public health to collect a DNA blood sample or genetic fingerprint from each newborn child. These would be put into a computerized DNA databank, with the identity of each newborn attached to the sample. How acceptable would you find this?

In response, 28% of a national sample considered this practice "very acceptable," with another 28% finding it "somewhat acceptable." In contrast, 16% found it "not very acceptable" and another 27% found it "not at all acceptable." The coercion implied in this survey question is universal (everyone would be tested) and without explicit consequences. Responses therefore might be more negative if coercion were applied more selectively and used as a precondition for receiving a specific good or service, like life insurance.

In contrast to the absence of surveys on the coercive aspects of genetic testing, there are several surveys on the public acceptance of mandatory HIV (or AIDS) testing--whether involving pregnant women, "high-risk" people, health care workers, or the general public. In these surveys, public acceptance of mandatory testing is fairly high, but HIV testing contrasts sharply with genetic testing inasmuch as HIV is communicable. It is therefore difficult to gauge public acceptability of mandatory genetic testing from studies of mandatory HIV testing.

Whereas coercion refers to the possibility that people might be forced to take a genetic test, control entails the privacy-related question of who would be entitled to know the results, whether the test was coerced or freely chosen. This issue of control was raised in a 1992 national survey conducted by Louis Harris and Associates (Louis Harris and Associates, 1997):

> In your opinion, if someone is a carrier of a defective gene or has a genetic disease, does anyone else besides that person, deserve to know that information?

This question is poorly worded since there is a large difference between carrying a defective gene and having a genetic disease. The wording problem aside, 57.1% of the sample stated that, yes, others deserve to know, compared to 41.0% who felt that no one else deserves to know. (The remaining 1.8% of the sampled answered that they weren't sure.) When asked about specific parties—employers, insurers, spouse/fiance, and other immediate family members—the percentages varied widely. Whereas 97.5% of the sample believed that a spouse/fiancé deserved to know about genetic information, the percentages for other immediate family members, insurers, and employers were 70.3, 57.8 and 33.5, respectively.

Several additional survey questions address issues of control or privacy but in a way that also raises the possibility that the disclosure of test results might have negative, discriminatory consequences. As a result, it is impossible to determine whether people are concerned about a loss of privacy per se or about the negative consequences that might occur when genetic test results are made available to third parties. For example, the 1995 Harris Poll referred to above contains a question about the possibility of being required to share genetic testing results with organizations making important decisions about an individual:

> If you had a genetic test, how concerned would you be that organizations that want to know the state of your health--such as health and life insurance companies, or employers--might require you to provide them with the test results so that they could decide whether to insure you or hire you?

Sixty percent of respondents indicated they would be "very concerned" about the required disclosure of genetic test results, with another 26% being "somewhat concerned." A similar level of public aversion to losing control of personal genetic information was expressed in 1993 Gallup Poll conducted in Great Britain (*Index to International Public Opinion:* 1993–94, 1995). The question posed was:

> Insurance companies could use genetic information from a blood test to target their policies, for instance, by ensuring that only those who will develop a disease pay a higher premium. Employers could screen the workforce, for instance, to ensure those at risk of cancer are not taken on to work with chemicals. In what circumstances, if at all, should insurers and employers be allowed to have genetic information?

Sixty-eight percent of respondents felt there were "no circumstances at all" under which genetic test results should be provided to insurers and employers, and additional 9% believed that disclosure of genetic information would be acceptable "only so long as a government-appointed body monitors the way information is used." Only 8% felt that insurers and employers should have this information "whenever they consider it necessary."

When people express concern about being forced to undergo genetic tests or share the results with third parties, they may actually be concerned with the potential consequences, such as discrimination by employers or insurers. It is best, then, for survey questions to separate issues of coercion, control, and consequences.

A question that focuses exclusively on the potential uses of genetic testing is contained in a Time/CNN poll reported in January, 1994. The poll found that 90% of respondents thought it should be illegal for insurance companies to use genetic test results in insurability decisions (Elmer-DeWitt, 1994). Whereas this poll dealt with a hypothetical situation, a study by Lapham et al. (1996) gauged self-reported discrimination in insurance and employment. The study's sample was a bit unusual, consisting of 332 people who: (a) had one or more family members with a genetic disorder and (b) were themselves members of a genetic support group. Respondents in the study reported high levels of discrimination, either against themselves or their affected family members. In the domain of insurance, 25% and 22% believed they were denied life and health insurance, respectively, because of genetic information. These rates of denial are far higher than the 1–5% typically cited by insurance industry representatives.

Balancing the view that genetic testing opens the door to extensive discrimination is a study of public opinion in Europe ("Europe Ambivalent on Biotechnology," 1997). Genetic testing was one of six biotechnologies about which national samples of adults in fifteen countries were asked. Using genetic tests to detect inheritable diseases was compared to: (1) introducing human genes into bacteria to produce medicines or vaccines, (2) transferring genes from plant species into crop plants to increase resistance to insect pests, (3) using biotechnology to improve food production, (4) developing genetically modified animals for laboratory research, and (5) introducing human genes into animals to produce organs for human transplants. Genetic testing was considered more useful, less risky, more morally acceptable, and more worthy of encouragement than all five of the other technologies. It is unknown whether these results would be replicated in the United States. Nevertheless, it is instructive to find that public opinion regarding genetic testing, while perhaps negative when taken in isolation, is more favorable than opinion about various types of genetic modification.

3.2. Individual Correlates of Public Opinion

Having overall measures of public opinion regarding genetic testing are useful in characterizing the public in general, but understanding sources of variation in opinion can be important as well. From a scientific point of view, the correlates of opinion provide clues to the reasons or mechanisms by which opinions are formed and change. For example, if opinions are most favorable among those who are also most knowledgeable about genetic testing, this might indicate that support for genetic testing will grow as information about this new technology diffuses within society. Alternatively, if opinions are most negative among respondents high in religiosity, then negative opinions are more likely to persist. From a practical, political standpoint, the correlates of public opinion suggest the social bases of support and opposition to genetic testing and thereby organizations which are most likely to participate actively in debates concerning it. For instance, if concerns about genetic testing are strongest among elderly respondents, then one can expect that sophisticated and well funded organizations like the American Association of Retired Persons might participate in policy debates regarding genetic testing. Conversely if concerns are greatest among the least educated segments of the population, resistance to genetic testing is less likely to be effectively expressed.

We have just seen that there are very few studies of public opinion regarding genetic testing. If bivariate relationships with characteristics such as age, gender, and education are reported at all, results are presented in a haphazard manner and multivariate relationships are not considered. In sharp contrast, studies of the correlates of public opinion regarding genetic engineering in agricultural production have been fairly sophisticated and may have some relevance to genetic testing. Both genetic engineering and genetic testing involve some level of intervention in the "natural scheme of things." Genetic engineering, for example, involves the creation of entirely new foods and the cloning of animals. Genetic testing provides people with insights into their future health status that formerly might only be provided by divine communication or, less reliably, by soothsayers and clairvoyants. Nevertheless, genetic engineering raises issues of health and safety, whereas genetic testing is more likely to raise concerns about privacy and discrimination. One must therefore be careful in extrapolating from the correlates of opinion regarding genetic engineering to those concerning genetic testing.

The most consistent findings in multivariate analyses of public concern about genetic engineering is that being female and having relatively low levels of income and education are associated with being more concerned about bioengineered foods (Florkowski et al., 1994; Grobe and Douthitt, 1995; Hoban et al., 1992). An intriguing question is whether knowledge of genetic engineering promotes more or less concern about it. In a study conducted in the state of North Carolina, greater awareness was associated with lesser concern (Hoban et al., 1992). This corresponds with the notion that knowledge promotes support for science and technology, and, conversely, that opposition to science and technology is rooted in ignorance. However, in a study conducted in Europe across time and in many countries, the conventional link between knowledge of genetic engineering and support for it was questioned (Europe Ambivalent...," 1997). Not only did concern and knowledge increase simultaneously between 1991 and 1996, but support was weakest in those countries where the use of biotechnology was most firmly established. Thus, the relationship between knowledge and concern about genetic testing is a worthwhile subject for empirical investigation.

3.3. Hypotheses

There is very little research from which to draw hypotheses about the correlates of public opinion regarding the coercion, control, and consequence issues raised by genetic testing, especially as these issues pertain to life insurance. Given this situation and our use of an all-female sample, we begin with a simple economic model that a woman's opinion is affected by her perceived ability to obtain life insurance. This ability, in turn, depends on demand and supply factors. Demand factors are those that affect a person's perception of the value of life insurance. For example, young, unmarried people without children are likely to view life insurance as less valuable than people who are older, married, and have children. Individuals with a family history of relatives dying at relatively young ages from genetic diseases will view life insurance as more valuable than people without such family history. An individual's personal health history should have a similar impact on life insurance demand, with poorer health associated with greater demand. Independent of family and individual health history, people with a relatively low subjective assessment of their longevity are likely to demand more life insurance than people who believe they will live to an old age. Of special relevance in this study, people who have been tested for a gene mutation like BRCA1 should have a stronger demand for life insurance than people who have not been tested at all, and those who have tested positive should have a stronger demand than those who tested negative.

Other demand factors are less obvious. A person's employment status likely influences the perceived value of life insurance to the extent that insurance payments more directly substitute for lost wages than lost home production or lost leisure time. Independent of employment status, income may serve as a proxy for financial assets worth protecting through the purchase of life insurance. Education may influence demand to the extent that it promotes awareness of risk-reducing and health-enhancing activities that promote life. Religious beliefs might affect the perceived value of life insurance since people who believe that a god is firmly in control of their fate might find life insurance less valuable than people who believe that human beings are able to affect their destinies. Finally, whether a person currently holds life insurance is a "super" demand factor inasmuch as it presumably is influenced by all of the demand factors already mentioned; moreover it represents something that might be threatened by the extensive use of genetic testing.

Supply factors influence the willingness of life insurance companies to offer coverage. As such, many of the same factors that influence demand for life insurance also influence supply. For instance, family and personal health history are presumably the two most important pieces of information used by insurers in making insurability decisions and also by consumers in seeking insurance. Similarly, a person's employment status is not just a demand factor but is likely to affect supply inasmuch as it affects the type, number, and price of insurance policies made available (e.g., group vs. individual).

While insurers have access to some of the same information that consumers presumably use in making insurance decisions, some demand factors are difficult for insurers to measure accurately or are unavailable to insurers. For example, insurers are generally unable to gauge a person's subjective assessment of longevity. Also, life insurers typically do not have access to any genetic test results.

4. METHODS

4.1. Samples

We explore the correlates of public opinion about genetic testing in relation to the practices of life insurance companies among a group of respondents composed of two subsamples. The first, "tested" subsample is drawn from a large kindred residing in Utah and Idaho whose members have been extensively studied at the University of Utah with funding from the National Cancer Institute. Many of these family members have been found to have a specific BRCA1 gene mutation on the long-arm of chromosome 17. This kindred, known as the K2082, is the largest family studied to date with a BRCA1 mutation. Women in K2082 who are BRCA1 gene mutation carriers are estimated to have an 85 percent lifetime risk of developing breast and/or ovarian cancer. As such, women who test positive face a substantially elevated risk of contracting a potentially fatal form of cancer during their lifetimes. Women in the kindred who test negative for the gene mutation have a similar risk of breast and ovarian cancer as women in the general population (since the BRCA1 mutation accounts for a small percentage of all breast and ovarian cancer).

In all, 410 adults (all Caucasian) in the K2082 were recruited for a large-scale, longitudinal study of the behavioral and psychological consequences of genetic testing. Recruitment of subjects began in October 1994 and continued until March 1997. If the individual agreed to participate in the study, a baseline survey was conducted by telephone. Next, the individual met with genetic and family counselors at the University of Utah Medical School to discuss the advantages and disadvantages of being tested for the BRCA1 gene mutation. Subjects were told that their test results would not be shared with any third parties, including insurers. If the individual still wished to be tested after the counseling, a blood sample was taken and the genetic test conducted. The individual then returned to the University of Utah to meet again with a genetic counselor to learn his/her test results and to receive non-directive counseling regarding the meaning of the results.

The post-test study protocol begins with a telephone interview one to two weeks after the individual receives his/her test results. The next interview occurs approximately four months after the test results. This is followed by one-year and two-year follow-up telephone interviews. The opinion data concerning life insurance as well as information about recent life insurance decisions are taken from the interview conducted one year after the receipt of genetic test results.

In all, 269 people in the K2082 have been tested. Of these, 195 have completed the one-year follow-up survey, 85 of whom are women between the ages of 18 and 55. (Women over 55 were excluded from this analysis because demand for life insurance drops off sharply—and prices rise steeply—after this age.) Thirty of these women tested positive for the BRCA1 gene mutation, and fifty-five tested negative. These women constitute the first, "tested" subsample in our analyses.

The second subsample of women in our study were interviewed as part of a NCI-funded project to study life insurance purchasing behavior among women varying in their experience with genetic testing and in their family cancer history. This study involved comparing the tested subsample with an "untested" subsample of women drawn from the general public in the state of Utah.

The process of obtaining the untested subsample began with a Utah Department of Health survey conducted in the spring and summer of 1996 via telephone interviews. Men and women in this state-wide, random sample survey were asked whether a woman (18 years old or older) in the household had a family history of breast and/or ovarian cancer among first-order (mother or sister) or second-degree (aunt or grandmother) female relatives as well as whether they were willing to participate in a follow-up survey.

The Utah Department of Health survey yielded a list of potential subjects for the current study identifiable through their telephone numbers. Although the tested sample had a few members living outside of Utah, the untested sample was confined for financial reasons to the seven most populous counties in Utah: Salt Lake, Davis, Utah, Weber, Cache, Tooele, and Summit. There were 1135 phone numbers available for contact, 421 of whom reported a family history of cancer. We oversampled women who reported having a family history of cancer in order to gain a better understanding of the potential impact of varying degrees of family history on life insurance attitudes and behavior.

In fall of the 1996 and the spring of 1997, research staff contacted prospective subjects by telephone to obtain their names, addresses, and permission to send them consent forms. At this point, they were told the purpose of the study and the approximate length of the telephone interview (20 minutes). Of the 421 possible subjects with a family history of cancer, 76 were never reached despite multiple phone calls at varying days of the week and varying times of the day and early evening, 53 phone numbers were disconnected, 37 declined participation, and 255 gave their names and addresses for the mailing of consent forms. Of these 255 people, 180 returned the consent form, 30 of whom declined to participate in the study. From the remaining 150 people, 147 completed telephone interviews were obtained.

Of the 714 people without a family history of cancer, 426 were never reached, 63 phone numbers were disconnected, 44 declined participation, and 181 gave their names and addresses for the mailing of consent forms. Of these 181 people, 136 returned the consent form, with 34 declining to participate further. This yielded 102 willing participants, from whom 99 completed interviews were obtained. Thus, combined with the 147 women reporting a family history of cancer, these 96 women constitute the 246 people in the untested sample. Taken together, the tested and untested subsamples yield an overall sample of 331 women.

4.2. Measures

The survey instrument contained four agree/disagree questions directly related to the life insurance issues raised by genetic testing. These questions constitute the dependent variables in the analysis that follows. One item addresses the possibility of coercion:

> If you were applying for life insurance and the company you approached required that you take a genetic test before they would sell you a policy, would you be willing or unwilling to do this?

A second question raises the issue of who should have access to test results:

> If people choose to be tested to determine if they have a genetic risk for certain diseases, then their life insurance companies should be allowed to know the results of the tests.

The remaining questions deal with the potential impact of genetic test results on insurance rates:

> Some people have a gene, called the BRCA1 gene, that increases their risk for breast and ovarian cancer....Life insurance companies should lower their rates for people they know do not have the BRCA1 gene.

> Life insurance companies should increase their rates for people they know have the BRCA1 gene.

In addition to these four policy questions, respondents provided information on their sociodemographic characteristics and health-related perceptions. The sociodemographic characteristics included: age, marital status, presence in the household of minor children, education, income, employment status, family cancer history, personal cancer history, personal cancer-related surgery history, religiosity, and, where appropriate, genetic test history, and genetic test results. The health-related perceptions included: relative health status ("Compared to other people you age, how would you describe your health?) and subjective longevity ("What do you think the chances are that you will to age 65...75...85 or more?").

5. FINDINGS

Table 1 provides descriptive data for all of the variables used in this study. It is important to note that the key dependent variables are dichotomous in nature and are skewed

Table 1. Descriptive statistics

	Mean	Standard deviation
Dependent Variables		
Should genetic test be required? REQUIRE (1=yes; 0=no)	.09	.28
Should life insurers know test results? KNOW (1=yes; 0=no)	.19	.39
Lower insurance rates if test negative? LOWER (1=yes; 0=no)	.24	.43
Raise insurance rates if test positive? RAISE (1=yes; 0=no)	.07	.25
Independent Variables		
Underwent genetic testing for BRCA1; TESTED (1=yes; 0=no)	.26	.44
Tested positive for BRCA1; POSITIVE (1=yes; 0=otherwise)	.09	.29
Tested negative for BRCA1; NEGATIVE (1=yes; 0=otherwise)	.17	.37
1st- and 2nd-order relatives with cancer history; FAMCANNM (number)	1.09	1.17
Age of respondent; AGE (years)	36.78	9.62
Education of respondent; EDUC (years)	13.92	2.00
Household income; INCOME (thousand of dollars)	45.88	24.08
Marital status; MARRIED (1=married; 0=otherwise)	.79	.41
Children less than 18 years old at home; NUMLT18 (number)	1.75	1.37
Full-time employment; FULLTIME (1=yes; 0=otherwise)	.36	.48
Personal cancer history; PCANHIST (1=yes; 0=no)	.12	.32
Probability of living until age 75;LIVETO75 (percent)	71.17	24.25
Religiosity; RELIGOS (1=deeply/fairly; 0=otherwise)	.81	.39
Currently have life insurance; LIFEINS (1=yes; 0=no)	.72	.45

toward one of the two possible answer categories. For example, only 9.0% of the respondents believed that life insurance companies should have right to make genetic testing a precondition for coverage (coercion issue). On the question of whether life insurers should be allowed to know the results of genetic tests (control issue), 19.1% of respondents answered in the affirmative. Nearly a quarter of respondents (23.8%) believed that life insurance rates should be reduced for people testing negative for the BRCA1 gene mutation (consequence issue), but only 6.5% indicated rates should be increased when a person tests positive. Overall, these results suggest that members of the general public are rather cautious about the use of genetic testing in life insurance underwriting.

Answers to the four questions about genetic testing and life insurance are not highly intercorrelated and can therefore be analyzed separately. Answers to the two questions about whether insurance rates should be lowered or raised depending on test results for the BRCA1 gene mutation have the highest Pearson correlation coefficient of any two items ($r= .40$, $p<.01$). Yet, of those respondents who believe that insurance rates *should* be lowered when a person tests negative for the gene mutation, 76.4% percent believe they *should not* be raised for people testing positive. Thus, the four questions measure related but different aspects of the debate about the role of genetic test results in establishing life insurance premiums.

What factors account for variation in opinion among members of the public. Our study differs from typical public opinion samples inasmuch as it contains a subsample for whom genetic testing is more than an abstract concept. Eighty-five of the sample members have undergone genetic testing, thirty of whom tested positive for the BRCA1 gene. We therefore begin the analysis of public opinion by comparing the responses to the four dependent variables among: (1) members of the 2082 kindred who have tested positive for the BRCA1 gene, (2) members of the kindred who tested negative, and (3) members of the general public with varying degrees of family history of breast and ovarian cancer. Table 2 summarizes these results.

Table 2. Opinion on questions regarding life insurance practices by groups

	Preferred life insurance industry practices (% Yes)			
Group N=329	Require tests	Know results	Lower rates	Raise rates
Tested: Positive n= 30	3.6	7.7	10.7	3.4
Tested: Negative n= 55	1.8	9.1	5.6	1.9
Family cancer history n=50*	14.3	23.4	20.0	4.1
No family cancer history n=194	10.6	22.5	31.9	9.0

* Two or more or first-order or second-order relatives with a history of cancer

The most striking difference within the overall sample is between women in the 2082 kindred and women in the general public. Among the women in the general public, however, the differences between women with and without a family history of cancer (defined as having two or more first- or second-order relatives with a history of cancer) are modest and inconsistent in terms of limiting life industry practices. There are some differences among the four groups in Table 2 in terms of sociodemographic factors, and these differences will be controlled in the multivariate analyses presented below. It is worth noting, however, that the women in the untested subsample who are defined as having a family history of cancer report more first- or second-order relatives with a history of cancer than even the people who tested positive for the BRCA1 gene mutation. This makes the differences between these two groups all the more striking.

To examine the correlates of public opinion in a multivariate fashion, we proceed in four steps. First, we examine only the effects of genetic testing and family history. Then, we explore these effects along with a summary measure of demand and supply factors, namely, whether a person is currently covered by a life insurance policy. Third, we examine the effects of genetic testing and family history in tandem with a variety of more specific demand and supply factors, such as age, education, and subjective longevity. Finally, given the importance of the tested-untested distinction in various analyses, we look at the role of independent variables only within the untested sample of respondents.

Given the dichotomous nature of the dependent variables, logit analysis is used. For any given level at which an independent variable is set, the logit equation predicts the natural logarithm of the odds that a respondent agreed with the particular policy statement. The odds of an individual agreeing with the statement are defined as $P_i/(1-P_i)$ where P_i is the probability that respondent i agreed with the question and $(1-P_i)$ is the probability that he/she disagreed. In the logit model, the estimated coefficients are cast so that a one-unit change in an independent variable produces a particular change in the natural log of the odds ratio, holding all other factors constant. Because this interpretation is cumbersome, odds ratios have been included in the tables below. Odds ratios constitute the odds that a respondent with a given value on an independent variable answers yes to an opinion question compared to the odds that a person without that value on the independent variable (holding other factors constant). So, in the case of test status, the odds ratio represents the odds that someone who tested positive answered yes compared to someone who tested negative.

Table 3 shows the results of a logit analysis using test status and family history of cancer as the only independent variables. Test status is a dichotomous variable that corresponds to the two major subsamples in the study--tested and untested. Family history of cancer is measured as a continuous variable; it is the number of first- or second-order relatives to have had breast and/or ovarian cancer. For all four opinion measures, test status has a strong effect whereas family cancer history has no effect. In considering the poten-

Table 3. Parameter estimates of the logit analyses for the probability of agreeing with a potential life insurance industry practice

Logit coefficient (Wald chi-square) odds ratio	Preferred life insurance industry practices			
	Require tests	Know results	Lower rates	Raise rates
Tested	-1.71	-1.22	-1.6	-1.25
	(5.04**)	(7.78**)	(12.91**)	(2.65*)
	0.18	0.29	0.19	0.29
FAMCANNM	0.07	0.11	-0.03	-0.00
	(0.16)	(0.75)	(0.04)	(0.00)
	1.07	1.11	0.98	0.89
Intercept	-1.32	-2.12	-0.85	-2.44
	(46.89**)	(66.44**)	(23.14**)	(65.90**)
	0.27	0.12	0.43	0.09
Overall chi-square	6.13*	8.43**	16.56**	3.13

* p<.05
** p<.01

tial effect of family history, recall that members of both the tested and untested subsamples have substantial variation in terms of the number of their first- or second-order relatives who have had breast and/or ovarian cancer. (In terms of statistical significance, one-tailed tests are used given the directional predictions for the various independent variables.)

Table 4 extends the analysis to include whether a person is currently covered by one or more life insurance policies. The results are similar to those in Table 3 except for the question of whether life insurance companies should have the right to know about genetic test results. Here, people with a life insurance policy are nearly three times (odds ratio = .35) as likely to want to deny insurance companies access to test results.

Given this small piece of evidence that demand and supply factors might influence opinions about life insurance industry practices, Table 5 shows the results of a multivari-

Table 4. Parameter estimates of the logit analyses for the probability of agreeing with a potential life insurance industry practice

Logit coefficient (Wald chi-square) odds ratio	Preferred life insurance industry practices			
	Require tests	Know results	Lower rates	Raise rates
Tested	-1.73	-1.32	-1.67	-1.28
	(5.20**)	(8.76**)	(13.31**)	(2.75*)
	0.18	0.27	0.19	0.28
FAMCANNM	0.07	0.10	-0.03	-0.01
	(0.16)	(0.64)	(0.06)	(0.00)
	1.07	1.11	0.97	0.99
LIFEINS	-0.40	-1.03	-0.33	-0.48
	(0.36)	(11.6**)	(1.29)	(1.06)
	0.68	0.36	0.72	0.62
Intercept	-1.84	-0.62	-0.60	-2.10
	(46.89**)	(5.13**)	(4.81**)	(24.65**)
	0.27	0.54	0.55	0.12
Overall chi-square	7.003*	20.22**	18.05**	4.23

* p<05
** p<01

Table 5. Parameter estimates of the logit analyses for the probability of agreeing with a potential life insurance industry practice

Logit coefficient (Wald chi-square) odds ratio	Preferred life insurance industry practices			
	Require tests	Know results	Lower rates	Raise rates
Tested	-1.79	-1.25	-1.77	-1.24
	(5.21**)	(7.61**)	(12.59**)	(2.44)
	0.17	0.29	-0.19	0.29
FAMCANNM	0.04	0.15	-0.06	-0.00
	(0.03)	(1.20)	(0.18)	(0.00)
	1.04	1.16	0.95	0.99
Age	0.03	-0.02	-0.00	0.03
	(1.67)	(0.81)	(0.04)	(1.02)
	1.03	0.96	0.99	1.03
Educ.	-0.36	-0.06	-0.02	-0.28
	(7.12)	(0.47)	(0.07)	(3.97*)
	0.70	0.94	0.98	(0.76)
Income	2.30×10^{-6}	-0.00	5.52×10^{-6}	0.00
	(0.04)	(1.68)	(0.58)	(5.97**)
	1.00	1.00	1.00	1.00
Married	-0.40	0.01	-0.01	-1.68
	(0.43)	(0.00)	(0.00)	(5.77**)
	0.67	1.01	0.99	0.19
NUMLT18	-0.26	-0.11	-0.18	0.10
	(2.12)	(0.90)	(2.52)	(0.26)
	0.77	0.89	0.83	1.10
Fulltime	0.66	-0.28	0.05	-1.68
	(2.14)	(0.69)	(0.02)	(0.13)
	1.94	0.76	1.05	0.83
PCANHIST	-1.99	0.12	-0.80	-0.03
	(3.27*)	(0.07)	(2.33)	(0.00)
	0.14	1.13	0.45	0.98
LIVETO75	0.01	0.01	0.00	0.01
	(1.65)	(1.17)	(0.42)	(0.30)
	1.01	1.01	1.00	1.01
Religos.	-0.40	-0.40	0.25	-0.24
	(0.57)	(1.10)	(0.40)	(0.15)
	0.67	0.67	1.29	0.79
Intercept	1.39	0.51	-0.83	-0.32
	(0.48)	(0.15)	(0.47)	(0.20)
	4.02	1.66	0.44	0.72
Overall chi-square	30.997**	19.03	22.60*	14.55

* $p<.05$
** $p<.01$

ate analysis that replaces life insurance holdings with a set of independent variables that past research has found to be associated with the likelihood of purchasing life insurance. Here again, there are just enough statistically significant results to pique one's interest. For the question of whether life insurance companies should be able to require genetic tests, people with higher levels of education and a personal history of cancer are most opposed to this possible requirement (other factors being held constant). A personal history of cancer has a similar but weaker association with opinion on whether people testing negative for the BRCA1 gene mutation should receive lower insurance rates. Specifically, people with a personal cancer history are less likely to favor such a discount. On the inverse question of whether people who test positive should have their insurance rates raised, personal cancer history does not appear to matter, but people who have higher levels of education and those who are married are *less* supportive of raising rates, while people with higher levels of income are *more* inclined to support higher rates.

Throughout the analysis to this point, whether a person is in the tested or untested sample is the only factor that consistently affects opinion. What happens when we examine the role of various sociodemographic, health, and perception factors within just the untested subsample? To avoid another lengthy table, and because the results differ little from those in Table 5, the findings are summarized here. On the question of whether life insurance companies should be able to require genetic tests, people with higher levels of education and a personal history of cancer are most opposed to this possible requirement, but older respondents are more favorably disposed to this possibility than younger ones. People who currently hold life insurance policies are more opposed to sharing genetic test results with insurers than people who do not hold a policy. Support for lowering insurance rates in response to negative test results (i.e., absence of gene mutation) is associated with having relatively few first- or second-order relatives with a history of breast and/or ovarian cancer. Finally, support for raising insurance rates in response to positive test results is higher among relatively wealthy respondents, unmarried respondents, and respondents with relatively low levels of education. Overall, the results are fairly similar regardless of whether the two subsamples are combined.

6. DISCUSSION

Taken in the aggregate, people appear to be very cautious about the use of genetic testing within the life insurance industry. A strong majority do not want life insurers to require genetic tests or know about test results. Not surprisingly, few people want insurance rates raised in response to a genetic test that shows a person has a high-risk gene mutation like the BRCA1, but neither do they want lower rates when a test shows the absence of such a gene mutation.

Looking within the overall sample, the only factor that consistently differentiates responses is whether a person has had a genetic test. Interestingly, though, there are no differences within the tested subsample depending on whether a test was positive or negative. Testing--with all the counseling and interviewing that goes with it in the tested subsample--appears to have an impact regardless of its outcome. This leaves us with an intriguing question: why does testing matter so much while test results matter so little?

Another intriguing "non-finding" is the extremely limited role of family or personal cancer history, whether in the tested or untested subsamples. Recall that 50 out of the 244 women in the untested subsample had two or more first- or second-order relatives with a history of cancer and another 85 women had one such relative. Similarly, there was substantial variation in the family cancer history of women in the tested sample, with 46 of 85 having two or more first- or second-order relatives with a history of cancer, 26 having one such relative, and 13 having none. Hence, one might have expected to find differences of opinion depending on family cancer history, whether within or across the subsamples. Yet, no such differences emerged.

It is possible that a different measure of family cancer history might have yielded different results. One possibility would be to confine the measure to first-order relatives. Another possibility is to include only relatives, whether first- or second-order, if the respondent was involved in their care when they had cancer. Both of these alternative measures of family history are possible given our data set, and future analyses will explore their potential usefulness. A third way of measuring family cancer history would be getting a sense of the emotional closeness between the respondent and any family member who experienced breast and/or ovarian cancer. Finally, one might relax the assumption that opinions are influenced by the cancer history of blood relatives. It is possible that watching a mother-in-law or sister-in-law struggle with breast and/or ovarian cancer might influence opinions even though there is no genetic link to the respondent.

Even if family cancer history is unrelated to opinions about life insurance practices, one might have expected that personal cancer history would be. Here, all forms of cancer except skin cancer were included in the measure. The resulting measure of personal cancer history is far more skewed than family cancer history, with less than 12% of the total sample having had cancer. Yet, this presumably strong factor was related to differences in opinion for only one of the four dependent measures (whether life insurers should be able to require genetic tests as part of the application process).

Along with the lack of differences in opinion according to test results and either family or personal cancer history, demand and supply variables had little explanatory power as well. A summary measure, whether a person currently is insured, was related to opinion on only one of the four questions--whether insurers should be allowed to know the results of genetic tests. For two of the four questions, none of the remaining sociodemographic and perception measures showed an association. For the questions about whether insurers should require tests and raise rates in response to positive test results, a few measures were statistically significant, but only in the case of education did the same factor show up for both questions. The direction of the association suggests that education creates skepticism about, rather than acceptance of, this new technology.

What should one make of so many non-findings? Perhaps it is premature to expect members of the general public to have stable and coherent views about the use of genetic testing in life insurance markets. It may be advisable to limit the analysis to people who describe themselves as being familiar with genetic testing. Our data set contains a question that allows us to distinguish among respondents who have heard a great deal, some, or not very much about the BRCA1 gene.

An alternative explanation for the surprising non-findings is that members of the general public may have opinions that do not fit comfortably within the dichotomous answer categories offered them in our study. Future research might offer a wider range of answer categories. And, of course, having more than one or two items each of opinion regarding the coercion, control, and consequence aspects of life insurance practices would enhance measurement quality.

Thinking more broadly, our basic model in which opinions about the role of genetic testing in life insurance practices are influenced by the same factors that affect life insurance purchases may need to be replaced with one that is less economic and more political in nature. To the extent that the opinion questions raise issues of public policy, differences among respondents may be based on their views of business and government. Are life insurance companies trustworthy? Are they excessively profitable? Is the government capable of implementing practices that effectively safeguard privacy or prevent discrimination? Answers to questions such as these may help explain differences in opinion within the general public about whether and how genetic test results should be used by the life insurance industry.

REFERENCES

Andrews, Lori (1997), 'Body Science', *American Bar Association Journal*, Vol. 83,April, pp. 44–9.

'Clinton Fighting Bias Based on Genetic Histories' (1997), *San Francisco Chronicle*, July 15, p.A3.

Elmer-Dewitt, Philip (1994), 'The Genetic Revolution,' *Time*, Vol. 143, January 17, pp. 46–54.

'Europe Ambivalent on Biotechnology' (1997), *Nature*, Vol. 387, June 26, pp. 845–7.

Florkowski, Wojciech J., Catherine Halbrendt, Chung L. Huang, and Lesa Sterling (1994), 'Socioeconomic Determinants of Attitudes Toward Bioengineered Products', *Review of Agricultural Economics*, Vol. 16, pp. 125–32.

Grobe, Deana and Robin Douthitt (1995), 'Consumer Acceptance of Recombinant Bovine Growth Hormone: Interplay Between Beliefs and Perceived Risk', *Journal of Consumer Affairs* Vol. 29, No. 1, Summer, pp. 128–43.

Herrmann, Sterngold, and Warland (in press), 'Comparing Alternative Question Forms for Assessing Consumer Concerns', *Journal of Consumer Affairs*, Vol. 32, No. 1, Summer.

Hoban, Thomas, Eric Woodrum, Ronald Czaja (1992), 'Public Opposition to Genetic Engineering', *Rural Sociology*, Vol. 57, No. 4, pp. 476–93.

Holmes, Eric Mills (1996/97), 'Solving the Insurance/Genetic Fair/Unfair Discrimination Dilemma in Light of the Human Genome Project', *Kentucky Law Journal*, Vol. 85, No. 3, pp. 503–663.

Index to International Public Opinion, 1995–96 (1997), Greenwood Press: Westport, CT.

Index to International Public Opinion, 1993–94 (1995), Greenwood Press: Westport, CT.

Lapham, E. Virginia, Chahira Kozma, and Joan O. Weiss (1996), 'Genetic Discrimination: Perspectives of Consumers', *Science*, Vol. 274, October 25, pp. 621–5.

Louis Harris and Associates (1997), *1992 Genetic Testing Survey*, Archival Study Number S912074, Institute for Research in Social Science: Chapel Hill, NC.

Schmitt, Frederick (1997), 'Insurers Oppose Arizona Genetic Test Law', *National Underwriter: Life & Health/Financial Services Edition*, May 19, pp. 3, 57.

Sterngold, Arthur, Rex H. Warland, and Robert O. Herrmann (1994), 'Do Surveys Overstate Public Concern?' *Public Opinion Quarterly*, Vol. 58, Summer, pp. 255–63.

5

GENETIC TESTING AND ADVERSE SELECTION IN THE MARKET FOR LIFE INSURANCE

Preliminary Findings for the BRCA1 Gene Mutation

Ken R. Smith,[1,*] Cathleen D. Zick,[1] Robert N. Mayer,[1] and Jeffrey R. Botkin[2,†]

[1]Department of Family and Consumer Studies
University of Utah
228 Emery Building, Salt Lake City, Utah 84112
[2]Department of Pediatrics
University of Utah
100 North Medical Drive, Salt Lake City, Utah 84113

1. INTRODUCTION

The human genome project is rapidly fueling the development of genetic tests for inherited diseases. Over 100 different genetic tests are currently available for a wide range of diseases including Huntington disease, hypercholesterolemia, sickle-cell trait, and selected forms of Alzheimer's disease, colon cancer, melanoma, breast and ovarian cancer. The U.S. Congressional Office of Technology Assessment has projected that there will be a tenfold rise in the number of genetic tests that are available during the 1990s (U.S. Congressional..., 1988). Increasingly, a variety of public and private interest groups are raising questions about what individuals and society should do with such genetic information. Within the life and health insurance markets, ethical concerns regarding genetic discrimination are clashing with the underwriting needs of insurance companies.

Consumers fear that if insurance companies gain access to genetic test results some insurers will use this information to deny coverage and/or raise rates to individuals who carry particularly serious gene mutations—even when the individuals in question are still asymptomatic. At the same time, insurance companies worry that if consumers are allowed to keep information regarding genetic test results to themselves, then those who test positive for disease-related gene mutations will be able to obtain high levels of insurance

* Phone: 801-581-5771; Fax: 801-581-5156; E-mail: smith@fcs.utah.edu
† Phone: 801-588-3640; Fax: 801-588-3642; E-mail: botkin@howard.med.utah.edu

Genetic Information, edited by Thompson and Chadwick.
Kluwer Academic / Plenum Publishing, New York, 1999.

protection at prices that are below actuarially fair values. In essence, each group worries that the other will exploit new genetic information to gain some advantage in the insurance market.

Both sides of the insurance and genetic testing debate have made assertions based on little data. In this paper, we capitalize on survey information from two groups of women to assess insurers' claims regarding consumer behavior. Specifically, we examine whether or not consumers exploit the information asymmetry created by maintaining consumer privacy with respect to genetic test results. The first group in our study is comprised of women in a large kindred whose members are at risk of carrying a specific mutation of the BRCA1 gene that dramatically increases their risk of developing breast and ovarian cancer. These women agreed to be tested for this mutation and they know their genetic test results. Their insurance companies do not have this information unless these women choose to reveal their test results to them. We separate these women into two sub-samples: those who tested positive for the mutation and those who tested negative. The behavior of both groups of women with regard to life insurance holdings four months after having been tested is compared to that of a group of women in the general public who have not undergone genetic testing. Comparisons across these groups allow us to assess the potential behavioral implications of asymmetric information in the life insurance market attributable to one particular genetic test—the test for a BRCA1 gene mutation.

2. BACKGROUND

2.1. Evidence of Adverse Selection

Adverse selection in insurance markets refers to the tendency for consumers who have an elevated risk of experiencing an event (e.g., being diagnosed with breast cancer) to use information, about which insurers are unaware, to obtain greater levels of insurance protection per dollar than if no information asymmetry existed. Several conditions must be met for adverse selection to occur: (1) there must be heterogeneity of risk among consumers, (2) insurers must not be able to detect (and therefore cannot appropriately price) this risk, (3) consumers must have access to information that allows them to predict their personal risk better than insurers, and (4) consumers must seek and obtain additional insurance coverage based on this information. These conditions, in combination, imply that high-risk consumers will have different demand curves than low-risk consumers.

While the role of adverse selection in insurance markets has been a topic of theoretical research within economics for over thirty years, empirical investigations that test for evidence of adverse selection have been rather limited. The empirical work that has been done generally demonstrates that individuals seek coverage that reflects their perceived level of risk but it does not document the extent to which consumers exploit any information asymmetries that may exist very well. Moreover, most empirical studies focus on the market for health insurance (e.g., Browne, 1992; Frank and McGuire, 1986; Marquis and Phelps, 1987) which differs from the market for life insurance in some key attributes that will be described later.

The two studies that test for adverse selection in the life insurance market rely on indirect evidence and reach different conclusions. Beliveau (1984) examines the relationship between the quantity of insurance purchased and coverage cost, using data from the Life Insurance Marketing Research Association, and finds that, holding age constant, the amount of insurance purchased is a positive function of its marginal cost. This willingness to pay a "quantity surcharge" suggests that people with higher perceived risks are selecting greater coverage even though it can only be obtained at a higher per unit cost. Cawley and Philipson (1996) examine the relationship between self-perceived risk and both the price and quantity of insurance purchased. In contrast to Beliveau (1984), they find little evidence of adverse selection and they conclude that "models of insurance exaggerate the superiority of the buyer's information relative to that of sellers," (p.27).

Pauly (1986) argues that the absence of empirical tests for the presence of adverse selection in the academic literature is in large part a function of the difficulty researchers have had in identifying a strong test for its presence. Indeed, Beliveau (1984) and Browne (1992) both note that their empirical tests provide necessary but not sufficient conditions to conclude that adverse selection exists. Insurance company representatives argue that the absence of evidence supporting their claims of adverse selection is simply an indication that insurance underwriters are doing their job well. That is, by gathering information from consumers on their family history, current health status, and risky behaviors (e.g., smoking), underwriters are able to place consumers in appropriate risk pools — pools that have different insurance prices associated with them — thus avoiding the potential economic losses that would occur if consumers engaged in adverse selection.

2.2. The Public Policy Debate Regarding Insurance and Genetic Testing

Despite the limited evidence that consumers engage in adverse selection, both insurance companies and privacy advocates are taking strong positions on the use of genetic test information in insurance underwriting. Insurance companies view adverse selection induced by keeping genetic test information private as unfair to them and to the low-risk consumers who, because of adverse selection, may pay unnecessarily high rates. In the extreme, insurance companies contend that adverse selection is a threat to the entire private insurance system. Consequently, the position of the American Council of Life Insurance is that it "would strongly oppose any proposed limitation or prohibition of insurers' right to have access to and to underwrite on the basis of genetic information or genetic tests" (American Council of Life Insurance, 1994:24).

Privacy advocates, in contrast, believe that genetic test results should be the exclusive property of consumers. They fear that test results will be used to deny insurance coverage or make it unaffordable and that this danger outweighs any potential danger of adverse selection. One such advocate described adverse selection as "not that big a factor," comparable to the unfortunate but controllable losses that occur due to shoplifting in the retail environment (Billings cited in Chase, 1992). Similarly, the Council for Responsible Genetics (1995), in testimony to the National Association of Insurance Commissioners Working Group on Genetic Testing, claimed that there is "no reliable evidence to suggest that consumers at risk for genetic conditions are gaming [the life and disability insurance] markets" (p.8).

In response to the growing controversy, policy makers have begun to enact legislation that will regulate who has access to genetic test results and how this information may be legally used. During the last congressional session, five bills were introduced that would have prohibited genetic discrimination with regard to health insurance and employment. At the state level, sixteen states have passed legislation regarding privacy of genetic test results and insurance (National Association of Insurance Commissioners, 1996). The intent of the state legislation varies considerably, however, depending on whether the law is directed at the health insurance or life insurance industries. In the case of the nine states where the statutes focus solely on health insurance, the laws prohibit the use of genetic test results to deny health insurance coverage and/or to establish a differential premium rate based on the results of genetic information. Similarly, the four states that have passed laws that deal exclusively with tests for the sickle-cell trait all stipulate that it is illegal to deny life or health coverage to someone solely because they have tested positive for this genetic mutation. Finally, three states (Arizona, Maryland, and Montana) have passed legislation directed at the use of genetic testing more generally in life and health insurance underwriting. These three statutes specify that an insurer may not deny coverage or charge differential premiums based on genetic test results *unless there is an actuarial justification for doing so based on the genetic trait in question*.

There are key differences between the markets for life and health insurance that may affect both firms' and consumers' behavior as well as policymakers' responses to the concerns that are being raised regarding the use of new genetic test information in these markets. First, the full price of life insurance is typically paid by the consumer even if it is obtained through an employer. In contrast, employers often bear a large portion of the costs of many health and disability insurance policies as part of their employee benefit packages. Second, the life insurance market contains considerable fluidity. That is, consumers generally have choice over a wide range of life insurance policies and they need not wait for an "open enrollment" period to change coverage. Thus, adverse selection induced by learning genetic test results, if present, could potentially reveal itself rather quickly. Finally, while consumers may have an incentive to reveal positive genetic test results voluntarily to their health insurers in certain situations (e.g., if a woman tests positive for a gene mutation that is associated with breast/ovarian cancer, she may want her health insurance company to know her test results, if in turn, the company would then cover the cost of a prophylactic double mastectomy—a radical and costly procedure that in a recent study was found to reduce the risk of breast cancer for women in high risk families (Haney, 1997)), consumers have no incentive to voluntarily share positive test results with their life insurance companies. A consumer who tests negative for a gene mutation may have an economic incentive to share his/her test results with his/her life insurance company voluntarily. If the company has been using family history information about a genetically linked disease to place the individual in a high risk pool, then the individual may be able to use information about his/her negative test results to gain a reduction in life insurance premiums for a given level of coverage. These three factors work in combination to create a situation where the potential for adverse selection in the life insurance market must be taken seriously. At the same time, because life insurance (as opposed to health insurance) is typically *not* viewed as an economic necessity, policymakers may be quick to respond to industry demands regarding their need to know genetic test results and to underwrite policies based on these results. As a consequence, it is imperative that we learn more about consumers' behavior in the life insurance market once consumers acquire information about their genetic status that is not currently available to insurance companies.

3. METHODS

3.1. The Data

We assess the impact of the potential information asymmetry in the market for life insurance created by genetic testing through comparisons across two groups. The first group is drawn from a large kindred residing in Utah and Idaho whose members have been extensively studied at the University of Utah. Many of these family members have been found to have a specific BRCA1 gene mutation on the long-arm of chromosome 17. This kindred, known as K2082, is the largest family studied to date with a BRCA1 mutation. Women in K2082 who are BRCA1 gene mutation carriers are estimated to have an 85 percent lifetime risk of developing breast or ovarian cancer. As such, those women who test positive face a substantially elevated risk of contracting a potentially fatal disease during their lifetimes. Women in this family who test negative for the gene mutation have the population risk for breast and ovarian cancer.

Four hundred and ten adult women and men (all Caucasian) in K2082 have been recruited into a longitudinal study funded by the National Cancer Institute (NCI) that examines the behavioral and psychosocial consequences of genetic testing (Botkin, et al., 1996). Recruitment of these subjects began in October 1994 and ended in March 1997, although testing will continue for several more months. When these individuals were initially approached, they were told that they were part of an extended family whose members have an elevated risk of developing breast and/or ovarian cancer. Each individual was then invited to participate in the study and given extensive informed consent documents to review.

If the individual agreed to participate in the study, a baseline survey was scheduled and completed by telephone. Next, the individual met with genetic and family counselors at the University of Utah Medical School to discuss the advantages and disadvantages of being tested for the BRCA1 gene mutation. Subjects were told that their test results were confidential and members of the research team would not share subjects test results with any third parties, including insurers. If, after counseling, the individual still wished to be tested, a blood sample was taken and the test conducted. The individual then returned to the University of Utah to meet again with a genetic counselor to learn his/her test results and to receive *non-directive* counseling regarding what the results mean. At this point, 250 study participants have been tested. But, because recruitment has occurred over an extended period, the tested kindred members are at various post-test survey stages.

The post-test study protocol begins with a telephone interview one week after the individual learns her/his genetic test results. The next interview occurs four-to-six months after learning the test results. This is followed by one-year and two-year post-test telephone interviews. Topics covered on the post-test surveys range from assessments of psycho-social functioning to life and health insurance purchasing decisions. Letters confirming upcoming interviews that include insurance questions also ask participants to locate life insurance documents in advance so that these documents might be consulted when answering specific survey questions.

Currently, 235 women and men in K2082 have been tested and have also completed the four-month follow-up survey. There are 117 women among these 235 who are between the ages of 18 and 55 and they form the subsample of tested individuals in our analyses. Among these 117 women, 79 of them (68%) have at least one first or second degree female relative who has been diagnosed with breast and/or ovarian cancer (the criteria we use to identify someone as having a "family history" of breast/ovarian cancer). Thirty-eight of the 117 (32.5%) tested positive for the BRCA1 gene mutation and 79 tested negative. All insurance information for these women has been taken from the four-month post-test interview.

The second group of Caucasian women in our study have been interviewed as part of an NCI-funded project that focuses specifically on life insurance purchasing behavior among women with a family history of breast/ovarian cancer. The sample of 169 women who have *not* been tested for the BRCA1 gene mutation was drawn from two sources: (1) women, age 18–55, who do not have breast or ovarian cancer but who had been referred to the University of Utah's High Risk Breast Cancer Clinic because they have one or more first-degree relatives who had been diagnosed with breast and/or ovarian cancer, and (2) a sub-sample of women, age 18–55, who participated in a state-wide health interview survey that was conducted in Utah during the summer of 1996. Women who participated in the state-wide, random sample survey, who said that they did not mind being re-contacted, formed the eligible pool of respondents. From that pool, we over-sampled those women who reported having at least one first or second degree female relative affected with breast and/or ovarian cancer. Among the 169 women in the non-tested sample, 99 (58.6%) are classified as having a family history of breast/ovarian cancer and 70 (41.4%) are not.

Women in the non-tested sample were recruited by telephone. If they agreed to participate in the survey they were sent a consent form which they had to sign and return. In the cover letter accompanying the consent form, they were also asked to locate insurance documents so that they could refer to them when answering questions during the interview.

3.2. Assessing the Potential for Adverse Selection

Standard economic models of insurance markets presume that consumers know more about their own risks than do insurers. Insurers respond to this information asymmetry by categorizing insurance applicants into risk pools based on observable characteristics that are highly correlated with actual mortality risk. Thus, when completing life insurance applications, consumers must typically provide information on such attributes as age, occupation, personal and family health histories, and personal smoking history. Consumers classified in the higher risk groups are then charged higher prices for a given level of coverage than are consumers in lower risk groups (or they are denied coverage altogether). The premium difference reflects the higher risk of mortality that members of these former groups face. In this context, life insurance company representatives argue that consumer knowledge of genetic predisposition to potentially life-threatening diseases must be shared with insurance companies so that policies may be appropriately underwritten.

The insurers' concerns regarding possible adverse selection associated with genetic testing stems from their belief that consumers who test positive for gene mutations that are associated with potentially fatal diseases will exhibit different price elasticities of demand than individuals who test negative and/or individuals who are not tested. And, if insurers do not have access to the same genetic test information, they will be unable to classify consumers into appropriate risk groups. In turn, this will lead some consumers to purchase higher amounts of life insurance at prices that are below what would be commensurate with their actuarial risk.

In the context of our study, a rather straightforward test for evidence of adverse selection is to compare the demand for life insurance among sample members who test positive, sample members who test negative and sample members who are not tested for the BRCA1 gene mutation. The strict test would require the simultaneous estimation of price and quantity equations. In this structural formulation, premium price per unit of insurance would be a function of the quantity of life insurance coverage and the vector of consumer characteristics known to the insurer to affect the risk of mortality (and hence the cost of insuring the consumer), (e.g., age, family and personal health history). The amount of life insurance coverage purchased, in turn, would be a function of the per unit price, individual characteristics that capture the consumer's risk aversion (e.g., income, education, number of dependents, risk attitude scales), and the individual's genetic test status. This two-equation system would also differentiate term insurance policies from life insurance policies that contain savings features. And, it would correct for the possible selectivity bias inherent in using data from only those respondents who report holding life insurance (and hence can provide price information).

Unfortunately, estimation of the structural equations model of the price and quantity of life insurance purchased is not yet possible because of the absence of insurance price and policy type information (which is currently being gathered from the members of K2082) and because of sample size considerations (which will lessen as more study participants get to the point where they participate in the four-month and one-year post-test interviews). Instead, we estimate a reduced form model that focuses on how knowledge of genetic test results influences the likelihood of having life insurance.

Standard measures of consumer risk aversion included in our reduced form model include age, the presence or absence of minor children in the home, current marital status, education, current employment status, and a scale (ranging from 0 to 100) that reflects the respondent's subjective assessment of the likelihood she will live to age 65. (These are fairly standard covariates perhaps with the exception of respondent's subjective assessment of her own mortality. This latter measure was also used with some success by Cawley and Philipson (1996) in their analyses of life insurance demand.) With the exception of age, the predictions regarding most of these variables are fairly straightforward (e.g., women with higher incomes have a greater economic loss to insure against and hence they should have a higher probability of carrying life insurance compared to women with lower incomes). Age is posited to affect insurers' assessments of mortality risk as well as consumers' levels of risk aversion. (Personal smoking history is another consumer characteristic known to the insurer that should affect insurance costs. We omit controls for personal smoking history from our estimating equation because over 85 percent of our sample identify themselves as devout members of the Church of Jesus Christ of Latter Day Saints (Mormons) and members of this church are prohibited from smoking. In Utah, where approximately 35–40% of the population identify themselves as devout Mormons, the adult smoking rate is the lowest of any state in the country (<16%) (Zick and Mayer, 1996).) If insurers capitalize on the fact that older women have higher mortality risks, then age should be positively correlated with premium price (omitted from this reduced form model) and negatively related to the probability of having life insurance. At the same time, older women may assess their relative need for insurance to be greater (e.g., they are more likely to have a home and want enough insurance to cover the remaining mortgage balance if they were to die), thus raising the probability that they will have insurance. On balance then, no unambiguous prediction about the effect of age can be made.

The two key covariates in our model that test for the presence of adverse selection are family history of breast and/or ovarian cancer and genetic test status. Family history raises the risk of mortality. But if the insurer has access to this information, the premiums charged to such high risk individuals will be adjusted upward and, on balance, this should result in family history having no impact on the probability of having life insurance. Similarly, testing positive for a deleterious BRCA1 gene mutation also increases the mortality risk. But, this information is known to the woman but not the insurance company. As a consequence, the potential for adverse selection exists. If we find a positive statistically significant coefficient associated with testing positive this would be supportive of the adverse selection hypothesis.

Three different reduced form models are estimated. The first model excludes all genetic test results information. The second model adds the genetic test result variables to the set of regressors. And, the third model interacts genetic test results with family history of breast/ovarian cancer. Estimation of these multivariate models is done using a probit estimation routine because of the dichotomous nature of the dependent variable.

4. RESULTS

Descriptive information is presented for the full sample and the K2082 and non-tested sub-groups in Table 1. The two sub-groups are similar on all of the key risk dimensions. The modest mean differences in income and the presence of minor children between the two samples are not statistically significant. Interestingly, the percentage of women having life insurance is moderately higher for those who have been tested compared to those who have not been tested, and, both fractions are somewhat above the average for all women in the United States. Nationally, it is estimated that about 65% of adult women carry life insurance (American Council on Life Insurance, 1994). The higher percentage of women holding life insurance in our sample (73% overall) may reflect the age restrictions we have put in place and/or the fact that we have over-sampled women with a family history of breast/ovarian cancer. It may also reflect the fact that while women in Utah have labor force participation rates that mirror the national average, they are more likely than their counterparts in other states to be married and to have minor children in the home—thus, increasing their need, on average, for insuring against the economic loss that would occur if they were to die prematurely.

Table 2 contains the multivariate probit parameter estimates of probability of having life insurance. Factors that are consistently associated with the presence or absence of life insurance include household income, age, and employment status. Employed women may have a higher probability of purchasing life insurance compared to non-employed women because of the potential loss in income that they are insuring against. At the same time, employment status may be picking up some price differentials as approximately 15 percent of all respondents report that their employer pays part or all of the premium for their primary insurance policy. Estimated coefficients associated with the presence or absence of minor children and marital status are statistically significant only in model 3. Education, and subjective assessments of longevity never reach conventional levels of statistical significance in any of the equations. These results are robust and parameter estimates vary little when alternative definitions of family history are used. Estimates obtained when

Table 1. Descriptive statistics: women, ages 18–55

	Tested sample (N=117)		Non-tested sample (N=169)		Combined samples (N=286)	
Variables	Mean	S.D.	Mean	S.D.	Mean	S.D.
Annual household income ($)	43,049	23,887	48,874	25,323	46,491	24,869
Has minor children (1=yes)	0.68	0.47	0.73	0.44	0.71	0.45
Currently married (1=yes)	0.81	0.39	0.80	0.40	0.80	0.39
Education (yrs.)	13.78	1.65	13.91	2.06	13.85	1.90
Chances of living to age 65 (0-100 scale)	83.25	21.30	87.01	14.80	85.34	18.40
Age (yrs.)	37.33	9.62	36.62	9.50	36.91	9.54
Currently employed (1=yes)	0.63	0.48	0.61	0.49	0.62	0.49
Family history of breast/ovarian cancer (1=yes)	0.68	0.47	0.59	0.49	0.62	0.48
Tested positive (1=yes)	0.32	0.47	—	—	0.13	0.34
Has life insurance (1=yes)	0.76	0.43	0.72	0.45	0.73	0.44
Number of life insurance policies held	1.09	0.86	1.17	1.01	1.14	0.95

Table 2. Probit parameter estimates of the probability of having life insurance: women, ages 18–55 (t-ratios in parentheses)

	Model 1		Model 2		Model 3	
Independent variables	Coefficients	Marginal effects[a]	Coefficients	Marginal effects[a]	Coefficients	Marginal effects[a]
Constant	-1.45	-0.44	-1.61	-0.48	-1.79	-0.53
	(-1.72)*		(-1.87)*		(-2.05)**	
Annual household income ($1000's)	0.02	0.006	0.0002	0.006	0.02	0.006
	(4.24)**		(4.37)**		(4.24)**	
Has minor children (1=yes)	0.23	0.07	0.25	0.07	0.28	0.08
	(1.05)		(1.16)		(1.30)*	
Currently married (1=yes)	0.31	0.07	0.28	0.08	0.36	0.10
	(1.25)		(1.13)		(1.40)*	
Education (yrs.)	-0.04	-0.01	-0.04	-0.01	-0.04	-0.01
	(-0.90)		(-0.87)		(-0.92)	
Chances of living to age 65 (0–00 scale)	-0.0009	-0.0003	-0.0005	-0.0001	-0.0004	0.0001
	(-0.18)		(-0.10)		(-0.08)	
Age (yrs.)	0.03	0.01	0.03	0.01	0.03	0.01
	(3.40)**		(3.30)**		(3.35)**	
Currently employed (1=yes)	0.40	0.12	0.39	0.11	0.39	0.11
	(2.13)**		(2.08)**		(2.08)**	
Family history of breast/ ovarian cancer (1=yes)	0.19	0.06	0.16	0.05	—	—
	(1.02)		(0.87)			
Tested positive (1=yes)[b]	—	—	0.31	0.09	—	—
			(1.09)			
Tested negative (1=yes)[b]	—	—	0.23	0.07	—	—
			(1.10)			
Family history and tested positive (1=yes)[c]	—	—	—	—	0.61	0.18
					(1.76)**	
Family history and tested negative (1=yes)[c]	—	—	—	—	0.34	0.10
					(1.12)	
Family history and not tested (1=yes)[c]	—	—	—	—	0.38	0.11
					(1.63)*	
No family history and tested positive (1=yes)[c]	—	—	—	—	0.40	0.12
					(0.72)	
No family history and tested negative (1=yes)[c]	—	—	—	—	0.76	0.22
					(2.01)**	
Chi-Square	63.36**		65.33**		68.51**	

* p .10, one-tailed test
** p .05, one-tailed test
[a] All marginal effects have been computed at the mean values for the independent variables.
[b] The omitted group in this sequence of dummy variables is those respondents who were not tested for the BRCA1 gene mutation.
[c] The omitted group in this sequence of dummy variables is those respondents who reported no family history of breast and/or ovarian cancer and who were not tested for the BRCA1 gene mutation.

family history is defined as (1) having a mother with breast/ovarian cancer, or (2) having two or more first or second degree female relatives with breast/ovarian cancer, are virtually identical to those presented in Table 2. These alternative estimates are available from the authors upon request.

Goodness of fit tests that examine the improvement in the equation's performance as we move from model 1 to the other two formulations, reveal that model 1 is preferred on statistical grounds. (The critical chi-square value (p =.05) is 5.99 with 2 degrees of freedom (df) and 9.49 with 4 df. The computed chi-square values are: models 2/1 (2 df) = 1.98, models 3/2 (2 df) = 3.16, models 3/1 (4 df) = 5.14.) That is, the addition of genetic test results variables (not interacted or interacted with family history) does not improve the overall goodness of fit of the estimating equation relative to the formulation that ignores genetic test results. This suggests that four months post-testing, there is little evidence of adverse selection among the women in K2082. Despite this finding, we elect to discuss the results of all three estimating models because the overall goodness of fit calculations may be influenced by the relatively small group-specific sample sizes used in the analyses.

Model 1 estimates reveal that a family history of breast and/or ovarian cancer does not affect the probability of a woman having life insurance. This is not a surprising result. If life insurance companies are effectively using family cancer history in the underwriting process, then the relatively higher premiums quoted to women with a family cancer history should serve to negate any possible adverse selection attributable to this characteristic.

In contrast, information regarding genetic test results is known by the consumer but not by the insurer. Thus, if women who test positive use this information asymmetry to purchase insurance at premiums that they know to be below what would be actuarially fair, we would expect to see a statistically significant positive coefficient associated with testing positive in model 2. While the estimated coefficient is positive, it does not reach conventional levels of statistical significance (p=.14, one-tailed test). Furthermore, the coefficient associated with testing negative is also positive, though not statistically significant (p=.14, one-tailed test), suggesting that going through the testing process, *per se*, may serve to increase the demand for insurance.

Model 3 contains the coefficient estimates when family history and testing status are interacted. Relative to women who are not tested and who have no family history of breast/ovarian cancer, three groups are significantly more likely to purchase life insurance: women with a family history who test positive (p=.04, one-tailed test), women with a family history who are not tested (p=.05, one-tailed test), and women with no family history who test negative (p=.02, one-tailed test). The positive coefficients associated with (a) having a family history and testing positive, or (b) having a family history but not being tested, are both supportive of the adverse selection hypothesis. The significant positive coefficient associated with having no family history and testing negative, however, is not consistent with the adverse selection hypothesis. As mentioned earlier, it may be that the testing process and the accompanying genetic counseling heightens the women's sensitivity to mortality risks. And, this heightened awareness may have its greatest impact on those women who had no family history of breast/ovarian cancer because they were likely unaware that they were in a high risk family prior to being enrolled in this study.

The absence of a statistically significant positive coefficient associated with testing positive but having no family history is also not consistent with the prediction of the adverse selection hypothesis. It is this group that has the most to gain by the information asymmetry (i.e., they are not currently identified as being in a high risk pool based on

Table 3. Crosstabulations of number of life insurance policies held by group membership (column fractions in parentheses)[a]

	Group					
Number of policies	Family history & tested positive (N=28)	Family history & tested negative (N=52)	No family history & tested positive (N=10)	No family history & tested negative (N=27)	Family history & not tested (N=99)	No family history & not tested (N=70)
0	6	15	2	5	26	22
	(.21)	(.29)	(.20)	(.19)	(.26)	(.31)
1	14	24	7	13	41	24
	(.50)	(.46)	(.70)	(.48)	(.41)	(.34)
≥2	8	13	1	9	32	24
	(.29)	(.25)	(.10)	(.33)	(.32)	(.34)

Chi-square = 7.86 (p=.64)
[a] Fractions may not sum to 1.00 due to rounding error.

their family history, yet they have tested positive). The small numbers of respondents in this group (N=10) may be partially responsible for the absence of any detectable effect, however. As more K2082 subjects are followed through the four-month (and one-year) interviews, we will want to take a closer look at this group.

The presence or absence of life insurance is a qualitative measure of consumer demand. The number of life insurance policies a woman has is another, more detailed indication of demand. In our sample, the number of life insurance policies held range from 0 to 9, but only 22 of the 286 women (7.7%) reported having more than two policies. Ordered probit estimates of models 1–3 using number of policies as the dependent variable did not converge probably because so few women had more than two policies. As an alternative, we present the simple crosstabulation of number of life insurance policies a woman has (collapsing all of those who have two or more policies together) by the testing/family history groups in Table 3. The associated chi-square test reveals that one cannot reject the hypothesis that the number of insurance policies held is equal across the testing/family history groups. That is, the figures in Table 3 do not support the adverse selection hypothesis at the bivariate level.

5. DISCUSSION AND CONCLUSIONS

A strong test of the adverse selection hypothesis awaits the completion of the on-going collection of price and quantity information and the addition of more K2082 sample members who will be completing the four-month (and one-year) post-test interviews in the near future. Nonetheless, the preliminary analyses done in this paper suggest that adverse selection in the life insurance market is not strongly evident among women who have learned that they carry the BRCA1 gene mutation. Only in model 3 did the coefficients on test results reach conventional levels of statistical significance and formal tests revealed that model 1 should be preferred over both models 2 and 3.

Why are women who carry this potentially life threatening gene mutation not exploiting their informational advantage in the market for life insurance? One possible answer is that positive genetic test results may simply serve to confirm something that many of these women *and* their insurers have known all along based on their family histories. If learning that they have tested positive simply confirms the suspicions of many of these women, it may have little behavioral impact on their life insurance purchasing behavior. Consistent with this argument is the research finding that women of all ages who test positive for the BRCA1 mutation experience little change in psychological well-being (as measured by anxiety, intrusion, and avoidance scales) between the pre- and post-test interviews because for many it only serves to affirm what they already suspected (Croyle, et al., 1997).

Alternatively, the reason we observe little evidence of adverse selection could be because our follow-up period of four months is too short. Behavioral changes regarding life insurance may only be observed with a longer follow-up period. To address this concern we re-estimated our three models using the non-tested sample and the 74 female members of K2082 age 18–55 who have already completed the one-year post-test survey. The estimates for all three models are quite similar to those obtained using the four-month post-test sample. The one exception is that the positive coefficient associated with having no family history and testing negative for the BRCA1 gene mutation is no longer significant using conventional statistical thresholds (p=.41). Thus, we have every reason to believe that the estimated behavioral response will not change as the follow-up period is lengthened.

Our conclusions regarding the possible threat of adverse selection in the life insurance market must be tempered, not only by the preliminary nature of the analyses done here, but also by the characteristics of the participants in study. The analyses make use of a somewhat homogeneous sample. All of the study participants are Caucasian women and the vast majority of them identify themselves as being active members of the Mormon Church. The impact of asymmetric information on the demand for life insurance may vary by gender, ethnicity, and personal life style (as proxied by religious affiliation). And, we focused on the testing for one gene mutation that is associated with two potentially fatal illnesses, breast and ovarian cancer. Learning the results of a test for a gene mutation that is associated with a non-life-threatening disease (e.g., the sickle cell trait), or learning the results of a test for a gene mutation that leads to the contraction of a disease that is always fatal (e.g., Huntington disease) may have a different behavioral impact. Finally, the study protocol required that the women who were tested have at least two meetings with a genetic counselor. The non-directive, educational content of these counseling sessions may have affected the post-test behaviors of these women in a number of ways — including, possibly, their life insurance purchasing behavior. It is unlikely that such counseling will be widely available as genetic testing spreads to other high-risk families and the general population.

Although the current work is preliminary and its generalizability may be limited, our findings suggest that policymakers should be very cautious as they move forward in the legislative process. They should view with some skepticism the insurance industry's contention that denial of access to genetic test results (and hence the denial of their use in underwriting) will threaten the industry's economic viability. Certainly, the results of this study bring this claim into question.

And, even if the use of genetic test information in the underwriting process where to help prevent industry losses, this possible benefit must be weighed against the costs of allowing such information to enter the public domain. Would broader access to test results ultimately discourage high risk individuals from being tested? Who would have access to the genetic test results and how would we insure that this information is used in a responsible manner? Already, the Council for Responsible Genetics has documented over 200 cases of coercion and/or genetic discrimination in employment and insurance (McGoodwin, 1997). And, in a recent survey of genetic support group members (an admittedly selective group), it was reported that 22% had been refused health insurance coverage, 25% had been refused life insurance, and 13% had been denied employment because of a genetic condition (Lampham, Kozma, and Weiss, 1996). Current legislative protections against genetic discrimination are weak and patchwork at best and there are no legislative provisions regarding coercion in testing (e.g., situations where an employer or insurer might require someone to undergo genetic testing before hiring or issuing an insurance policy). At the federal level, the 1995 revision to the compliance manual for the Americans with Disabilities Act clarifies that genetic discrimination in employment should be considered under this act. This provision has not yet been tested in the courts, however. The 1996 Health Care Reform Act also provides some protection within the health insurance domain by guaranteeing health insurance portability across jobs. But, it says nothing about the price an employee must pay for such insurance if an individual's medical records reveal that s/he has tested positive for a gene mutation associated with a serious disease. Discussion of the wider range of public policy issues associated with genetic testing from the states' perspective can be found in a report put together by the Council of State Governments (1993). Wider access to genetic information is certain to spark the need for detailed articulation of what constitutes legitimate and ethically responsible uses of such tests and the genetic information they reveal.

ACKNOWLEDGMENTS

The research reported in this manuscript was supported in part by NCI grants R03 CA70669 and R01 CA63681. Diana Lane, Jean Nash, and Debra Dutson provided invaluable survey management support for this project.

REFERENCES

American Council of Life Insurance (1994), 'The need for genetic information in risk classification', Washington, DC.

Beliveau, B. (1984), 'Theoretical and empirical aspects of implicit information in the market for life insurance', *Journal of Risk and Insurance*, Vol. 51, pp. 286–307.

Billings, P.R. cited in Chase, D. (1992), 'In the battle over genetic testing, critics must be reconed with', *Council Review*, Vol. 17:9–10.

Botkin, J.R., Croyle, R.T., Smith, K.R., Baty B., Lerman C., Goldgar D., Ward, J., Flick, B., and Nash, J. (1996), 'A model protocol for evaluating the behavioral and psychological effects of BRCA1 testing', *Journal of the National Cancer Institute*, Vol. 88, pp. 872–882.

Browne, M.J. (1992), 'Evidence of adverse selection in the individual health insurance market', *Journal of Risk and Insurance*, Vol. 59, pp. 13–33.

Cawley J. and Philipson, T. (1996), 'An empirical examination of information barriers to trade in insuamce', working paper, University of Chicago.

Council for Responsible Genetics (1995), *Testimony to the National Association of Insurance Commissioners' Genetic Testing Working Group*, March 12.

Council of State Governments (1993), *Advances in genetic information — A guide for state policy makers*, Iron Works Pike: Lexington, KY.

Croyle, R.T., Smith, K.R., Botkin, J.R., Baty, B., and Nash, J. (1997), 'Psychological responses to BRCA1 mutation testing: Preliminary findings', *Health Psychology*, Vol. 16, pp. 63–72.

Frank, R.G. and McGuire, T.G. (1986), 'Adverse selection in insurance markets: A selective survey', in Dionne, G. (ed.), *Contributions to Insurance Economics*, Kluwer Academic Publishers: Boston, MA, pp. 97–140.

Haney, D.Q. (1997), 'Double mastectomy radical, effective', *Wisconsin State Journal*, April 14, p. 1A.

Lapham, E.V., Kozma, C., and Weiss, J.O. (1996) 'Genetic discrimination: Perspectives of consumers', *Science*, Vol. 274, pp. 621–624.

Marquis, M.S. and Phelps, C.E. (1987), 'Price elasticity and adverse selection in the demand for supplementary health insurance', *Economic Inquiry*, Vol. 25, pp. 299–313.

McGoodwin, W. (1997), 'Biotechnology and the consumer.' Esther Peterson Consumer Policy Forum Speech at the American Council on Consumer Interests' annual conference, April 5, and forthcoming in *The Journal of Consumer Affairs*.

National Association of Insurance Commissioners (1996), *Report of the Genetic Testing Working Group to the Life Insurance (A) Committee*, June.

Pauly, M.V. (1986), 'Taxation, health insurance, and market failure in the medical economy', *Journal of Economic Literature,* Vol. 24, pp. 629–675.

U.S. Congressional Office of Technology Assessment (1988), *Mapping our genes: How big, how fast?* OTA-BA-373, U.S. Government Printing Office: Washington DC.

Zick, C.D. and Mayer, R.M. (1996), 'Consumption in the land of milk and (home-baked) cookies', in Heaton, T.B., Hirschl, T.A., and Chadwick, B. (eds.) *Utah in the 1990's — A demographic perspective*, Signature Books: Salt Lake City, UT, pp. 139–154.

6

GENETIC ENGINEERING AND GERMAN HEALTH INSURANCES

Thomas Uhlemann

Institute of Medical Sociology
University of Hamburg, Germany

1. INTRODUCTION

Technological innovation quite often results in far reaching changes within the matrix of society which forms a meaningful base for individuals or groups. Technological implementations often lead to a practice where the individual is burdened with the incurred costs. These costs, however, are not necessarily of a financial nature but range from difficulties in the assessment and the decision making process, responsibility in areas of high risk behaviour, to stigmatisation and discrimination.

A number of methods, especially diagnostic techniques based on molecular genetics, have been developed among others under the collective name of genetic engineering. Their development heralds a promise which is directed at two of the most fundamental aspects of human existence: health and security. Certainty with regards to health status and the elimination of illness and suffering minimalised risk is promised, thus suggesting the engineering of health and with it creating an illusion of far reaching control of life.

Innovations in genetic engineering indicate new ways for its application. The innovations therefore set new genetic standards and create a shift in the medical definition of health and illness. New illnesses are continuously diagnosed and an increasing number of people no longer comply with the "new norms". The "normal" turns into a deviation and those who were classed as "healthy" are now classified as "so-far-without-illness". The subsequent genetic deviation of health-norms may not only result in considerable problems for the individual but also affect society as a whole. Genetic know-how will be of importance to health and social politics when technical developments lead to the implementation of technology, thus posing the question of whether there is a need for legal empowerment within society.

At the same time covetousness is stimulated among those interest groups who deal with the economic aspects of health care. These are, apart from producers and distributors of tests and medics from various disciplines, primarily health insurance companies.

Genetic Information, edited by Thompson and Chadwick.
Kluwer Academic / Plenum Publishing, New York, 1999.

The following discourse focuses on the question of how the development of genetic engineering is received by the German insurance companies.

2. PROBLEM SITUATION

Genome tests for the diagnosis of hereditary diseases and genetically conditioned sensitivities towards environmental influences are increasingly available. In the USA, a well developed market for tests already exists as does the competition between various (private) suppliers. Numerous test-kits are freely available. Suppliers advertise, irrespective of the small practical value of many tests, by highlighting the tests' sensitivity and reliability.

This situation has lead to open discussions in Germany regarding the risks of this new technology. It has also resulted in state-lead activities evaluating economic and social consequences and the need for legislation. Finally, discussions were held among economic sectors where a broad application of genetic methods have already been or are likely to be used in the near future. Primarily, the farming industry, food, chemical industry and the pharmaceutical industry should be mentioned in this context, where a considerable variation in the application is possible.

Health insurances have also been discussed by various parties as possible areas for the application of gene tests from an early stage. However, the following questions arise:

a. How far would these new diagnostic key technologies change the financial planning of health insurances?
b. Would a need for social legislation arise?
c. What public health changes could be implemented?

3. PUBLIC DEBATES

Public discussion in Germany regarding the potential implementation of genetic tests within the health insurance system is embedded in the general debate surrounding the chances and risks genetic engineering entails and the current reform of the health system. Not only are various political parties and authorities involved, but also members from groups endeavouring to adopt an explicitly critical approach to research paradigms are involved in setting the agenda for these discussions.

The insured and applicants do not appear as independent protagonists although they as customers and contractual partners of insurance companies are in the end the deciding target group for state politics and insurances alike. Further, they are only indirectly involved and are, at best, represented by individual interest groups such as political parties and unions etc., who are represented in the self administrative organisation of the social health insurances.

The topic has been predominantly left to scientific and legal experts from industry and public administration, representatives from the private insurance market as well as organised medical professions. Questions on the use of genome testing have long been discussed in the European framework and competition regarding the responsibility and use of genetic diagnostic techniques (application and termination) exists within the various areas involved in the medical profession in Germany. Yet this discussion passes—with some exceptions—the social insurances by. Opportunities for influencing present discussions on the future of health insurances are therefore not taken.

Genetic engineering does not only evoke hopes for the future and fears in the above mentioned areas but also in the area of health insurances. The public has far reaching positive expectations regarding the opportunities genetic diagnosis offers and regarding the potential of somatic gene-therapy.

These expectations have been fuelled on the one hand by genetic researchers and politicians and on the other hand by the vast advances genetics has made. However, these somewhat unrealistic expectations regarding the application of genetic methods in a medical-curative context are in opposition to the considerable scepticism about genetic engineering as a whole.

Normative and informative aspects and the danger of possible cuts in service and increasing demand for additional contributions dominate the public discussion. The following will illustrate the main concerns.

3.1. Stigmatisation Due to Genetic Deviation

Insurance companies could be tempted to link their services to very restrictive behavioural standards. Pregnant women could for instance be asked to take genetic pre-natal diagnostics or to participate in other early examinations or to take courses on prevention. Based on recent examples in the USA, where in individual cases insurance companies refused the expected back payment of costs because parents had decided to have their child although the prenatal findings had shown the foetus to be disabled (Fromme and Raabe, 1997), there is fear among the insured in Germany that exclusion from insurance services or other exclusions may be introduced. Not participating in such diagnostic or preventative programmes might therefore entail lower payments by the insurances companies in case treatment is needed. This is comparable to the system in dental care where a so called Bonus-System is in operation. This form of economising is continuously promised to the insured by the medical councils and by state institutions and it also forms part of the government's endeavours to cut costs in health.

The Rule Book V of the Social Welfare Legislation does with its §1 indeed allow an adaptation of services according to appropriate criteria or the plan for financial sanctioning in the case of inappropriate behaviour and failure to co-operate.

Pessimistically this could mean that genetic risks involved say in a pregnancy, are in reality considerably less severe than the insurance companies let on. This would therefore result in the company refusing to take on potentially occurring costs if the child were to be diagnosed at an earlier stage with a genetic disorder. Not using any form of contraception or refusal to carry out an abortion turn into risk behaviour. Whether this can be agreed to in in a social context needs to be assessed, considering whether society would be burdened with avoidable costs incurred by individuals.

Using this as an example, disabled groups illustrate the apparent normative character of genetic diagnostics. They point out the danger of the general exclusion of genetically different humans and also the similarity of present cost-benefit analysis to those in the 20s and 40s. Genetic diagnostics could serve as a scientific justification for a new political movement in Eugenics.

Considering society as a whole, pre-natal tests could lead to euthanasia without governmental and legal guidelines, because a large number of parents, pressured by considerable social conformity might use negative test results as a reason to abort their unborn children who quite often are capable of life.

3.2. Care Obligations and Inappropriate (Abnormal) Behaviour

The thought that it may be in the interest of the insurance companies to broaden the term of "inappropriate behaviour" due to the establishment of special health related care obligations linked to genetic pre-dispositions is increasingly plausible. This becomes even more plausible when considered in the light of intensifying competition between individual health insurance companies, between social insurances and private providers and in the light of the effort of all insurance companies to improve their cost-structure which entails binding "low risks" and excluding "high risks". Social insurances also follow this logic since on the one hand their income through contributions has stagnated and on the other legal provisions do not allow a rise in contributions nor do they foresee it being linked to the inflation rate of gross income.

Critics of the implementation of genome analytical knowledge warn that private providers will hardly turn a blind eye to the new possibilities of differentiation whereas the social health insurances provide at least some protection from access to the individuals' health information. The insured fear that private health insurances will make contractual completion dependent on genetic tests. Here genetic tests would be selection criterion for the potentially insured and for those in need of paying obligatory risk contributions. The realisation of contractual agreements with health insurance companies could perhaps be on a voluntary basis, be legitimised by the existing general contractual freedom and be bound to an introductory genome test; coverage of cost incurred through illnesses due to genetic irregularities could be sought from supplementary insurances. These fears are fuelled by individual life insurance companies who introduce items regarding genome analysis in their questionnaire for potential customers.

3.3. The Transparent Human Being

The public is repeatedly asking whether—and if so to what extent—governmental or private interest groups are compiling a database on health related information in Germany and would this openly invite misuse. Health insurances in particular might be interested in databases in order to draw further distinctions between the risk groups their clients fall into. Risk differentiation based on genetic criteria could already take place due to the information supplied when completing a contract; it would, however, be in the interest of health insurance companies to have the right to question extended.

Further reservations exist with regard to the possible violation of the individual's privacy. Generally speaking, the vision of a far-reaching investigation into individuals' personalities entails a violation of basic human rights which is described as a negative futuristic vision. This prognosis is often coupled with a fear of general risk selection, which would consequently lead to the exclusion of larger population groups from protection against health costs.

4. GOVERNMENTAL POLITICS—POLITICAL DISCUSSION AND NEED FOR ACTION WITHIN JURISDICTION

Methods of genome analysis and its possibilities, dangers, further development and application substantiated governmental concerns in the mid to late '80s.

In order to deal with "Chances and Risks in Genetics" a working group was set up which was also concerned with the implications of genetics for health insurances.

Possible benefits for health insurances, legal regulations of the framework required for the application of genetic processes and questions emanating from this were also discussed in the German parliament.

Reflecting the fears voiced in public the working group of the parliament recommended that:

> in order to guarantee the protection of the applicant and prevent genetical screening potentially offensive to common decency, guidelines have to be drawn up by the health insurance companies as part of the supervision of insurances which consider the working group's developed principles.
> If a restriction of the implementation of genetic analysis cannot be reached this way, a change in the Rulebook concerning contracts between Insurances and Applicants needs to be considered. (Enquete-Kommission des Deutschen Bundestages 1990)

However, the Federal Government did not see a need for intervention regarding the implementation of genome analytical knowledge or procedures in the insurance sector.

5. THE POSITION OF HEALTH INSURANCE COMPANIES

Opportunities and possibilities for the implementation of knowledge and procedures derived from genome analysis are also being discussed within insurance markets. But both here and in public discussions distinctions have to be made between the social insurances (or comparable insurances) and private insurances companies. In comparison to the private sector, discussions within the social insurances are not as advanced. There has not yet been a "final formation of opinions". Representatives of the most significant insurance companies and their associations proved to be generally more defensive due to the rather frequent and critical public discussions.

Questions on the future links of gene technology with health insurances evoke comments which most commonly deal with technological advance in general and more specifically, Germany's ability to compete in the global market.

The question of whether substantial ethical responsibility and adherence to meticulous safety precautions can be guaranteed with the future implementation of genetic diagnosis is voiced as an additional reservation. Prevention especially is viewed as a potential use for genetic analysis; lifestyle changes might help slow the manifestation of a (partially) genetically evoked disease or possibly restrain its progress.

The social insurances in general do not yet pursue set policies regarding the implementation of genome analytical procedures; in future they will be—along with financial complications due to the increased demand for genetic testing—mainly affected indirectly.

The report by the German parliament's working group points towards other possible consequences the genetic procedures could have on the social insurances:

> Genome analysis can intensify the already existing tendency of private insurance companies to exclude risk patients. After the analysis they will be able to attract low risk patients with low premiums while risk patients are left to the Social Insurance.(Enquete-Kommission des Deutschen Bundestages 1990)

Whereas the social insurances (or comparable insurances) are not yet unified in their approach to the application of genetic procedures, the private health insurance companies have a much more accentuated and stringent argumentation.

It is here that the different interests between the private and the social insurances, and the private companies' involvement in public debate become apparent since the private insurances find themselves under suspicion of implementing genome based knowledge in order to carry out risk-differentiation. The publicly expressed fears are, of course, not acknowledged but dismissed.

6. THE CONCEPT OF INSURANCE

The insurance companies' main argument refers to circumstances that are based on a generalisation of individual's health costs. Protection against unknown risks is part of an insurances' agreed service and therefore the main concept of an insurance; a differentiation within the group of the insured according to different risk characteristics would therefore deem the axiom of actuarial mathematics inapplicable and undermine the concept of insurance.

The associations of leading private insurance companies are therefore generally disputing that genetic testing forms part of the rationality for the underwriting of specific risk-estimations.

Obviously the parliament's working group does not agree with the above: instead it vaguely recognises possibilities for genetic testing's future use. At present genetic tests are not regarded by private health insurances as profitable, but this could change in the future.

Health insurance companies need accurate data on their customers' current health status and anticipated illnesses in order to carry out cost- and risk calculations. Information on the genetic (pre-) disposition of the insured would allow a prediction of expected morbidity for the relevant clients of previously unimaginable accuracy.

A split within the groupings of the insured into certain risk categories could—to pick up on the argument the insurance companies made earlier—only undermine the axiom of actuarial mathematics, especially if the insured informed themselves on their individual, identifiable health predispositions and consequently developed their own insurance strategies.

It can therefore be said that the above mentioned fears of the insurance companies result in the following question: Is it possible to maintain ignorance among the insured as a foundation for insurance or is detailed information held by the insured going to lead to an improper use of this knowledge, according to their understanding and the meaning of the rules concerning contracts with insurances (§16). This could result in the insuring of someone with bad risks because they have not complied with the obligatory disclosure of relevant information.

It is conceivable that health insurance companies will also offer cut-rates or other enticing conditions, comparable to life and car insurances etc., for certain characteristics since they are interested in keeping the low-risk customers.

However, DNA analysis plays a minor role in current health insurance practice. Until now health checks prior to taking out insurance included the overall known health history of the applicant but did not include—with some exceptions—the obligation to carry out certain medical examinations. Further there have not been any contractual conditions that were better for some individuals, such as special premiums or reductions in contributions, if a comprehensive (voluntary) examination was carried out prior to taking out the contract. Applicants are obviously legally obliged to pass on information on examinations, previous illnesses and treatments etc.

This law applies particularly to aspects relevant to health issues and includes insights gained through genetic analysis. No differences are made between conventional and genetic diagnostics. Applicant and insured alike have to release the previous health insurance companies and medical doctors of the pledge of confidentiality. The insurance companies can therefore check the applicant's details on previous illnesses and received treatments.

Private health insurances, unlike the social ones who are bound to contractual obligations, can refuse a contract or demand higher contributions. However, the unrestricted ability of private health insurances to identify aspects that do not only reflect current health status but also have a bearing on the future status is legally problematic. How far reaching is the obligation for disclosure of such information, in the light of any existing conflicts with various human rights laws?

After having taken out personal insurance the question arises whether the obligatory pre-contractual report has been completed by the insured. If this is the case, then genome analyses, which were carried out after the insurance was taken out neither results in the exclusion from benefits for certain illnesses nor in the right for special benefits.

7. CONCLUSION

Presently DNA tests are neither requested nor used when health insurances are taken out in Germany. However, most notably the private health insurances have investigated the possibilities for and the future implementation of genome analysis and any implications it would have for the health insurances market. The focus has been both the increasing accuracy, economic viability and the availability of diagnostic genetic tests and the demand on behalf of the insured.

The implementation of genetic testing is presently not included in the insurances' risk calculation. It needs to be investigated whether insurances should have the right to access their clients' genetic information. Although this implies that at present there is no apparent economic use (benefit) for genome analysis, it is possible that in the near future it could become profitable within the private insurance sector to do so.

The social health insurances and the Commission of the German Parliament see a new danger of further risk distinctions based on genetics, since individual companies will make use of every possibility to embrace so-called "low risks" and exclude "high risks" in their cost calculations. Considering the imminent deregulation of the insurance sector, discounted premiums and risk-money within the private sectors affect the member structure and structure of contributions of the social insurances. A bonus/ high risk premium system devised by the private health insurances on the basis of (voluntary) genome analysis would increase the difference in risk categories between the clients of social and private health insurance companies. Basic principles of solidarity of the German social security system would be endangered if such a far reaching genetically argued risk distinction came to fruition.

The Europeanisation of the insurance landscape that is resulting in increasing interconnection of large insurers in Europe, the shift of insurance companies from national markets to the European market and the increasing opportunities for the insured to select health insurance freely, calls for an international perspective when assessing the national situation.

REFERENCES

Bundesministerium für Bildung, Wissenschaft, Forschung und Technologie (Hrsg.) (1995), 'Bildungs- und forschungspolitische Schwerpunkte 1995', Bonn.

Deutscher Bundestag (Hrsg.) (1994), 'Bericht des Ausschusse für Forschung, Technologie und Technikfolgenabschätzung (20. Ausschuß) zur Technikfolgenabschätzung (TA)/Genomanalyse'. Bundestags-Drucksache 12/7094, Bonn.

Enquete-Kommission des Deutschen Bundestages, W.-M. Catenhusen, H. Neumeister (Hrsg.) (1990), 'Chancen und Risiken der Gentechnologie'. Dokumentation des Berichts an den Deutschen Bundestag, Frankfurt-New-York.

Fromme, H., Raabe, K. (1997), 'Noch werden Daten für Risikoprüfungen nicht genutzt'. In: *Ärzte-Zeitung*, Nr. 93, Jg. 16, S. 18.

GenEthics News (1995), 'US panel proposes ban on genetic discrimination in insurance'. In: *GenEthics News*, November/December 1995, 3.

Gill, B. (1991), 'Gentechnik ohne Politik', Frankfurt.

IKK-Bundesverband (Hrsg.) (1995), *'Die Krankenversicherung'*, 11, Bielefeld.

Institut für System- und TechnologieAnalysen (1992), 'Perspektiven der Anwendung und Regelungsmöglichkeiten der Genomanalyse in den Bereichen Humangenetik, Versicherungen, Straf- und Zivilprozeß'. Im Auftrag des Büros für Technikfolgenabschätzung des Deutschen Bundestages, Bad Oeynhausen.

Pieper, A. (1994), 'Der Rücktritt vom Versicherungsvertrag wegen Verletzung der vorvertraglichen Anzeigepflicht'. In: *Versicherungsmedizin*, 49, Heft 2, S. 33–34.

Rosenbrock, R. (1994), 'Die Gesetzliche Krankenversicherung am Scheideweg'. In: *Jahrbuch für Kritische Medizin*, Bd. 23: Gesundheitskult und Krankheitswirklichkeit, Hamburg, S. 189–205.

Sachverständigenrat für die konzertierte Aktion im Gesundheitswesen (1995), 'Gesundheitsversorgung und Krankenversicherung 2000. Kurzfassung und Empfehlungen'. Sondergutachten 1995, Bonn.

Sahmer, S. (1995), 'Genomanalyse und Krankenversicherung'. In: *Versicherungsmedizin* 47, Heft 1, S. 5–9.

Schulz-Weidner, W. (1994), 'Genomanalyse und Sozialversicherung in der Bundesrepublik Deutschland'. In: Centro De Direito Biomedico, Universidade De Coimbra/Institut für Arzt und Arzeneimittelrecht, Universität Göttingen (Hrsg.), *Genome Analysis. Legal Rules—Practical Application. Reports of the Workshop 11th-14th June 1992.* Coimbra 1994, S. 265–306.

Schulz-Weidner, W. (1992), 'Genomanalyse und gesetzliche Krankenversicherung'. In: AOK-Bundesverband (Hrsg.), *DOK*, 1–2, Bonn, 23–29.

Schulz-Weidner, W. (1992), 'Genomanalyse und gesetzliche Krankenversicherung'. In: AOK-Bundesverband (Hrsg.), *DOK*, 3, Bonn, 68–74.

Steinvorth, U. (1995), 'Genomanalyse—Lasten, Rechte und Pflichten des Gebrauchs genetischen Wissens'. In: *Z Gerontol Geriat* 28:378–381.

Verband der privaten Krankenversicherung e.V. (Hrsg.) (1992), *PKV Publik*, 4/92, Köln.

Wiese, G. (1994), 'Die Entschlüsselung des menschlichen Genoms—Gegenwärtiger Rechtszustand in der Bundesrepublik Deutschland und Diskussionen über gesetzliche Neuregelungen'. In: Centro De Direito Biomedico, Universidade De Coimbra/Institut für Arzt und Arzeneimittelrecht, Universität Göttingen (Hrsg.), *Genome Analysis. Legal Rules—Practical Application. Reports of the Workshop 11th-14th June 1992.* Coimbra 1994, S. 385–443.

7

SELLING SOULS

Ethical Theory and the Commercialisation of Genetic Information

Martin Johnston

Department of Health Studies
University of Central Lancashire

My attitude towards him is an attitude towards a soul. I am not of the opinion that he has a soul
—Wittgenstein, 1978, p. 178

1. INTRODUCTION

The prospect of an increasing range of genetic screening tests, with ever improving accuracy, raises numerous concerns. Whilst the scientific optimist may point to numerous advantages in terms of enhancing quality of life, reducing the incidence of serious genetic illnesses and the like, numerous writers (Post,1992, Terrenoire, 1992, Chadwick, 1993, Boss, 1994, Hepburn, 1996) have pointed to a number of difficulties associated with such technologies. A key concern relates to the control of genetic information pertaining to a particular individual following such screening. Clearly it is not possible to deal with this concern in any complete sense in a paper of this size, nonetheless, by focusing on one particular issue—the use of genetic information by insurance companies—this paper aims to sketch the general territory in which discussion may prove most fruitful. It will be argued that by exploring the metaethical foundations of our concerns, we may go some way towards addressing the problems.

2. THE NATURE OF THE CONCERN

Detailed and accurate information concerning the likelihood of illness and the longevity of particular individuals would clearly be of tremendous advantage to the insurance industry. Such information would offer an unprecedented degree of accuracy in determin-

Genetic Information, edited by Thompson and Chadwick.
Kluwer Academic / Plenum Publishing, New York, 1999.

ing the likely size and timings of any future payouts on the part of individual companies, which in turn would allow a far greater degree of control in what is an essentially risky business. That the insurance industry should take a keen interest in developments in genetic screening should come as no surprise then. But among the possibilities here two major concerns emerge. Increased knowledge concerning the future well being of a given individual could lead to the prospect of particular individuals being considered to be too high a risk to be worth insuring, or the establishment of a sliding scale of premiums dependent upon one's genetic constitution, where those identified as being more likely to develop particular conditions pay more. Essentially the danger is that the use of such information will ultimately result in discriminatory practices. Now there may well be occasions where discrimination between individuals, and groups, is legitimate, and equally discrimination may be viewed as either positive or negative, the question then becomes one of where decisions to not insure, or to charge higher premiums on the basis of genetic information lie in this scheme of things.

Though by no means a complete account, a key feature of positive discrimination is that it confers some advantage upon those discriminated. Facing higher premiums or the impossibility of obtaining insurance cover does not fit well with this view, to have to face this possibility is then quite clearly a disadvantage, and hence discrimination on these grounds is quite clearly negative, from the individual's perspective. Of course it is important to note that those who test as 'genetically sound' may well gain some advantages in terms of lower premiums; this type of issue will be dealt with more fully below. Much of the ethical argument concerning the advantages of genetic screening stress the possibility of the subsequent increase in autonomy that this information may make possible (Hoedemaekers, ten Have and Chadwick, 1997), but any kind of negative discriminatory reaction on the part of the insurance industry is likely to offset any such advantage.

This brings us to issue of the legitimacy of discriminatory practices. At the heart of this are the broadly complementary notions of merit and desert. Both of these fail to make sense without some account of responsibility and choice. Where an individual makes particular lifestyle choices, smoking tobacco for example, which have a negative influence on their health they share some responsibility for those outcomes. The extent of this responsibility is the subject of a great deal of debate, nonetheless, I think it is clear that the smoker is in some sense complicit. Equally, though it may prove practically difficult, it makes sense to ask the smoker to desist, and perhaps even make the performance of particular medical procedures contingent upon just such an attempt. Similarly, were an insurance company offering health cover to charge higher premiums of smokers, there would not be an in principle objection. Smoking is something the individual has control over, and consequently can take appropriate measures to remove themselves from the discriminated against group, not only does this result in the removal of the higher premiums but it also produces a degree of health gain in itself. The same type of case can be made for any number of high risk activities. Though there is a need to work out extensive details in all such cases, the fundamental point comes down to this; it is possible to identify some cases where individuals, or groups, *deserve* to be treated differently, even when such treatment works to their disadvantage.

The situation is radically different when we come to consider an individual's genetic constitution. It would be absurd in the extreme to suggest that individuals are in any sense responsible for their genetic constitution, or that they could be asked to alter it at will. (Though clearly there is an issue of concern here for prospective parents in terms of additional pressures being brought to bear against bringing children into the world who are anything less than perfect, and as gene therapy techniques are developed then individuals may find themselves pressurised into undergoing treatment.) Insofar as this is the case then, there is nothing like the notion of the smoker deserving unequal treatment; remove responsibility and choice, and we lose merit and desert. Any justification of discriminatory practices on the part of insurance companies on the grounds of genetic constitution must therefore lie elsewhere. It is in the exploration of such possible justifications that we must turn our attention to moral theory.

3. CONSEQUENTIALISM AND INSURANCE

Given the failure of merit and desert to legitimate discrimination on the grounds of genetic constitution some further justification is required. Attempts could be made to justify such practices in terms of their consequences, and indeed consequences may well be included alongside desert and merit in the previous example. In these terms the advantages provided by genetic information in terms of the increased accuracy in the predictions of insurance companies may lead to the improved efficiency of said companies. We then need a further account of the social utility of the insurance industry in order to identify the overall benefits which would result from such improved efficiency. Whilst the detail may be lacking, for present purposes we may concentrate on key issues of the compensation of individuals for loss or damage, and the significant contribution insurance companies make in terms of general business investment. On these grounds alone the insurance industry is clearly socially useful, and in some cases necessary. Improved efficiency of the industry would then appear to open up the prospect of a range of advantages.

But an important omission in this sketch is another, arguably more fundamental, purpose of insurance companies; to make money for shareholders and underwriters. Now there need be nothing wrong—although this is debatable—with the generation of profit *per se*, but we do clearly recognise limitations on the types of practices which are considered morally sound in the pursuit of profit. The selling of drugs such as heroin and cocaine is indeed profitable, but few would argue that current drug dealing practices are ethical. So, in considering the possible advantages accrued by discriminatory practices in the insurance industry it is important to note that some of these result in individual gain for relatively small numbers of people.

Equally it may be argued that improved efficiency is a necessary feature of any business venture. Current thinking in the libertarian mould tends to incorporate a kind of social Darwinism into economic theory, whereby only the strong survive. Improved efficiency is a must for any company if it is to prosper and this is simply a hard economic fact which must be faced devoid of sentiment. It may be desirable on broad humanistic grounds to avoid any kind of discriminatory practices in one's business dealings, but if one's competitors adopt them to their advantage then this requires that one follow suit or go under. There is something of the 'selfish gene' in all of this where the continued survival of the company comes to be seen as an end in itself, any further consequences become secondary to this prime directive of survival.

Thus we arrive at an account which attempts to justify discriminatory practices in terms of the consequences which result from such a course of action. Presumably all of those consequences identified, and indeed others, may play a part in attempts to justify discrimination on the grounds of genetic constitution, but the real question is whether any, some or all of them will prove sufficient. In order to address this issue it will be necessary to consider notions of value generally, paying particular attention to the idea of 'harsh economic realities'.

4. FACTS AND VALUES, ECONOMICS, AND ETHICS

The preceding section made reference to the idea of harsh economic realities. What appears to be at stake is a tension between what we would like to do and what we are able to do. Now whilst this tension is commonly dealt with in ethics in terms of ought implying can, that is, it makes no sense to require a particular type of conduct unless it is possible for the individual under the obligation to perform said act. In present terms the question concerns just how far it is appropriate to see economic restrictions as providing real obstacles to ethical conduct. It is my contention that the main issue of difficulty is a reliance on too sharp a version of the fact/value distinction. Classical conceptions of the distinction tend to rest on the idea that facts are in some sense objective and independently verifiable, whereas values are inherently subjective and essentially matters of individual preference. Whilst there are external conditions which allow us to settle disputes concerning matters of fact, nothing of this form will help when dealing with disputes involving values. Such differences have been said to characterise the sharp discontinuity between science and ethics. In terms of the present discussion then, economics, as a science, presents us with facts concerning the world whilst ethics is perhaps reduced to no more than wishful thinking. The classical fact/value distinction becomes an economics/ethics distinction and the issue to be addressed is just how marked this distinction is.

In the case of the physical sciences Putnam (1992) offers a sustained critique of the legitimacy of the fact/value distinction, pointing to the inclusion of criteria such as coherence, completeness and simplicity as grounds for theory selection. Such notions entail an essentially value based element. Insofar as this is the case then, science cannot be viewed as entirely value free. What is lost is the idea of science being primarily concerned with establishing the Archemedian point or 'view from nowhere', as Nagel (1986) describes it. What we arrive at then is a view of science as an essentially constructive and creative endeavour, grounded in a human perspective. It is a short move then to apply such thinking to economics. In the absence of any complete account as to why economics, *qua* science, is different from any other science it is legitimate then to postulate that economics too is essentially value infused.

Resistance to such a view is perhaps based upon what could be described as 'the bogey of relativism', the idea being that if we give up on the notion of objective facts then everything comes down to matters of personal preference and anything goes. But this is by no means a necessary outcome of recognising the value component of science, rather it follows from an arguably mistaken view of the nature of value. Various writers have identified accounts of the nature of moral value which whilst not involving a reduction to objective facts nonetheless admit the possibility of notions of truth being applicable to moral discourse (Darwall, Gibbard and Railton, 1992). A full account of such approaches is not possible here, but a selection of particular themes will be of use for present purposes.

Blackburn (1984) identifies the essentially projective nature of moral evaluations, that is, moral properties are not to be found in things themselves but are based in our interactions with them. In many respects this amounts to saying that moral properties supervene upon other more mundane properties of things, events and so on. Now whilst the precise nature of the supervenience relationship may elude us, this is not to say that it is arbitrary, in fact, if moral evaluations are to have any action guiding function, then the relationship needs to be particularly strong. But even without the full detail of the relationship, we nonetheless arrive at an approach which begins to introduce a notion of truth into moral evaluation based on how we come to ground our moral discourse whilst recognising the constructive nature of the whole endeavour. (But notice this constructive element is consistent with Putnam's remarks concerning the physical sciences.) If we then augment this approach with Gibbard's (1992) account which places normative judgement in a naturalistic context, the tendency to engage in normative discussion being seen as essentially the result of evolutionary processes, we can develop the notion of truth in ethics further. A central tenet of Gibbard's account is the idea that co-operation and co-ordination between human beings has significant advantages in terms of survival, in these terms then we can consider the respective benefits of differing moral perspectives in terms of their ability to enhance co-operation and co-ordination. The 'truth' of our moral assertions then becomes measurable in terms of this enhancement.

What all of this amounts to is that we can have a conception of ethics which is not that removed from our conception of science. Putnam (1992) employs the notion of warrant in both cases, rather than attempting to operate with some absolute notion of objective truth with which our assertions correspond, the real task is to provide sufficient grounding or warrant for those assertions. This is by no means a straightforward task, and will involve a range of elements which cross both sides of the classical fact/value distinction, but the difficulty of the task is not sufficient grounds to give up. With this broad sketch in mind then, we can return to the use of genetic information by insurance companies.

5. GENETIC INFORMATION AND INSURANCE

We are now in a position to relocate the concern identified above regarding discriminatory practices based upon genetic constitution. The situation is no longer one of a choice between moral values and economic facts, rather, to use Gibbard's terminology, we are dealing with competing systems of norms. On the one hand we have a system of norms which stresses the value of notions such as efficiency and prudence, whilst on the other are those which take humanitarian concerns as primary. Certainly economic considerations have a part to play in our deliberations, but not as some sort of factual trump which should lead us to set aside our ethical concerns. It is here that Wittgenstein's remark cited at the beginning of the paper has particular relevance. There is value in adopting an attitude towards a soul in our dealings with one another irrespective of the factual basis concerning the existence of souls. Such an attitude is likely to enhance co-ordination and co-operation in human affairs to an extent which is not achievable when we set aside concerns for humanity in favour of the generation of profit. Indeed it is surprising that we should be prepared to risk the potential benefits of genetic screening simply because the resulting information may be misused by the insurance industry, yet this type of concern is surprisingly common in the literature (Post, 1992, Ledley, 1994, Hepburn, 1996). But if my arguments are at all plausible then there is nothing inevitable about genetic information being used in this way. The introduction of new, or extension of existing, privacy laws would be one way of avoiding unethical discriminatory practices. Whilst such intervention in the free market may be viewed with abhorrence by some, allowing unrestricted access to genetic information, or allowing insurance companies to require that prospective clients undergo genetic screening, opens up the possibility of the introduction of eugenics in the name of free enterprise (Ledley, 1994) which represents a far greater degree of interference in the lives of individuals. As Ledley (1994) points out, it is not the technology itself which presents the threat, but rather the uses to which it is put.

We are dealing with a matter of choice, many of the dangers associated with genetic screening will only be realised if we allow them to be, and it would be evasive to hide behind a screen of economic necessity and consequently abrogate our responsibility in making choices. Economic considerations do not by themselves provide sufficient warrant for possible discriminatory practices and it is difficult to see how they could be adequately augmented. In many respects the basic issue comes down to whether or not it is morally acceptable to make profit from another's misfortune; I would suggest that it is not. Clearly the choices are our own, we may choose to enter into the business of 'selling souls', but let us be clear about why we choose to do so. To allow the use of genetic information by insurance companies as a basis for discriminatory practices is to take a particular stance on the matter, it is not to be neutral or amoral, rather it is to endorse a particular normative system. It is my contention that such a position is not sufficiently warranted and that what we stand to lose is of far greater significance than any mere economic gains.

REFERENCES

Blackburn, Simon (1984), *Spreading the Word*, Oxford University Press: Oxford

Boss, Judith A. (1994), 'First trimester prenatal diagnosis: earlier is not necessarily better', *Journal of Medical Ethics*, Vol.20, No.3, September

Darwall, Stephen, Gibbard, Alan and Railton, Peter (1992), 'Toward *Fin de siecle* Ethics: Some Trends', *Philosophical Review*, January

Chadwick, Ruth F. (1993), 'What counts as success in genetic counselling?', *Journal of Medical Ethics*, Vol.19, No.1, March

Gibbard, Alan (1992), *Wise Choices, Apt Feelings; A Theory of Normative Judgement*, Oxford University Press: Oxford

Hepburn, Elizabeth R. (1996), 'Genetic testing and early diagnosis and intervention: boon or burden?', *Journal of Medical Ethics*, Vol.22, No.2, April

Hoedemaekers, Rogeer, ten Have, Henck and Chadwick, Ruth (1997), 'Genetic screening: a comparative analysis of three recent reports', *Journal of Medical Ethics*, Vol.23, No.3, June

Ledley, Fred A. (1994), 'Distinguishing genetics and eugenics on the basis of fairness', *Journal of Medical Ethics*, Vol.20, No.3, September

Nagel, Thomas (1986), *The View From Nowhere*, Oxford University Press: New York

Post, Stephen G. (1992), 'Huntington's Disease: prenatal screening for late onset disease', *Journal of Medical Ethics*, Vol.18, No.2, June

Putnam, Hilary (1992), *Realism With a Human Face*, Harvard University press: Cambridge, Massachusetts

Terrenoire, Gwen (1992), 'Huntington's Disease and the ethics of genetic prediction', *Journal of Medical Ethics*, Vol.18, No.2, June

Wittgenstien, Ludwig (1978), *Philosophical Investigations*, Basil Blackwell: Oxford

8

THE ETHICS OF 21ST CENTURY BIOINFORMATICS

Ethical Implications of the Vanishing Distinction between Biological Information and Other Information

Dick Holdsworth

STOA (Scientific and Technological Options Assessment)
Directorate-General for Research, European Parliament, L-2929 Luxembourg

1. THE CHRONOLOGICAL PERSPECTIVE

The title of this paper was chosen to highlight the chronological perspective which is needed when considering current developments in genetics and related scientific areas, and the moral and ethical implications of these.[1]

There is no law of nature which says that significant events have to happen on the century's turn, and yet sometimes they do. One of the most striking instances occurred in 1900, with the rediscovery of Gregor Mendel's classic paper on the genetics of the pea-plant by de Vries, Correns and von Tschermak. As is well-known, the paper, Versuche über Pflanzenhybriden, was originally published in the Proceedings of the Natural Science Society of Brünn, or Brno, in 1865. Between 1865 and 1900 the paper was not lost in complete obscurity. There were some references to it. But it was only in 1900 that its full intellectual impact was detected.

Nearly one hundred years on, a scientific and technological revolution as important as the discovery of the laws of genetics itself is in progress. The difference between what is happening now and what happened then is that the present revolution does not involve one single momentous discovery by an individual researcher. It involves an event or process of a different kind. What we are experiencing is the confluence of two intellectual currents: information technology and molecular biology.

Even now, there is no single name for this revolution. It has no single geographical or intellectual locus. If you wish to search for it, the key-words you might use could include bioinformatics, combinatorial chemistry, protein sequencing, the protein folding problem, high-throughput screening, rational drug design and, some would add, nanotechnology.

Genetic Information, edited by Thompson and Chadwick.
Kluwer Academic / Plenum Publishing, New York, 1999.

Those who work in the area of technology assessment have a professional bias against the easy use of terms like 'breakthrough' or 'revolution' when talking about the growth of knowledge in science and technology. The usage is so often inappropriate, if not entirely meaningless. Nevertheless there is a need to signal the fact that something extraordinary is happening in the area I have set out to discuss here. Perhaps later I shall find a more exact cliché.

2. THE SCOPE OF THE ETHICAL DISCUSSION

The practice of bioinformatics and the scientific and technological activities associated with it, like every significant human undertaking, does raise certain moral issues. Those which are at present of most immediate concern relate to the ownership, authenticity, publication and use of information. In particular, serious attention has had to be given to the issue of whether researchers who succeed in determining a DNA sequence (perhaps part of the genetic code of a pathogen) have an obligation to make it immediately available to other scientists. The elucidation of these problems in turn raises a number of second-order, or ethical questions. There are a number of ways of approaching these. The elucidation of these problems in turn raises a number of second-order, or ethical questions. There are a number of ways of approaching these.

The approach that will be adopted here is determined by the newness of the subject-matter. This is why the chronological perspective is important. To speak about the 'leading-edge' of knowledge sounds fine until we remember that it is also the brink of ignorance (Holdsworth, 1995, p.143). When we apply our knowledge, we have a duty to realise that we may also be applying our ignorance. We have a duty to remember that thought, but not to be paralysed by it. Morally, what we have to do is to devise strategies for proceeding to apply knowledge that we have reason to believe will yield benefits, while taking precautions against the risks entailed by our own ignorance. Ethically, we have to find grounds for validating, or not validating the strategies.

3. ACCOUNTABILITY

One example is the strategy of accountability: of making the actors in a situation accountable for what they do. This is not necessarily the same thing as making them managerially or politically responsible or making them liable in law. At its most fundamental, it can be understood as the imposition of a responsibility to communicate, upon request, a narrative of the actions undertaken in the context of the circumstances that obtained at the time (Holdsworth, 1994). This strategy assumes that in certain cases action will have been taken. It also tacitly assumes that the action involves risk. It is a strategy which may well result in the ascription of blame, but its first consequence is not that but to provide the information necessary for society to learn by its mistakes. To the extent that it assumes that we act from behind a veil of ignorance it could be thought of as Rawlsian. To the extent that it entails an evolutionary, trial-and-error view of the growth of knowledge, it can be thought of as Popperian. As a combination of these two elements, it implies a social contract: a contract between society at large and the actors at the leading edge of knowledge.

4. SECOND-ORDER ACCOUNTABILITY

But those who strike the social contract are also to be held to account. This is what might be called second-order accountability. The narrative that supports their actions will be nothing less than history itself. Just as an operator in an industrial plant who is asked to explain a sequence of events which led up to an accident will be expected to have shown 'situation awareness', so the same thing is expected of those who strike the social contract on our behalf. They will be expected to show evidence that they had grasped what was going on around them: that they had done their best to understand the context of their actions—the context within which their actions assumed meaning. They will be expected to have shown 'situation awareness' in the flux of human history.

This paper, then, will present a brief attempt to situate the phenomenon of bioinformatics in the context of a wider historical and conceptual account, of the sort that we could justifiably be asked at a later date to provide if we, as citizens of the society in which the phenomenon arises, were asked to show that we had taken pains to understand it and to position it in the moral universe that we deem ourselves to inhabit.

When dealing with a new phenomenon, I suggest that this step should be prior to the issue-by-issue analysis of the specific moral and ethical questions to which the phenomenon gives rise.

5. BIOINFORMATICS

Bioinformatics as it is known today is a phenomenon of the 1990s. There is a temptation to define bioinformatics as the use of electronic computing to analyse and interpret gene-sequences which have been yielded by gene-sequencing research in the laboratory. However, this misses some key elements of the phenomenon. It is not simply the computer that is the defining tool: it is a system comprising computer, databases, on-line networking and specialised software. That is why it is a phenomenon of the 90s: it takes the form it does today because of the present availability of large amounts of sequencing data in databases which are linked into the World Wide Web. There are other points, but we shall come to those in a moment.

The history of bioinformatics must itself take its place in a sequence of events. It was logically preceded by the development of 'robust and generally applicable sequencing methodologies', which Primrose (1995, p. 11) dates to 1977. Following that development, interest grew in the possibility of sequencing whole genomes, and a start was made on viral genomes. Success here emboldened scientists to think in terms of sequencing the entire human genome. Primrose reminds us of the difference in scale between these two types of task. The size of the phage λ genome, sequenced by Sanger et al. in 1982, was 48.5 kb (kilobase); the Epstein Barr virus DNA sequenced in 1984 by Baer et al. was 172 kb, but the size of the human genome is three billion base pairs (3×10^9 bp).

Before taking this line of discussion further, an observation should be made concerning protein sequence data. Burks (1997, p. 12) points out that:

> Protein sequence data became available as a databank resource well before nucleic acid sequences because amino acid sequences were—at that time—easier to determine directly than were nucleic acid sequences.

Gradually we can build up a picture of the sequence of steps that had to be taken before the present situation was reached. Another step along the way was made when the technology of the CD-ROM was introduced into molecular biology as a data distribution medium. The advantages and disadvantages of this are discussed by Fuchs and Cameron (1997, pp. 46–48). The two disadvantages they mention both relate to time. One point concerns the slower speed of access to data offered by CD-ROM by comparison with magnetic disk. The other point is no doubt more significant. They put it this way:

> Production cycles of two or three months make CD-ROM more suitable for periodic releases of databases every few months than for rapid updates. For many researchers this time lag is unacceptable and more immediate access to latest information, for example by using computer networks, is required...

Before exploring that idea, we need to have a clearer idea of the kind of knowledge that bioinformatics can yield.

A researcher who recently entered the field has put it this way (Thonnard, 1996, p. 6):

> molecular biologist[s] rely more and more often on computer aided sequence analysis tools... [For] example, they need computer programs to compare DNA and protein sequences, to search for coding regions in DNA sequences, to predict the secondary and tertiary structure of DNA, RNA and proteins.

Let us consider a specific example, by taking a brief look at recent research on Treacher Collins Syndrome.

6. TREACHER COLLINS SYNDROME

It so happens that it was in the same year that Mendel's research came to light, in 1900, that E. Treacher Collins described the essential features of the syndrome which is today known by his name. Treacher Collins Syndrome has clinical manifestations that include abnormalities of the ears with hearing loss, anomalies of the facial bones and cleft palate. The 'Identification of the complete coding sequence and genomic organization of the Treacher Collins Syndrome gene' has recently been reported by a team based at the University of Manchester (Dixon et al., 1997). In their article, the authors report isolating the complete nucleotide sequence of the TCS gene, referred to as *TCOF1*, and the derived amino acid sequence of its predicted protein product, known as Treacle. They explain that in their research they turned to the use of a number of bioinformatics programs:

> As initial database sequence comparisons failed to show any strong homologies between *TCOF1* and previously identified genes, gene families, or motifs of classic importance. (p. 227)

This helps us to understand the nature of the task for which bioinformatics programs are designed. They must search through millions of bytes of information seeking out regularities in the data that may prove to yield homologies.

Such programs would be useless if there was not a large store of gene-sequence data already stored in a form accessible to computers, and if this store was not being continuously fed with new data from laboratory research. To someone working in the field now, that must seem an obvious statement. But until quite recently it would by no means have been obvious that the scientific research community would organise its activities in this way. It is not just that computing tools are rather convenient for doing genomics and protein sequencing. Rather, these and related disciplines have re-organised themselves around the bioinformatics paradigm.

This point can be brought home by dwelling for a moment on the training theme, which helps us to get a feel for the fact that we are not just talking about a research methodology or an information technology system, but about a human culture. In order to see how someone is inducted into that culture I return to the case of the new entrant to the field whom I have just cited.

7. A COURSE IN BIOCOMPUTING

A researcher such as the one I have just quoted (Thonnard, 1996), who is at the Laboratoire de Biologie Moleculaire of the Universite Catholique de Louvain in Belgium, learns the skills by taking part in an on-line interactive biocomputing course organised by the University of Bielefeld in Germany.[2] She reports (p. 6) that:

> The main components of the course were the Hypertext Coursebook, on-line sessions using the Electronic Conferencing system BioMOO, and the hands-on visualization of some important concepts, like sequence alignments.

More specifically, the topics in the coursebook on the Web include topics such as: pairwise sequence alignment; weight matrices for sequence similarity scoring; Fasta/Blast exercises; multiple alignment; the mathematical basis of molecular phylogenetics; genetic algorithms and protein folding.

Be it noted in passing that Fasta and BLAST are what are known as heuristic algorithms, or database search programs for finding optimal alignments between sequences. There is an interesting, if brief discussion in Altschul (1997, pp.152–153) of questions of optimisation in such programs. In each case he refers to a trade-off between speed and sensitivity.

Returning to the training course, it is reported that the student works at her own pace on the hypertext coursebook, but is enabled to interact with other students and an instructor in the on-line conferencing environment, BioMOO. We are told that MOO stands for 'MUD, Object-oriented'. A MUD is an on-line interactive communication tool. The initials originally stood for Multi-User Dungeon, since the tools were first developed for 'dungeon and dragon' role-playing games. Now, in BioMOO, they can permit six groups of six students and an instructor to meet once a week over a two-month period to discuss course chapters and assignments.

The student whose experiences we are considering has an interesting comment on biocomputing tools as such. She writes (pp. 6–7):

> The first programs for sequence analysis were mainly running on UNIX systems and required cryptic commands. Currently, more user friendly programs are available. Therefore, when biologists want to use tools for computer aided sequence analysis, they do not necessarily need to be expert in mathematics and computer sciences. However they still face a difficult task because the numbers of tools is large and even exploding: there are a large number of databases (DNA sequences, gene maps, protein sequences, protein 3-D structure, specific databases,...) a lot of programs (these are able to perform various tasks, on the basis of different principles), various ways to access the tools (one can use [a] terminal session in a UNIX machine, install programs on his personal computer, access tools via the Web).

In spite of the difficulties,

> All the participants agree that the course allowed them to acquire some skill in biocomputing but also meet other interesting people and, moreover, have a lot of fun together.

It was good that the author placed this human comment alongside the technical ones. This helps us in the task of positioning the phenomenon in our moral universe. Science as fun deserves to take its place among the alternative characterisations, often psychologically less convincing, that have from time to time been served out to the public: science as lofty calling, science as ineffable mystery and so on.

Also, it does us good to be reminded, by implication, of changes in the moral universe of the scientist. Not long ago the idea of user-friendly on-line science for non-mathematicians whose hands-on research is tapping a keyboard would have been regarded as shocking. Perhaps it still is. Should it be?

8. AUTOMATION IN MOLECULAR BIOLOGICAL RESEARCH

In order to discuss automation in molecular biological research, one must try to understand something of the magnitude of the tasks involved.

Primrose (1995) offers some interesting data. Discussing genome sequencing strategies, he refers to two alternatives: complete genome sequencing and the alternative strategy of confining the analysis to "complementary" DNA (cDNA). The latter implies limiting the task to the sequencing of expressed genes, using partialcDNA sequences, known as ESTs: expressed sequence tags.

Primrose points out that complete genome sequencing is "very labour intensive" (p. 88). In fact this is a point he mentions twice in his book, laying a realistic and therefore welcome, but unusual emphasis on the occupational psychology of research. Elsewhere he writes:

> It has to be realized that, to be carried out effectively, genome sequencing must be managed by skilled scientists. It is also repetitive, boring and labour intensive. The two aspects are not compatible and can lead to a high staff turnover rate (p. 11).

An important factor is the ratio between the quantity of raw sequencing data and that of confirmed gene sequence. Primrose points out that while gene density is very high in prokaryotes and lower eukaryotes at around 900 genes per Mb, this comes down to 200 in the nematode and 10 to 20 in human DNA.

He also compares sequencing rates as between manual and automated sequencing. He reports a case where manual sequencing proceeded at an average rate of 25 kb per person per year. By contrast, the figure for a 24-channel fluorescent sequencer is about 2.7 Mb per machine per year.

However, Primrose says that even with the development of automation and multiplexing there are still limitations on the method. One is the temperature limitation on acceleration of the rate of electrophoresis. Primrose points out that the simplest way to increase "separation speed is to increase the electrical potential but this significantly increases the amount of heat produced". This causes distortions.

Here is another trade-off: between the rate of work (or power) and the rate of generation of waste heat. Perhaps it is not entirely irrelevant here to contrast what happens in the laboratory with what happens in the living cell. The robots have not yet improved on the economy of nature.

However, it is important to understand that it is not only at the point where sequencing data is analysed—at the level of bioinformatics—that automation and information technology intervene.

9. MAPPING THE DISCIPLINES AROUND BIOINFORMATICS

Did space allow, it would also be helpful to map the situation of bioinformatics among the cluster of disciplines that surround it. For present purposes, a brief indication will have to suffice.

The common object of attention is the process whereby, in the cell, the construction of the biological macromolecules, of proteins, is effectuated from the nucleic acid templates, and of the interaction of these macromolecules with each other and with compounds occurring naturally or as the result of pharmaceutical research.

Genomics helps to identify genes responsible for illness. For the pharmaceutical industry this is only the start of a longer process, which has recently been described in the following way (Stevens, 1997):

> To understand how genes interact is to understand how life develops and unfolds, what can go wrong, and what tools we have at our disposal to intervene if things do go wrong so we can direct fine strategies that address these deficits', says Hoffman-La Roche's Drews. 'All of our drug therapy today is directed at about 500 targets—molecular sites in cells where drugs effect their actions. Now genomics will clearly enlarge this by at least an order of magnitude. There will be 5,000 to 10,000 targets identified through genomics, which means there is a tremendous opportunity for new-drug discovery'.
>
> As drug targets are identified by genomics, other technologies kick in to help find drugs to interact with targets. First, robotics and automation are applied to the gene-finding task itself. Combinatorial chemistry allows generation of libraries of millions of chemical compounds that can be synthesised, and then directed at drug targets to test interaction. Miniaturization of samples and automated robotic handling for high-throughput screening, allows hundreds of thousands of samples to be evaluated daily.

Bioinformatics applies high-powered computer manipulation to help understand correlations, draw conclusions, and generally make sense out of the incredible mountains of data process generate.

A brief glimpse of combinatorial chemistry at work was provided by a recent article in Chemistry Week (Cookson, 1997), which discussed work within the wide/ranging R&D programme of BASF at Ludwigshafen, a research centre with a total of 7,000 R&D staff. The article said BASF's work on combinatorial chemistry started three years ago in the main laboratory, where the company's biotechnology research is based. It said a multidisciplinary project team began work in 1994. It explained that:

> They are developing an ever-expanding toolkit of synthetic reactions for preparing compounds with potential biological activity, and finding new ways to automate the process of making and then separating the chemicals.

The article reported the head of research planning at BASF, Dieter Jahn, as saying that there had "already been some 'hits' from testing the products of combinatorial chemistry in high-throughput screening for pharmaceuticals, though it is too soon for any of the candidates to have moved into clinical trials".

Another statement in the article was of particular interest. It quoted Jahn as stating that "the biggest challenge is how to manage the exponential growth in information, which is most obvious today in the life sciences but will affect the whole of science and technology".

The researchers in the pharmaceuticals sector can indeed generate huge numbers of compounds by combinatorial chemistry, working through computer libraries of data to identify promising candidates. The possibilities are so huge that they must be constrained in some way so that the process can be optimised. This is done by adducing the findings from other disciplines, including research into the structure of macromolecules having the active sites that are the targets for the disease/blocking drugs that the pharmaceuticals sectors seeks to develop.

There is another approach, called rational drug design, which involves using the structural information about the stereospecific and electrostatic properties of the target molecule to build up a conception of the type of compound that might be devised to act upon it.

10. THE INFORMATION EXPLOSION

Protein modelling, drug design and combinatorial chemistry all use computer information/processing to a high degree. In this respect, they are the direct counterparts in their various spheres of bioinformatics.

Many of the commentators already quoted here have stressed the huge quantities of information being generated in and by their respective disciplines. The interesting question is whether the generation of information is proceeding on the basis of positve feedback resulting from the interaction between the disciplines, or whether the interaction is serving to optimise, and therefore constrain the process.

If the positive/feedback model is the correct one, then what we are talking about is an information explosion. At last the exact cliché has been found. There is evidence that the positive/feedback model is the correct one.

Speaking to the BioIndustry Association at The Science Museum in January 1997, George Poste, responsible for Research & Development at SmithKline Beecham, said

> The rise of genomics, combinatorial chemistry, automated screening and bioinformatics have completely re-engineered the drug discovery process in less than five years, bestowing vital competitive advantage on companies who captured these skills ahead of others.
>
> History may well record the century from 1950 to 2050 as one of the most remarkable in the expansion of human knowledge, driven by inter-dependent advances in biology and computing.
>
> It is impossible to convey to those outside of the technical community the staggering speed and scale of research progress in molecular biology, genetics and computing. It is both frightening and exhilarating.

Exhilarating it may be, but if the correct model for the process whereby our information is growing is indeed the positive-feedback one, then the corollary is that the process is *not* being optimised.

11. HALF THE CENTURY FROM 1950 TO 2050

As far as the century from 1950 to 2050 is concerned, we only know half the story so far. How well have we understood it? Let me conclude by thinking about the history of the concept of information during that time.

The origins of information theory are traced to a paper published by Claude Shannon in 1948. Shannon had originally gone to the Bell labs to work on the carrying capacity of telegraph lines, so in the earliest days there was no organic link between the emergent disciplines of information theory and of electronic digital computing. There are many surprising facts about the history of the concept of information, and that is one of them. Another is that it has proved possible for at least two books to have been published on the history of computing which did not discuss Shannon. They are Shurkin (1984) and Campbell-Kelly and Aspray (1996).

Electronic computing had admittedly already got under way by 1948. It was on 14 February 1946 that the ENIAC machine, which had been developed to perform ballistics calculations for the US Army, was first shown to the public at the Moore School of Electrical Engineering at the University of Pennsylvania in Philadelphia (Hughes, 1996).[3]

These developments were approximately contemporary with the origins of molecular biology. Thuillier (1975, pp. 14–15) has divided the early history of molecular biology into three phases. First, there was what he calls a 'romantic phase' beginning in 1935 with the reflexions on the gene of Max Delbruck. Then he sees a 'dogmatic' phase, lasting for a decade after the discovery of the double helix by Watson and Crick in 1953. This set the central dogma on the functions of DNA and RNA. It was also the time when Francois jacob and Jacques Monod extended the theory by their work on messenger RNA. The period after 1963 is seen by Thuillier as an 'academic' period, charcterised by the consolidation of the scientific position of molecular biology.

It was during the 1960s, as numerous texts attest, that the term 'information' became fully absorbed into the vocabulary of molecular biology, whether in the context of the nucleic acids or in that of proteins.

At the end of that decade, precisely in 1970, Francois Jacob published the book he called 'a history of heredity', under the title *La logique du vivant*, or 'the logic of the living'. That was the same year that Monod published *Le hasard et la necessité* (*Chance and necessity*). The two of them, with Andre Lwoff, had won the Nobel Prize in 1965, principally for work published in 1961.

In *La logique du vivant*, Jacob meditates (p. 270) on the statement by Schrodinger in *What is life?* (1944) that:

> Life seems to be orderly and lawful behaviour of matter, not based exclusively on its tendency to go over from order to disorder, but based partly on existing order that is kept up.[4]

He comments that:

> It is the concept of information which, in the middle of this century, opens an access to the analysis of this order and of its transmission.

He then discusses the relation between information and thermodynamics, and goes on:

> This isomorphism of entropy and information establishes a link between the two forms of power: that of doing and that of directing that which does. In an organised system, whether living or not, it is the exchanges, not only of matter and energy but of information, which unite its elements. An abstract entity, information becomes the locus where different types of order are articulated.

These remarks of Jacob's are made in the context of a section of his book headed 'The macromolecules'.

12. INFORMATION IN MOLECULAR BIOLOGY AND ELECTRONIC COMPUTING: A CHRONOLOGY OF KNOWLEDGE

We see, then, that molecular biology developed at around the same time as electronic computing and information theory. Though there has been no space here for a thorough historical survey, some hints have been picked up which suggest that these were two parallel processes which indeed remained parallel. The admittedly selective quotes given here have perhaps suggested that molecular biologists were more aware of information theory than the exponents of electronic computing were aware of molecular biology. In many ways, this was because information technology could not yet provide the special tools for molecular biology which have now been yielded, and also because molecular biology was not yet ready to use such tools.

If you pick up a book about computer simulation in genetics published in the early 1970s, such as Crosby (1973) you find its subject matter is still closer to the simulation of population genetics than to the modelling of DNA. As we have already seen from Primrose (1995), it was not until 1977 that 'robust and generally applicable sequencing methodologies' were developed, and even then the modern bioinformatics techniques of gene discovery which we were discussing earlier were still years away.

The result has been that although the development of information/processing by electronic computer proceeded contemporaneously with progress in research into biological and biochemical information processing, their trajectories, though they sometimes touched each other, were never unified.

It can be argued that this is precisely what is happening now, and that this fact helps us to understand the information explosion which is currently being experienced in the sectors that have been described in this paper. It could also be taken as evidence that what has been seen so far is only the beginning of an even greater process.

13. THERE ARE NOT TWO KINDS OF INFORMATION

One of the tasks left uncompleted as we reach the half-way mark in the century from 1950 to 2050 (notice how we have shifted the time-frame since the beginning of this paper: perhaps the turn of the century is not such a significant milestone after all) is that of defining what 'information' means. Although we have a formula that permits us to quantify information and therefore to use it in scientific and engineering calculations, we lack an ontology of information.

A necessary element in the ontology of information will certainly prove to be that which is encapsulated in the distinction between the signal that is transmitted in an information process and the channel that vectors the signal. A channel can take a number of alternative states. The signal is an ordering of such states. It is not a particular state or states, but an ordering of the states.

Information is signals. This conception of information enables us to see that there are not two kinds of information: that of electronic computing and that of molecular biology. In each of these two contexts, we find signals that are expressed by orderings of the states of physical substrates. Viewed in this light, the nature of the substrate, silicon or carbon, is irrevelvant, even if, in real-world situations, there are contingent reasons for its relevance.

Therefore, if it sometimes seems that there is more than one kind of information, it is because 'information' is a term like 'colour', which has one sense but a plurality of references.

However, information processing systems do not function in isolation. They exist to do work, and what is special about the biological information-processing systems that we have been considering is the astonishing rate at which they work: in a word, their huge power.

That is why we need to understand them. That is why the prospect of human agents coming into possession of that power alerts our ethical awareness.

NOTES

1. The views expressed in this paper are personal and do not represent the opinion of the European Parliament. At the same time, if I may do so without making them responsible for my opinions and short-comings, I should like to thank those who have helped me in various ways during the preparation of this paper, especially Jean-Marc Egly, Bob Harper, Robert Herzog, Carlos Martinez, and Shoshana Wodak.
2. Virtual School of Natural Sciences—BioComputing Division (VSNS-BCD), University of Bielefeld. The home page is at 'http://www.techfak.uni-bielefeld. de/bcd/welcome.html'.
3. The article by Hughes cited here reported the celebrations, attended by Vice-President Al Gore, of the 50th anniversary of ENIAC at the Moore School, University of Pennsylvania.
4. The citation is from Schrödinger (1944), p.69 (not p. 68, as stated in Jacob, 1970). Translations here are by the present author.

REFERENCES

Altschul, Stephen F. (1997): 'Sequence comparison and alignment', Bishop and Rawlings (eds.), 1997, pp. 137–167.

Bishop, M.J. and Rawlings, C.J. (eds.) (1997), *DNA and protein sequence analysis—A practical approach*, IRL Press, OUP, Oxford.

Burks, Christian (1997): 'Molecular biology databases', Bishop and Rawlings (eds.), 1997, pp. 1–30.

Campbell-Kelly, Martin and Aspray, William (1996), *Computer—A history of the information machine*, BasicBooks, HarperCollins, New York.

Cookson, Clive (1997), 'R&D combining brainpower', *Chemical Week*, 24 September 1997, supplement, pp.11–12.

Crosby, Jack L. (1973), *Computer simulation in genetics*, John Wiley & Sons, London.

Dixon, Jill; Edwards, Sara J.; Anderson, Isobel; Brass, Andrew; Scambler, Peter J., and Dixon, Michael J. (1997): 'Identification of the complete coding sequence and genomic organization of the Treacher Collins Syndrome gene', *Genome Research*, 7, pp. 223–234.

Fuchs, Rainer and Cameron, Graham N. (1997): 'EBI databases and tools', Bishop and Rawlings (eds.), 1997, pp. 45–58.

Holdsworth, Dick (1994), 'Accountability: The obligation to lay oneself open to criticism', Chadwick, Ruth (ed.): *Ethics and the professions*, Avebury, Aldershot, pp. 42–57.

Holdsworth, Dick (1995), 'Ethical decision-making in science and technology', Almond, Brenda (ed.), *Introducing applied ethics*, Blackwell, Oxford, pp. 130–147.

Hughes, Samuel (1996), 'Taken aback by ENIAC', *The Pennsylvania Gazette, Alumni Magazine of the University of Pennsylvania*, March 1996, pp. 20–25.

Jacob, Francois (1970), *La logique du vivant*; Gallimard, Paris.

Mendel, Gregor (1865), 'Versuche über Pflanzenhybriden', *Proceedings of the Natural Science Society of Brno*, 1865, pp. 3–47.

Poste, George (1997), 'Address by George Poste, President, Research & Development, SmithKline Beecham Delivered to the BioIndustry Association, The Science Museum, London, England', January 14, 1997, BioIndustry Association, London.

Primrose, S.B. (1995), *Principles of genome analysis—A guide to mapping and sequencing DNA from different organisms*, Blackwell, Oxford.

Schrödinger, Erwin (1944), *What is life?—The physical aspects of the living cell*, Cambridge University Press, Cambridge.

Shannon, Claude E. (1948), 'A mathematical theory of communication', *Bell System Technical Journal*, 27, July and October 1948, pp. 379–423 and 623–656.

Shurkin, Joel (1984), *Engines of the mind—A history of the computer*, W.W. Norton & Company, New York.

Stevens, Tim (1997), 'The gene machine', *Industry Week*, August 1997, pp. 168–175.

Thonnard, Joelle (1996): 'The VSNS-BCD biocomputing course', BEN News, Belgian EMBnet node biocomputing quarterly newsletter, Vol. 4, No. 3–4, Winter 1996, pp. 6–7.

Thuillier, Pierre (1975): 'Comment est née la biologie moléculaire', *La Recherche en biologie moléculaire*, Editions du Seuil/La Recherche, Paris.

9

THERE IS NOTHING SPECIAL ABOUT GENETIC INFORMATION

Søren Holm

Department of Medical Philosophy and Clinical Theory
University of Copenhagen
Blegdamsvej 3, DK-2200 Copenhagen N, Denmark

For in much wisdom is much grief: and he that increaseth knowledge increaseth sorrow
—*Ecclesiastes 1: 18*

1. INTRODUCTION

One often encounters the idea that there is something special about genetic information, that it is somehow different from, or more important than other health related information we can get about ourselves or other persons. There are conferences and research projects dealing with genetic information that implicitly reaffirm this idea, and it is also prevalent in the public debate where genetic tests are generally viewed with a mixture of awe and suspicion. But is there really something special about genetic information, or are we just mislead by the fact that genetics is the latest in a long range of scientific fields which have at different times captured public attention?

My contention in the present paper is that there is nothing special about genetic information which distinguishes this kind of information from other health related information in any morally or politically relevant way. This view has implications in a lot of areas, but here I will confine myself to look at the regulation of the use of genetic and other health related information in connection with employment decisions. The sooner we rid ourselves of the idea that there is something special about genetic information, the sooner will we be able to deal constructively with the large issues raised by the use of all kinds of health related information in employment decisions.

If we do not rid ourselves of this idea we will for a long time be busy regulating an ever increasing range of genetic information. The invention and practical perfection of the DNA-chip will revolutionize genetic testing (Butler, 1997; Hacia *et al*, 1996; Stimpson *et al*, 1995), and we have no chance of catching up if we resort to piece-meal legislation.

Genetic Information, edited by Thompson and Chadwick.
Kluwer Academic / Plenum Publishing, New York, 1999.

2. THE FALSITY OF GENETIC ESSENTIALISM

Genetic essentialism is the idea that the essence, the nature of a human being is defined by its genes (Cranor, 1994). This is often expressed in the statement that "we are our genes". If genetic essentialism is true then there seems to be a fairly straightforward argument for the claim that genetic information is special. It would be information about the very essence of a person, whereas other non-genetic information would only be about accidental attributes. We would still need to be able to distinguish between genetic and non-genetic, but if we could do that, genetic information would surely be special.

We do have reason to doubt whether we can make the distinction between genetic and non-genetic information. Is information about bloodtype for instance genetic or non-genetic? This is a problem which I will not pursue in this paper, but the term genetic information should generally be read as in scare-quotes pending further analysis as to whether it is a useful category.

There is no doubt that genetic essentialism plays a large role in the public perception of genetics (Nelkin and Lindee, 1995), but there is similarly no doubt that genetic essentialism is false. The ideas embodied in genetic essentialism are not new, similar ideas lay behind the classic eugenics movement (Pernick, 1996; Friedlander, 1995; Larson, 1995), but the great developments in human genetics, and the public portrayal of these developments have again given these ideas a measure of scientific respectability.

With regard to physical characteristics it may be true that the genetic endowment (itself a tendentious essentialist term) of a given person sets the outer limits for what the person can maximally achieve in height, physical strength etc.. But whether these limits are reached is obviously a result of the environment the person has encountered during intrauterine and postnatal life. The maximal limits are only reached if the person has lived in very favourable environments. This is for instance evidenced by the fact that the mean height of Northern European populations is still slowly increasing, and by the still existing differences in mean height between social classes even in affluent European countries. It is of course also the case that a single catastrophic illness or a single accident may prevent a person from reaching his or her physical maximum, even if they live the rest of their life in optimal surroundings. The actual physical development of a given person is thus always controlled by a complex mixture of genetic and environmental determinants.

When it comes to non-physical characteristics the picture becomes much more complicated. There is still debate about whether there is one underlying mental ability which reflects itself in some kind of general intelligence, or whether there is a range of separable and not closely connected abilities (a range of different intelligences).[1] Whether or not this controversy is resolved we can, however, safely affirm that the outer limits of a person's achievement in the intellectual field are not set by genetics alone. Most would probably concede that Archimedes and Leonardo da Vinci were geniuses of the first rank, but a school child in a good modern school surpasses the knowledge of either of these two geniuses a good number of years before compulsory schooling is ended. Our intellectual achievements are a complex result of genetic, environmental, and general cultural influences, and to fix one's eye on the genetic component is simply an unwarranted reduction.

Another version of genetic essentialism, sometimes called "genetic determinism", claims that our genes are our destiny. According to this version of essentialism it is our genes which ultimately control how our lives will go, and especially how they will end. There is again something correct in this idea, but there is more which is wrong. It is true that certain genes do have a large influence on the life of specific people (e.g. the genetic mutations leading to severe genetic defects), but for most people their life is not controlled by their genes in any strong sense. The genes may play a role, but the roles played by the inanimate environment, and by the social environment are much larger. It is also true that genes may make certain causes of death more likely for a given person, but a predisposition to heart disease is not enough to substantiate a claim that heart disease thereby becomes the destiny of a person. It is worth remembering that it is only very recently that most people have lived long enough to contract heart disease. Previously the same genes were there, but they did not in any way influence the destiny of the person. That destiny was determined by hunger, infection and numerous other environmental factors. Yes, even today it is the case that I may die tomorrow due to infection or accident even though my genes "predict" that I ought to die of heart disease some years hence.

Connected to the idea of genetic essentialism is the idea that genetic information is special because it describes the deepest level of human biology: a level which has been hidden from mankind until very recently, and which still requires extensive technological intervention to be brought out into the open. Genetic information is in the body, in the cells, and it requires an act of boundary breaking to get it out.

It is a very ancient human idea that the hidden is at the same time valuable and potentially dangerous. Our folk tales abound with this notion, and it may play some role in our assessment of genetic information. Just like it was a fact that the trolls and the ogres lived under the ground and could become angry if men started mining for ore or gems (Ingemann, 1992), and that mining was therefore both profitable and dangerous, it is now an accepted fact that genetic information lives under the skin and the effects of acquiring it are unpredictable. This is true in a certain sense, but it does not give us an argument for treating genetic information in a special category. Many non-genetic forms of health-related information can also be acquired only through bodily invasion and with the risk of detecting something unpleasant and/or dangerous.

3. GENETIC INFORMATION IS...

A claim that genetic information is special compared with other health related information does not have to rely on genetic essentialism. It can also be based on a further claim that there is some other feature of genetic information which makes it different. A number of such features have been proposed in the public debate. Some have pointed out that genetic information is predictive, some that it tells us something about conditions which can be transmitted to offspring, some that it is informative about other persons, and some that it is specific in being very sensitive.

In the following I will briefly aim to show that for any of these allegedly special features of genetic information we can find other non-genetic health related information which shares the feature.

The first claim which can be brought forward is that genetic information is predictive. First, it is worth pointing out that a lot of genetic information is non-predictive. One could look into the genes to discover the colour of the eyes of a specific person, but this information would not be predictive in any meaningful sense of that word. We could obviously "predict" the colour of the eyes, but since this is directly observable it would not be much of a prediction. A lot of health related information which is non-genetic is highly predictive. Knowing that a person has cancer of the stomach is highly predictive of death in the not too distant future. Similarly knowing the LDL-cholesterol level or the homocystein level in the blood of a person can be very predictive of the person's risk of coronary heart disease.

The second claim is that genetic information is special because it tells us about constitutional features which can be transmitted to the offspring of the person. This feature of genetic information can also be found in a range of non-genetic information. Social class is for instance transmissible, as is level of education; and recent studies indicate that mothers "transmit" certain dispositions or non-genetic risk factors to their fetuses through the intra-uterine environment (Dennison *et al*, 1997; Barker, 1997; Barker, 1995).

The third claim is that genetic information is not only informative about the person in question but also about other persons, for instance other family members. This claim is true in some cases, but it is also true for a number of non-genetic cases. A few examples should again suffice: The birth of a child with the stigma of congenital syphilis is highly informative about its parents, and the finding that one member of a family suffers from low-grade lead poisoning is indicative of the same condition in other family members.

The fourth claim that genetic information is particularly sensitive paradoxically receives its force from the popular belief in genetic essentialism. Because genetic information is believed to reveal the very nature of a person it is also considered as very sensitive, and its disclosure as very problematic. There is, however, non-genetic information which is equally or more sensitive in the public sphere.[2] HIV-status is a very obvious example of such a piece of information, and others could be psychiatric illness, or inclinations toward special kinds of sexual practices (cf. the British Spanner case (*R v. Brown et al*, 1994)).

But could it not be the case that although no specific feature is unique or even special for genetic information, the combination of all four features mentioned is unique and does make genetic information into something special?

This could be the case, but it happens not to be. One of the examples mentioned above, congenital syphilis, is at the same time predictive, transmissible to offspring, informative about others, and sensitive in the requisite sense. Three of the features happen to co-occur in the very common case of smoking (predictive, transmissible to offspring, and sensitive (at least in some countries)).

It could be objected that the examples offered above are examples of persons already suffering from a disease, whereas genetic information is predictive of future disease and that it is this feature which makes genetic information special. This objection is, however, not valid. If some persons for instance have a very high diet induced LDL-cholesterol level they are not ill or diseased, but they do have an increased risk of future ischaemic heart disease. This shows that pure prediction of future risk can also be found outside of the area of genetic information.

4. THE FUTURE REGULATION OF THE USE OF HEALTH INFORMATION

If genetic information is no different from other kinds of health related personal information we can say that the use of all such information should be regulated in the same way, but we still have to decide how it should be regulated.

Here there seems to be two opposing models. The first of these is the traditional model in employment where the employer is allowed to seek any health information he or she finds relevant for the employment decision. Regulation following this model seeks to restrict frivolous or discriminatory collection and use of health information, but the underlying assumption is that if the information is relevant it can be legitimately collected.

The second model is the relatively new model often applied to genetic information where the employer is not allowed to seek this information, or where stringent restrictions are put on exactly what information can be sought.

We have established that the same model should be used for all kinds of health information, but should it be one of these two models, or one in between?[3]

In the present paper it is impossible to give this question a full analysis.

The main argument pointing towards the first model is that employment contracts are contracts. They are binding legal agreements entered into after negotiation between two (or more) parties, and each party has a right to set its own terms during the negotiation process leading to the contract. The state should uphold such contracts, and should perhaps prohibit contracts entered into under duress, but apart from this the content of the contracts and the negotiation process is a matter for the parties to decide on without state interference. If one of the parties requires health information from the other party this is quite legitimate, since the other party can just withdraw from the contract negotiation if it is unwilling to supply this information.

The first model tacitly presupposes a more or less equal relation between the parties to an employment contract. But this presumption is very naive and most often false. Employers very often have much more power than the job applicant (this is even implied in the term "applicant"). Employment markets are far from ideal markets but are characterised by large power-differentials between the participants in the market.

Talk about power and social classes have gone out of favour since the final fall of Marxism. It is, however, important to realise that although the specific Marxist concept of the class-struggle was deeply flawed, the idea that some people and some companies have and use power to further their own aims is basically a very sound idea (on the general analysis of power see for instance the papers in Steven Lukes' "Power" (Lukes, 1986)). Even neo-classical economics predict that powerful cooperations will emerge through competition in the market and the result of economies of scale.

The fact that there are power-differentials in a given market is not always ethically important. It may not be ethically important whether or not there are large power-differentials between buyers and sellers in the market for Elvis memorabilia or vintage Barbie dolls. The existence of such power-differentials is, however, important in those markets where everybody has to participate. In modern Western societies there are a number of markets where everybody have to participate, and the employment market is one of these. In such markets the powerful can illegitimately exploit the powerless who cannot withdraw from the market. There is therefore good reason to regulate the kind of contracts which can be made, and in this case, the kind of health information which may be asked, sought and used.

The work leading up to the 1996 Danish law on health information and employment was started because Danish politicians and labour unions feared the newly accessible genetic information and its impact on the employment market in the future (Lov nr. 296, 1996). It was their idea to implement new, stringent, and specific regulations for genetic information, but during the work in the preparatory committee it quickly became evident that the distinction between genetic and non-genetic information could not be upheld. The final legislation is therefore not based on such a distinction but covers all health-related information an employer desires to get in connection with an employment decision. The basic principle of the law is that the employer is prohibited from seeking such information, and from using it if he should obtain it without seeking it. Exemptions can be granted in specific circumstances by the Minister for Employment and Industrial Relations, who must first ask the opinion of a newly formed Health Information Council consisting of representatives of employers and labour unions, as well as representatives from a number of ministries, the Danish Medical Association, and the Danish Council of Ethics. To date only 5 exemptions have been granted, and of those 4 have been given to off-shore companies for health checks aimed at ascertaining whether off-shore employees are fit to perform their auxiliary functions as fire-men etc. in the cases of an emergency off-shore.

This law is thus an example of a law operating without a distinction between genetic and non-genetic health-related information, and with a very stringent set of regulations on the use of such information.

The main problem with the Danish law is that the provision for enforcing the strict regulations are fairly weak. There is no central authority whose task it is to investigate possible violations of the law and prosecute such cases in the courts. The enforcement of the law is left to the individual employee, or in reality to a labour union acting for an individual employee to bring a case against the employer for illegal collection of health data. At present only one such case has been brought to court. The case involved a person with a chronic disease who had been asked at an employment interview whether her disease was in a stable phase. The court awarded her substantial damages, but unfortunately the case has not been reported officially, so no account of the reasoning behind this verdict can be found.

With stronger provisions for enforcement the Danish law would give better protection for the vulnerable party in employment decisions, and this would enable a more unequivocally positive verdict concerning this law.

5. CONCLUSION

This paper has argued for two conclusions. First, it is argued that there is nothing intrinsically special about genetic information compared to other kinds of health related information. For every allegedly special feature or special combination of features of genetic information it is possible to find non-genetic health related information which shares these features. This has implications for legislation in a number of areas where we wish to control health related information (employment, insurance, health registers etc.).

The second part of the paper explores these implications in connection with regulations of employment decisions and shows: 1) that we need to regulate the use of health related information in this context because of the power differentials between employers and employes, and 2) that a uniform regulation of all health related information is preferable to a specific regulation of genetic information. The recent Danish law on health information and employment is discussed, and it is concluded that it is a useful model for such regulation, although stronger enforcement provisions ought to be recommended.

NOTES

1. One specific example of this debate and its implications is the debate over Herrnstein and Murray's "The Bell Curve" (Herrnstein and Murray, 1994; NIH-DOE Working Group, 1996).
2. There is an underlying very difficult question: what features make a specific piece of information sensitive? There are probably no pieces of information that are sensitive in and of themselves. A piece of information becomes sensitive in a specific context, and loses its sensitivity in other contexts. Within an association of HIV-positive persons HIV-status is, for instance, not sensitive.

3. This is not intended to deny that some specific items of information might be more sensitive than others, and that this should reflect itself in legal regulation. The statement is only intended to deny that genetic information falls in a special sensitive category just by virtue of being genetic.

REFERENCES

Barker, D.J. (1995), 'The fetal and infant origins of disease', *European Journal of Clinical Investigations*, Vol. 25, p. 457–63.

Barker, D.J. (1997), 'Fetal nutrition and cardiovascular disease in later life', *British Medical Bulletin*, Vol: 53, p. 96–108.

Butler, D. (1997), 'Truce is likely in battle over 'DNA-chip' patent rights', *Nature*, Vol. 387, p. 221.

Cranor, C.F. (ed.) (1994), *Are Genes Us? The Social Consequences of the New Genetics*, Rutgers University Press: New Brunswick, NJ.

Dennison, E., Fall, C., Barker, D. (1997), 'Prenatal factors influencing long-term outcome', *Hormone Research*, Vol. 48(Suppl 1), p. 25–9.

Friedlander, H. (1995), *The Origins of Nazi Genocide: From Euthanasia to the Final Solution*, University of North Carolina Press, Chapel Hill, NC.

Hacia, J.G., Brody, L.C., Chee, M.S., Fodor, S.P., Collins, F.S. (1996), Detection of heterozygous mutations in BRCA1 using high density oligonucleotide arrays and two-colour fluorescence analysis', *Nature Genetics*, Vol. 14, p. 441–7.

Herrnstein, R.J. and Murray, C. (1994), *The Bell Curve - Intelligence and Class Structure in American Life*, The Free Press, New York.

Ingemann, B.S. (1992), *Fjorten Eventyr og Fortællinger (1820–1864)*, Borgens Forlag, København.

Larson, E.J. (1995), *Sex, Race, and Science: Eugenics in the Deep South*, John Hopkins University Press, Baltimore, MD.

Lov nr. 296 af 24. april 1996 om brug af helbredsoplysninger mv. på arbejdsmarkedet.

Lukes, S. (ed.) (1986), *Power*, Blackwell Publishers, Oxford.

Nelkin, D. and Lindee, M.S. (1995), *The DNA Mystique: The Gene as a Cultural Icon*, W.H. Freeman and Company, New York.

NIH-DOE Working Group on Ethical, Legal, and Social Issues (ELSI) (1996), 'Statement: 'The Bell Curve'', *Journal of Medical Ethics*, Vol. 22, p. 190.

Pernick, M.S. (1996), *The Black Stork: Eugenics and the Death of "Defective" Babies in American Medicine and Motion Pictures Since 1915*, Oxford University Press, New York.

R v. Brown et al. [1994] 1 A.C. 212.

Stimpson, D.I., Hoijer, J.V., Hsieh, W.T., Jou, C., Gordon, J., Theriault, T., Gamble, R., Baldeschwieler, J.D. (1995), 'Real-time detection of DNA hybridization and melting on oligonucleotide arrays by using optical wave guides', *Proc Natl Acad Sci USA*, Vol. 92, p. 6379–83.

10

HIGH SPEED GENETIC TESTING TECHNOLOGY AND THE COMPUTERIZATION OF THE MEDICAL RECORD

Curtis R. Naser

Fairfield University

1. INTRODUCTION

The Human Genome Project—a three billion dollar, fifteen year effort to map and sequence the human genome[1]—was begun in 1990 and is presently estimated to be ahead of schedule. The base-pair by base-pair sequencing of the entire three billion bases of the human genome is within reach within the next five years. The phenomenal success of the project in meeting and surpassing its goals is largely due to advances in sequencing technologies and informatics. Indeed, the standard genomic sequencing technologies are largely automated and computer assisted (Gibbs, 1995). In addition, the analysis of raw genetic sequence, and its integration with maps of the various chromosomes is largely a project in informatics, the results of which are freely available over the World Wide Web. The purposes of this massive, industrial strength science project, are manifold. Besides providing the building blocks for answering questions about the ontogeny of the human organism, how the human species has diverged both from its evolutionary precursors as well as how individual races and ethnic groups have diverged genetically[2], the Human Genome Project promises to revolutionize the practice of medicine and is already revolutionizing the strategies of biomedical research (Watson, 1992).[3] It is this latter promise that is probably the most compelling reason that the United States Congress was willing to invest $200 million a year in this project.

Even as the Human Genome Project enters the final phase of sequencing, having already produced high resolution maps of each human chromosome, the fruits of this project have begun to be applied to medical practice in the rapidly growing field of medical genetics. There already exist genetic tests for a number of human diseases, including Huntington's disease, breast cancer, and cystic fibrosis to name just a few of the more well known conditions. Mendelian Inheritance in Man, an online catalogue of human disease genes lists over 4000 gene locuses (OMIM, 1996). Although the Human Genome Project has yet to give rise to any new treatments of disease, early diagnosis of genetically related diseases has aided individuals in family planning, and the early management of their genetic diseases.

Genetic Information, edited by Thompson and Chadwick.
Kluwer Academic / Plenum Publishing, New York, 1999.

Genetic testing, however, is a double edged sword, offering both benefits and burdens. Some genetic diseases, such as Huntington's have no treatment, and while early genetic diagnosis may be an aid to family planning, knowledge of the disease may be a heavy burden to bear. Indeed, there is good evidence that a majority of individuals with a family history of Huntington's disease have opted not to seek the knowledge the genetic test offers (Decruyenaere et al., 1993; see also Huggins et al., 1990). Other conditions, such as the genetic predisposition to certain cancers may leave one with the unfortunate choice of prophylactic surgery such as double mastectomy for breast cancer or colostomy for colon cancer, or in either case, the anxiety of the vigil, waiting for the inevitable appearance of the cancer. Such testing is complicated in some cases by the incomplete penetrance of many disease genes, leaving patient and doctor to weigh probabilities rather than certainties. Prenatal testing also has the consequence that because treatments for genetic conditions are so rare, the only options when a genetic disease is detected in the fetus are abortion or carrying the fetus to term. In addition, there are social consequences of genetic testing, ranging from denial or discontinuation of health insurance, life insurance and employment, to social stigmatization and discriminiation.

Because medical genetics is so loaded with risks, geneticists have been very careful to formulate exhaustive protocols for counseling prior to, during and after genetic testing. Patients need to be educated about each specific test: what the test can and cannot determine, the risks of the disease, the social implications of the genetic knowledge, and the psychological burdens of the knowledge.

Though these efforts are quite laudable, even if still controversial (see Clarke, 1991), the very technology which has given rise to the disciplines of medical genetics and genetic counselling, threatens to undermine them. Currently, most genetic tests are done at a central laboratory and can take several days to a week to complete. The tests are labor intensive and expensive. Until recently, most such testing laboratories would work only with certified genetics counselors in a medical genetics practice, thus guaranteeing that patients would be fully informed of all the risks and benefits of the test results as well as receive counseling after the test. But genetic testing is still a cottage industry which is now entering its adolescence. Just this past year, for instance, the Genetics and IVF Institute in Fairfax, Virginia broke ranks with two other testing labs which had agreed not to make the BRCA1 genetic test available directly to the public because of uncertainty about the risks posed by the gene. For $295 the Institute offered the test to anyone, arguing that it was any woman's right to have the test (Kolata, 1996; Silverman, 1995). The industry's adolesence is now upon us.

This is just the tip of a very large iceberg, the hidden outlines and mass of which I hope to at least partially reveal in this essay. Genetic testing and sequencing technologies already exist which have the potential of putting a complete genetics testing laboratory in every physician's office, and maybe even in the local drugstore along side other over the counter medical testing. The stakes are quite high, not only because human genetics has rapidly become very big business (Anderson, 1993; Greely, 1995; Stone, 1993), but also because the problems already associated with genetic testing will only get worse unless adequate solutions can be found. Let us take a look at one of these newer technologies which promises to transform the practice of medical genetics.

2. SEQUENCING AND TYPING STRATEGIES

Standard methods for genetic testing employ what is called the "polymerase chain reaction" or PCR. In order to test for the presence of a particular gene or allele's of a gene, a sample of DNA is isolated by means of two "probes"--each a 10 to 20 base long strand of DNA complimentary in sequence to DNA sequences at each end of the target gene. As the sample DNA is heated up, the two complimentary strands of DNA that form the double helix separate. When cooled again, the short probes are able to "hybridize" or bond to the target strands of DNA (one probe for each complimentary strand at opposite ends). Then as the sample is incubated at a moderate temperature, polymerase enzyme initiates the formation of a complimentary strand of DNA starting at the probe. The cycle is repeated twenty to forty times and the result is that the target DNA has been amplified exponentially. The sample is then placed on an electrophoretic gel and even single base pair differences in length of the target gene can be resolved visually on the gel.

Sequencing of genes employes the same electrophoretic gel technique but the amplified DNA is first divided into four separate solutions and "digested" by enzymes that specifically cleave the DNA at one of the four base pairs that make up the molecule and a flourescent tag is then added. The resulting solutions are then passed through an electrophoretic gel, and the sequence of the DNA is reconstructed by summation of the differing lengths of DNA. All standard sequencing machines employ computer assisted reading and analysis to reconstruct the sequence.

2.1. Gene Chips

An alternative strategy has been developed by several companies, including Affymetrix of Santa Clara, CA, Hyseq of Sunnyvalle, CA and Synteni of Palo Alto, CA (Wade, 1997). This stategy employs what are called "oligonucleotide probe arrays". The idea is that rather than using a single probe and applying it to a solution of DNA, a large number of probes are fixed at one end to the surface of a silicon chip approximately one centimeter square. A large number of different probes of predetermined sequence can be so fixed in a predetermined spacial array or grid (Fodor et al., 1991; Pease et al., 1994). Sample DNA is amplified and digested by enzymes into lengths equal to the probes' lenghts (10 to 30 bases) and labled with a flourescent tag (sometimes the DNA is first converted into RNA). The resulting samples of DNA then hybridize to those probes whose sequences they match. A microscope is then able to read the resulting hot spots of flourescence when illuminated by a laser of appropriate wave length. A computer is employed which has been programed with the sequences and their positions on the chip. Thus a flourescent mark on the chip indicates that a piece of the sample DNA has hybridized to those probes and therefore, the sequence of that particular sample is known (Kozal et al., 1996). The intensity of flourescence can also be used as an idicator of the quantity of sample sequences present. This is important, for instance, in assessing total gene expression of cells by measuring which genes have been "switched on" and so have been expressed in the form of messenger RNA. The intensity of flourescence correlates directly with how much messenger RNA in each case has been produced (Lockhart et al., 1996).

Affymetrix is currently producing chips for research purposes which have been divided into a grid of 40,000 squares, each 50 microns square. Consequently, a single such chip can contain approximately 40,000 different probes. In order to increase the accuracy of the chips, which they have demonstrated to be equivalent to standard sequencing techniques, a great deal of redundancy is built into the probes. Presently, they are able to detect the expression of up to 6000 genes by analyzing expressed cellular RNA (Lockhart et al., 1996). Affymetrix claims to be able to produce chips of 20 micron resolution giving the chip a total of 250,000 sites and they claim 10 micron chips are possible with a total of 1,000,000 sites (Kozal et al., 1996). Affymetrix anticipates commercial production of chips capable of monitoring the expression of 50,000 genes (Wade, 1997).

Such chips can be used to test for particular sets of genes, by using probes specific to those genes and their alleles. Alternatively, it is possible to employ such probes to do raw sequencing (Drmanac et al., 1989). Reaction times vary from an hour to overnight depending upon the procedure and the necessity of amplifying the sample DNA prior to hybridization. Scanning time is quite short: fifteen minutes for a 20,000 probe array.

While Gene Chips presently remain tools for researchers, the long term goal will be to produce a fully automated system for clinical genotyping and sequencing. While this goal still remains on the horizon, it is being actively pursued by Affymetrix and other companies. Affymetrix presently has a $31 million grant from the U.S. Department of Commerce to develop an automated Gene Chip machine for the doctor's office.

3. IMPLICATIONS

Though oligonucleotide probe arrays are not yet on the drug store counter top nor even in the doctor's office, we are not in the genetic Kansas anymore. The pace of genetic technologies is quickening, and we have not yet figured out how to handle the implications of the clinical genetics we already have available to us. Will we be able to use this technology for the good of all concerned? Or will the imperative "a technological can do implies a commercial must do," rule this new technology as it seems to rule so many others? Let us examine some of the benefits and risks of this new technology.

3.1. Benefits

Genetic testing is presently labor intensive, time consuming and expensive. High speed testing, by whatever means, will bring genetic information to the physician and patient much more rapidly and at lower cost.

From a purely medical point of view, we should recognize that once the necessary research has been completed, highspeed genetic testing and analysis will be an unparalleled diagnostic tool at the disposal of the physician. It will not be long before a physician will be able to get a complete genetic analysis of each patient, indicating genetic predispositions to diseases. Such information will form the basis of long term preventive recommendations and treatment. Further, as our understanding of the genetic bases of drug action increases, physicians will be able to tailor their treatments and prescriptions to the peculiar genetics of their patients. Likewise, as research progresses in infectious diseases, the genetic bases of susceptibility will become clearer. Thus immunizations and prophylactic treatments can be targetted to those who are susceptible. Speedy and efficient genetic diagnosis will be a necessary part of this process.

From the point of view of research, such technologies are already speeding up our ability to map, sequence and develop definitive diagnostic tests for disease genes. In addition, entire genomic RNA expression chips will become available to monitor the expression of genes in each type of cell as an organism develops from egg to adult. This research will ultimately prove invaluable in the study of developmental disorders as well as contribute profoundly to our basic understanding of ontogeny. And, as more and more genetic information winds its way into the medical record, such medical records will become invaluable sources for genetic epidemiology, containing both genetic information and medical history along with pertinent demographic data.

So researchers and clinicians alike are quite excited by the prospects of the increasing geneticization of medicine, which will only be accelerated as high speed genetic testing equipment becomes more available and inexpensive. But these benefits are not univocal. Let us now turn to how such a development both impacts on old problems and raises a host of new ones.

3.2. Old Problems Made Worse

3.2.1. Health Insurance. At its inception, the Human Genome Project was recognized to raise many profound ethical, legal and social issues. Detailed knowledge of genetics impacts upon our self understanding and our evolutionary history, the relationships between races and diverse ethnic groups, as well as our understanding of disease. Genetic testing has manifold implications, not just for how diseases are recognized, but also for how they will be treated, and just as important, how society responds to these diseases and the knowledge of them genetics affords. As a result, the architects of the Human Genome Project argued for and won a commitment of 5% of the overall budget to the study of the Ethical, Legal and Social Implications (*ELSI*) of the Human Genome Project.

One of the first areas recognized to be impacted by increased understanding of genetics and the develpment of genetic testing techniques is health insurance. A 1993 subcommittee of the ELSI working group—Task Force on Genetic Information and Insurance—published a report entitled *Genetic Information and Health Insurance* (Task Force on Genetic Information and Insurance, 1993). This report analyzed the nature of insurance underwriting and the impact of genetic testing.

The basic problem is this: insurance companies have an interest in minimizing risk. They seek knowledge about the nature of the risks of the prospective insured in order to make a prudent judgment about whether that risk should be assumed and if so, at what cost. There are, of course, many limits to that knowledge, including the costs of gathering it. Insurance companies rely on a combination of actuarial tables, medical history and physical examination to make these judgments. On the other hand, individuals seeking insurance have an interest in hiding their risks. In addition, individuals who are generally healthy may not seek health insurance coverage, whereas individuals who suffer from diseases or conditions have a clear interest in getting insurance coverage. The predictable result is that those most in need of insurance will have the most difficulty obtaining it. And when they are able to obtain it, the insurance company may exclude coverage for the very thing that presents the greatest risk to the insured—a pre-existing condition or predisposition to disease.

Enter genetic testing. Genetic testing promises to provide highly specific information about present and future health problems, what George Annas has called a "coded future diary" (Annas, 1993). Such information would obviously be a boon to insurance underwriters who could use this very detailed predictive information to minimize their risk by excluding some candidates from insurance coverage altogether, excluding specified genetic conditions from coverage for others, or increasing insurance premiums for specific risks on still others. Genetic information, however, can work against the interests of the insurance company if an individual learns of specific genetic risks and seeks coverage on the basis of that knowledge but does not share that knowledge with the insurance carrier. The result is the problem of self selection, or what the insurance industry calls "adverse selection," which works to overburden insurance pools with higher risks.

The conclusions of the ELSI task force on Genetics and Insurance was that because of these conflicting interests, a rational policy on insurance in an age of genetics was impossible. Unless there is fundamental reform of the U.S. health care system whereby everyone is guaranteed basic health coverage, the contradictions inherent in this system of health insurance will be radically exacerbated by genetic testing. Insurers will use this knowledge to exclude those of high risk, thus obviating the medical advantages of the testing in the first place. Likewise, individuals may seek genetic knowledge of themselves or their children, in order to make decisions about whether to carry insurance and how much to buy.

Presently insurers generally do not require the genetic testing of applicants for the simple reason that it is too expensive: the costs of testing large numbers of individuals exceeds the amount of money the insurance companies would save by excluding those of high genetic risk. If an individual, however, has undergone genetic testing, insurance companies will often demand that information, though if the testing has been done "off the record," the individual may be able to refuse that information, at the risk of lying to the insurance company.[4]

High speed, inexpensive genetic testing, however, radically tips the precarious balance our society presently enjoys between the competing interests of insurers and insured. When genetic testing becomes cheap enough to be cost effective for insurers to demand, they no doubt will demand it, thus making what the task force recognized as an already irrational system worse. The insurance companies argue that they cannot selectively exclude themselves out of existence, and that an equilibrium will be achieved (Pokorski, 1994; but see also Hudson et al., 1995), but this may well be at the cost of a lack of coverage for a number of devastating genetic conditions which insurers will simply refuse to underwrite. The consequence will be that some of the medical benefits of genetic testing will be lost because individuals will not have the means to pay for the treatment of them.

This is a social disaster waiting to happen. Whether high speed genetic testing becomes a routine part of insurance underwriting and medical practice in five years or fifteen years, unless the underlying contraditions of health care financing and insurance are resolved, the inevitable result will be an economic boon to the insurers and a corresponding loss—often devastating—to the individual person.

3.2.2. Employment. The above nightmare is not confined only to those who must purchase individual insurance policies. Large group insurers and those who self-insure—almost exclusively employers—may also use genetic testing to exclude individuals of high risk. Employers may also have an incentive to exclude individuals of specific genetic risk if it is probable that they will lose that employee to health problems and disability in the foreseeable future. Some businesses invest a great deal of resources in the training of their employees. If they are able to determine in advance who will be around long enough for them to recoup their investment, then employers will have an incentive to use genetic testing for that purpose, provided that the testing is cost effective. High speed genetic testing will eventually make such practices cost effective. The result may be a class of people who are able bodied, but who are genetically future disabled and may not be employable, or may not be able to find employment in their chosen profession.

Some employers may also seek to use cost effective genetic testing in order to minimize the risk of health problems which arise in certain work place conditions, such as exposure to toxins. As the genetic basis of toxicology becomes better understood, it will be possible to discriminate between individuals who are more or less susceptible to particular toxins in the work place. It may be cheaper for the employers to weed out their susceptible employees (or future employees) than to clean up the work place of the toxins. Thus again, some individuals will find their ability to realize their professional goals thwarted by a form of genetic discrimination.

Another possibility exists that corporations involved in environmental pollution detrimental to human health may argue that the effects of those pollutants are a result of the genetic predisposition of those who are affected, rather than the result of the pollution itself, thus seeking to absolve themselves of liability. The same argument might be constructed in the case of cancers related to the use of tobacco products.

We presently have little or no effective public policy to deal with these problems, yet the development of genetic testing technology proceeds apace. Must it? Must we develop and then market these technologies before we solve the problems that they will inevitably create? Will we simply wait to see what happens and who gets hurt, before we, as a society, are willing to take action to solve these dilemmas? Considerations of social justice would lead us to proactively address these problems before they become worse. The ELSI Task Force on Genetics and Insurance, writing its report in 1993 at the height of the Clinton Health Care Reform effort, looked forward to its successful completion and felt confident that these problems could be headed off before they become worse. Unfortunately, their optimism was misguided.

The result presents an interesting lesson. Market forces have since penetrated heavily into health care in the form of managed care. It was recognized early on that the introduction of market forces to health care would lead to compromises in patient care. Managed care advocates argued otherwise, contending that managed care would make medicine more efficient. We have, in fact, seen both predictions come to pass. Health care costs have been at least temporarily held down and there is good evidence of compromise of patient care in a number of arenas (Green, 1996a). As a result, society has responded piecemeal to these compromises, with legislation being drafted to limit gag rules on physicians and to force insurers to offer greater lenghts of stay in hospital after certain procedures.

If the history of managed care is a guide, then we can only expect change in public policy when a sufficient number of individuals are sufficiently harmed to mobilize our representatives into action. The unfortunate consequence, however, is that not only are individuals sacrificed in the process but that the resulting policies are cobbled together piece meal. The result is an unsystematic morass of regulation.

In the area of genetic testing, I suspect that it would be naive to expect that those who are contributing to the "progress" of the human genome project and the translation of its knowledge into practical and clinical applications, such as high speed genetic testing technologies, would be willing to slow down their efforts until society can be convinced of the need for proactive legislation. There simply is too much money at stake. Too many scientists have invested their careers and money in the commercialization of genetics and biotechnology, too many corporations are invested heavily in the same, for us to be able to slow the tide of genetics that is presently flooding our society. It will most likely be only in reaction to the injustices and harms that result from the application of genetics to social institutions such as insurance and employment that we will see any movement to ameliorate the most egregious harms that are to come.

3.3. New Problems

While the impact of genetics upon insurance and employment have been recognized from the start of the rise of contemporary genetics, there are a number of areas just below the surface in which high speed genetics creates new problems. These areas include its impact upon genetic counseling and a host of difficulties raised by the convergence of this technology with the increasing computerization of the medical record.

3.3.1. Genetic Counseling. As noted above, the present process of genetic testing involves intensive counseling of testees and sometimes their families who may be impacted by the testing as well. The development, however, of high speed, comprehensive and inexpensive genetic testing, while a boon to the practice of medicine, may make it all but impossible to effectively counsel patients. Each genetic disease has its own unique set of circumstances, impacting the individual and family in very different ways. Strong social stigma may be attached to some genetic conditions, and the interpretation of genetic testing is complex and requires a great deal of knowledge and background. Patients need a great deal of education to be able to assess the need for and consequences of genetic testing.

Yet when we multiply the need for this education for each genetic condition several thousand fold in testing for multiple genes at a time, and when we place the testing apparatus in the hands of individual doctors and counselors, how will we be able to adequately educate and counsel patients about these tests? There is already evidence that many physicians ordering diagnostic genetic tests do not fully or even adequately understand the results, let alone adequately prepare their patients for testing, even when the testing is restricted to conditions of their specialty (Giardiello et al., 1997).

The prospect of easily accessible genetic testing, perhaps one day marketed directly to the consumer, that is fast and inexpensive may lead to too much information too fast. Without adequate counseling and interpretation by qualified physicians and counselors, individuals may be overburdened with information that they are simply not competent to understand. The possibility for misinterpretation without proper guidance is great, and could lead to imprudent decisions by patients and physicians who are not trained to interpret the results of genetic testing. While I expect that along side of high speed genetic testing computerized interpretation of its results will develop, a consistent level of expertise amongst health care providers necessary to adequately interpret and act upon genetic testing results outside of the specialty of medical genetics will be long in coming. Genetic testing technology is rapidly outstripping the competence of health care providers to interpret it and high speed genetic technologies will radically exacerbate this problem.

3.3.2. Genetic Testing and the Computerized Medical Record. This is perhaps the most disturbing result of the development of this new genetic testing technology. Because the gene chips produced by Affymetrix and others rely upon the computer to analyze the raw data coming from the reading of the chip itself, the results of the testing are presented in computer code of the sequences which have matched. Depending upon the construction of the oligonucleotide probes on the chip, this code may be the entire sequence of a gene being tested, or it may indicate the specific allele(s) of perhaps a large number of specific genes being tested. Hence the computerized results will contain either raw sequence, or the specific sequences of small portions of a large number of genes, in addition to the immediate interpretation of these sequences in terms of the presence or absence of specific known alleles of the specified genes. While the results can be reviewed directly on screen using the software provided by the company, or printed out, it is unlikely that this computerized genetic information will then be discarded. Indeed, when such testing technology is finally approved for clinical use, whether in commercial testing laboratories, hospital cytogenetic laboratories or in the physician's office, the computerized code representing the specific genetic sequences tested will inevitably be ported into a computerized medical record.

There are a number of advantages to this arrangement. First, it allows the physician or other specialist to access the raw sequence, thus enabling them to apply new knowledge to genetic information which has already been gathered. Specific alleles, for instance, may indicate how an individual will respond to different therapies developed in the future. Furthermore, by retaining genetic information in the computerized medical record, this information is more easily shared amoungst consulting physicians, or transmitted to remote sites to which the patient may travel or relocate. Furthermore, by including detailed genetic information in the medical record, it will be possible for researchers later to cull such vast and detailed records for associations between specific genes or alleles and the medical history and demographic information included therein. Such would be an exceedingly powerful database for a wide variety of epidemiological research.

But the sword cuts in several directions. While a boost to the practice of medicine and the enterprise of research, the inclusion of large quantities and highly specific genetic information in the medical record can also be a hazard to the patient in a number of respects.

We have already discussed the impact of high speed genetic testing on insurance and employment. I will only point out here that the inclusion of this information in a computerized medical record further exacerbates these problems by making detailed genetic information readily available to insurers and employers, who already have a variety of ways of accessing the medical record. High speed genetic testing results coded in a computerized medical record simply hands insurers and employers genetic information on a silver platter, or rather, on an aluminum disc.

Equally troubling, however, are the issues of privacy and confidentiality. Medical records, and especially computerized medical records, have long been recognized as veritable sieves. Such records are routinely accessible to large numbers of individuals at hospitals, whether they have a need to know or not. In addition, patients are routinely required to authorize the release of their medical records to insurance companies as a condition of coverage, and these insurers in turn hand patient records over to such agencies as the Medical Information Bureau and other data clearing houses for sharing with other commercial interests.

Aside from the use of these records by insurers and employers for purposes of minimizing their risks and costs, is there good reason to decry the openness of the medical record? There are two responses to this question. First, other *harms* may befall individuals whose medical records have been made accessible to other individuals.

3.3.3. Embarrassment of Disclosure. Some medical and genetic information is highly personal and its disclosure may be an embarrassment to the individual and what is embarrassing or sensitive is itself a matter of personal feeling. Employees of a hospital, for instance, must bear the risk that their fellow employees with access to medical records may browse their records, simply out of curiosity. While some individuals may have the means to get their treatment else where, many employees will be forced by their insurance to get treatment in specific hospitals, and perhaps in the very hospital in which they work.

This problem, however, is not limited simply to employees of a health care institution. The large numbers of individuals who work in health care and have access to medical records exposes the general public who use these hospitals and facilities as well to the prying eyes and curiousity of electronic voyeurs.

While it is unpredictable what specific information an individual will find sensitive and thus rather not have disclosed to her neighbors or acquaintances, there is some genetic information that will be more sensitive than others. Particularly sensitive will be that information pertaining to personality and intelligence. Though we presently know very little about the genetic bases of personality, intelligence and psychiatric disorders, this knowledge will come in time. And just as we do not routinely discuss or disclose our IQs because we feel that that is personal information, so we may anticipate that individuals will have a strong interest in maintaining the confidentiality of such genetic information.

3.3.4. Genetic Reductionism. There is a danger that as we develop a more and more sophisticated understanding of genetics, that we will fail to recognize the limits of genetic knowledge. Dorothy Nelkin and Susan Lindee (1995) have already chronicled the penetrance of genetic ideology and metaphor into our popular culture. As the advancement of genetics continues, we can only expect an increasingly genetic ideology to envelop us. While genetic scientists and physicians will admit that genes are not the complete story of the human being, the increasing reliance on genetics enforces this erroneous stereotype. The danger is that individuals may be categorized on their basis of the genes, rather than on the basis of who they are and what they have accomplished or can accomplish. The increasing reliance on genetics leads us to make all sorts of judgments on the basis of things that are completely out of the control of the individual—his or her genes—to the exclusion of what is in his or her control—namely, the capacity to adapt and overcome whatever limitations may have been placed upon the individual by biology. Before high speed genetic testing becomes a routine part of our lives, we need to address society's tendency to oversimplify complex information and problems. What we need is vastly better education of the public about the meaning and limits of genetics.

3.3.5. Patient Trust in the Health Care System. It is perhaps a cliche that a physician has an obligation to maintain the strictest confidentiality in relationships with patients. The wisdom of this principle, expressed first by Hippocrates[5] has never been seriously questioned. There are, however, a number of circumstances in which confidentiality is overridden by higher needs. These include consultation with other providers about patient care. Thus these other providers enter into a similar relationship with the patient. While the circle of those who are privy to confidential information is thus widened, it remains within the scope of the principle. But in addition to provider access to medical records which is a necessary condition of care, insurance companies, billing personnel, hospital records and informatics personnel, QA personnel and sometimes government and accrediting agencies all have access to the medical record. Patients have simply had to accept this less than optimal state of affairs as a condition of getting the medical care they need. The addition of detailed genetic information to the medical record, however, raises the stakes for the patient, as such information can harm their economic and employment interests in other domains. Given the large number of individuals and agencies that have access to the medical record, and given the conflicting interests some of these may have vis-a-vis the patient's interests, it would be irresponsible to place such detailed genetic information in the electronic medical record until adequate safeguards to protect this information have been established. The fate of the model Genetic Privacy Act (Annas et al., 1995), which sought to do just that, does not leave one sanguine about the prospects of meaningful reform in this area. The openess of the medical record is already of concern to many patients. The addition of information that has the potential to work against their own interests in that record can only serve to undermine patient trust in the health care system.

3.3.6. Harms vs. Wrongs. Even if the possible harms due to the disclosure of genetic information can be avoided by appropriate public policy, there remains the fact that patients expect that their records will remain confidential. Simply eliminating the possibility of harm due to disclosure does not eliminate the *wrong* or disservice to a patient whose records are open to perusal by so many individuals and agencies. Alexander Capron has elaborated on this distinction by arguing that a person who enters your house while nobody is home and neither takes nor disturbs any of your possessions, has not *harmed* you by removing or damaging any of your property, but has *wronged* you by both trespassing on your property and by violating your privacy (Capron, 1991).

The inclusion of detailed genetic information in the medical record leaves individuals open to such violations of their personal privacy, by exposing information which taken in its totality is unique to that individual and details a part of that person's past, present and future medical history. While genetic information is by no means the whole story of the individual human being, it is a significant enough part of the individual that the individual should expect that that information only be disclosed to those people who have a genuine need to know in order to advance that individual's own interests. Such information is simply too sensitive to leave open to perusal and use by as large a cohort as presently has access to the medical record, whether or not such use constitutes an actual harm to the patient.

3.3.7. Commercial Interests in the Medical Record. The introduction of managed care has increased the level of bureaucratic access to the medical record, not only to review individual patient care, but also for the purposes of reviewing provider practices which the managed care organization (MCO) may want to encourage or discourage. For instance, MCOs regularly review the patient records of treating physicians in order to assess the frequency with which physicians order various tests, treatments and prescriptions. While the MCO may argue that this activity is not different from standard quality assurance (QA) reviews, the purposes of the MCO review are at least two fold: 1) to indeed assess the quality of care delivered, but 2) to evaluate the practices of the provider in light of predetermined economic goals. Physicians who routinely order more diagnostic tests than average are encouraged to limit such orders. If they fail to, their future participation in the plan and consequent ability to treat current patients may be threatened and ultimately terminated. Although patients have of necessity (some might say, by coercion) signed over to the MCO the right to review their medical records, few patients actually understand that their records are being used for a purpose other than their own medical interests, namely, the advancement of the MCO's own economic interests.

There are other purposes as well to which a computerized medical record may contribute, particularly one which contains so much detailed genetic information as high speed testing will make possible. The Consumer Advocate for the City of New York, Mark Green, published a report in 1996 detailing the practices of *pharmaceutical benefits managers* (PBMs) in the development of formularies and the use of patient records to modify physician prescribing habits to conform to the set of drugs the PBM and MCO have decided to cover (Green, 1996b). Unfortunately, the choice of preferred drugs is in many cases based upon the amount of money a pharmaceutical manufacturer is willing to pay the PBM, or in some cases whether such manufacturer owns a controlling interest in the PBM. The report details several cases in which the pharmaceutical manufacturer has been given access to patient records for these purposes. One can only imagine to what other ends our medical records are put as well, once these corporations have a copy of our medical records. As more detailed genetic information becomes a part of these records, they will increasingly become more and more valuable as a research tool. Will PBMs and their manufacturer partners limit the use of these records to just their marketing potential—an activity that is already questionable enough—or will they not go on to further exploit our records and the genetic information they hold to further advance their commercial interests through basic biomedical research? For instance, a detailed genetic profile of patients taking a particular drug would serve a pharmaceutical manufacturer in discovering which genetic types are more or less susceptible to that drug's specific action, as well as side effects. While such information may also be of use to the patient, should they not have some say in whether their records and genotypes are used for this purpose?

Unfortunately, the situation is worse than that. It is routine for MCOs, pharmaceutical corporations and data management firms such as Equifax and IMS America to *sell* medical records back and forth (Kolata, 1995). The inclusion of detailed genetic information will not only make these records more valuable as a research tool, but they also obviate any pretense at the protection of individual confidentiality, even if the records have been stripped of personal identifiers. This is because the amount of genetic information is already specific enough to uniquely identify an individual. Indeed, it has even been proposed that a person's social security number be replaced by a unique genetic sequence of their DNA as a personal identifier (Dahm, 1997). Thus, in principle, anybody into whose hands such records fall may be able to identify the individual person even if their name and address has been stripped from the record.

There is another domain in which "anonymity" becomes highly problematic: the use of "anonymous" tissues and cells in research. As soon as a sufficient amount of genetic information is incorporated into the medical record, the idea of an "anonymous" tissue sample will become an oxymoron, since one need only test the genotype of the specimen and then search the medical record databases for the corresponding unique sequence or genotype. Since a great deal of genetic research is conducted on anonymous tissue samples precisely because it is believed that the source of those tissues cannot be harmed because they are anonymous, the impending loss of anonymity will suddenly create a host of human research subjects suddenly "at risk" but for whom no institutional protections exist since the research is often conducted outside the purview of the institutional review boards whose charge it is to protect such subjects. The research is outside of their purview precisely because in the U.S. the Federal Regulations exempt such research from IRB review and consequently from any process of informed consent on the part of the sources of those tissues.[6]

3.3.8. Research Interests in the Medical Record. It is now-a-days difficult to discuss the conduct of biomedical research outside of the context of commercial interests.[7] The above scenario raises deep problems about individuals' right to privacy, who "owns" the medical record and hence who can profit from it, and what obligations are owed to patients. High speed genetic testing, as I have argued will make these records even more valuable and we can expect large numbers of interests to line up at the electronic feeding trough in the future. It is becoming increasingly difficult to distinguish between pure "research" interest in these electronic medical records and the resultant commercial interests since scientists themselves have discovered their own financial interests in exploiting the knowledge they produce, even if at public expense.

Nevertheless, if we can abstract for a moment from these commercial interests, we find behind them a genuine interest on the part of biomedical research to exploit the value of genetics and the wealth of information genetic testing will produce in the electronic medical record, for the public good. After all, if knowledge is commercially valuable, it is ultimately so because that knowledge will contribute to the practice of medical and/or public health. Thus researchers claim that apart from the commercial exploitation of medical records, they should have access for the purpose of generating generalizable knowledge of benefit to all. Thus if we clear away the obfuscation of market interests, we find at bottom a fundamental conflict between individual patients' expectations of, if not rights to, the privacy of their medical record against society's interest in exploiting the information contained therein for the benefit of all. It is fundamentally a value judgment how we settle this conflict.

This raises one of the more interesting questions of how we evaluate problems raised by complex new technologies. Typically, the ones who best understand these technologies are the ones who seek to exploit them, for whatever purpose. The individuals or groups, however, who may have an interest—whether as a matter of right or a matter of protection from harms—and in this case individual patients, generally have neither the background, experience nor knowledge to fully appreciate the possibilities of such technologies and how they may be exploited. Thus, who is going to speak for those who essentially have no representation in such decisions? Is it sufficient for the managers of managed care, the scientists and the CEOs of various biotechnology firms to set their own standards for protection of patient confidentiality and privacy, or is this tantamount to simply letting the fox guard the chicken coop?

3.3.9. The Darker Side. On August 21, 1996, the New York Times carried a story reporting that New Jersey law enforcement officials arrested 12 alledged members of the Genovese crime family on charges of racketeering and conspiracy for infiltrating Tri-Con Associates, a health care corporation which "arranged and managed group medical, dental and optical programs for employers and unions with networks of health-care providers" (Raab, 1996). In addition to submitting false claims, inflating fees and laundering money—usual mob activities—prosecutors were very worried that because Tri-Con had access to patient records, these records might be used for the purpose of blackmail and extortion. Robert Buccino, of the Organized Crime Bureau of New Jersey, was quoted, saying "They did have access to medical records, and it could be used for extortion and other criminal purposes." To my knowledge, they found no further evidence of such activities, but I suspect that if not at Tri-Con, we will see other examples of this in the future.

The possibility of individuals gaining access to medical records, and particularly medical records replete with detailed genetic information, for the purpose of exploiting that information for personal gain and against the interests of the patient is a scary prospect. For here, not only are patients' rights to privacy and confidentiality violated, but their interests may be directly harmed in a variety of ways. We do not need a take over of an MCO by the mob for this to happen. We need only individuals who have access to medical records to exploit this access.

4. CONCLUSION

High speed genetic testing, in whatever form it eventually comes to our doctors' offices, promises to radically transform the practice of medicine. It can be used to great medical benefit of all persons. At the same time, however, the social conditions in which we presently live are not ripe for exploiting this technology without causing as much if not more harm to the very individuals we seek to medically benefit. Until we can resolve the underlying contradictions in health care financing and access to medical care; until we can come to some consensus about the limits of using genetic information in employment; until we can establish a national policy on the privacy and confidentiality of electronic medical records; until we can empower individual patients to control the uses of their medical records; and until we can safeguard electronic medical records from their unscrupulous use by criminal interests; the volumes of genetic information that will be liberated by these technologies promise to harm as much as they promise to help.

Can we responsibly develop and use these genetic technologies before we solve these underlying problems, or will we only be forced to solve them when wide spread genetic testing hits the fan of commercial exploitation? I am not particularly optimistic. Inspite of the commitment of 5% of the Human Genome Project's funds to the study of the ethical, legal and social implications of genetics, the problems genetics raises are still with us. Has ELSI simply been ethical window dressing, absolving researchers, physicians and corporations of their responsibility to use new technology wisely? If we cannot translate even the basic recommendations of the ELSI Task Force on Genetics and Insurance into a coherent policy, then what purpose does this ethical reflection serve in advance, if we will only wait until a large enough number of persons are harmed by the introduction of genetic technology to routine medical practice?

In establishing ELSI, scientists and policy makers sought the advice of a broad range of disciplines to help advise and formulate policy in advance of what was forseen to be the revolutionary impact of genetics on all of our lives. That advice has been forthcoming in spades. Much remains controversial, but the road ahead is fairly clearly charted, if we have the will to address these issues now. But if we fail to have the will to make the changes necessary to make genetic technologies safe for society, what then? What responsibility do the scientists and physicians who have profitted enormously from this three billion dollar project have to the public who has paid for it and yet stands to equally and simultaneously be harmed (and helped) by it? But alas, so many of the scientists have lost their impartiality, have they not, by investing in development of the technology their work has made possible? It would be unrealistic for us to hope that the scientists who have so invested in these projects would, as a matter of principle, refuse to complete them until our social conditions have been transformed sufficiently to make these technologies genuinely safe for the public. Technology appears to have an inevitability about it that seems to drive itself into our lives and transform our society and self understanding. But is it really the technology that is the driving force, or is not just the thirst for profits that really fuels the engine of technological change?

NOTES

1. For a general history of the Human Genome Project see Cook-Deegan, 1994. For a general account of its purposes, see Gilbert, 1992; Koshland, 1989; and Watson, 1990.
2. This is largely the subject of the Human Genome Diversity Project, an international effort to collect DNA samples from every ethnic population around the globe.
3. The social, scientific and even moral worth of the project has also been called into question on a number of fronts. See, for instance, Hubbard and Wald, 1993; and Lewontin, 1992.
4. This leads to the unfortunate consequence in genetic testing and counseling that many individuals will seek testing "off the record," traveling to distant clinics and paying cash for the testing and counseling in order to insure that the results of the tests do not become a part of their medical record. Counselors typically inform prospective testees that they may be required to provide the results of any genetic tests to insurers and employers, if asked. The purpose of seeking the testing "off the record," however, is to precisely avoid getting caught if they choose not to divulge the results of genetic tests. This places genetic testors and counselors in the position of at least aiding in consumer fraud.
5. "Whatever, in connection with my professional service, or not in connection with it, I see or hear, in the life of men, which ought not to be spoken of abroad" (Edelstein, 1943).
6. See 45 CFR Part 46.101(b)(4) and 46.102(f).
7. For a small sampling of the problem, Fisher, 1996; Wadman, 1996; Anonymous, 1996; Anonymous, 1996; Hilts, 1997; Fersko and Connolly, 1992; Annas, 1990; McNally and Wheale, 1996; Anderson, 1993; and Stone, 1993.

REFERENCES

Anderson, C. (1993), 'Genome Project Goes Commercial. [News]', *Science*, Vol. 259, No. 5093, 15 Jan, pp. 300–302.

Annas, G.J. (1990), 'Outrageous Fortune: Selling Other People's Cells', *Hastings Center Report*, Vol 20, No. 6,, pp. 36–39.

Annas, George J. (1993), 'Privacy Rules for DNA Databanks: Protecting Coded 'Future Diaries'', *Journal of the American Medical Association*, Vol. 270, No. 19, 17 November, pp. 2346–2350.

Annas, George J.; Glantz, Leonard H.; Roche, Patricia A. (1995), '*The Genetic Privacy Act and Commentary*' Boston University School of Public Health, Boston.

Anonymous (1996), 'Gene Donors' Rights at Risk', *Nature*, Vol. 381, p. 1.

Anonymous (1996), 'Capitalizing on the Genome [editorial]', *Nature Genetics*, Vol. 13, No. 1, May, pp. 1–5.

Capron, Alexander M. (1991), 'Protection of Research Subjects: Do Special Rules Apply in Epidemiology?', *Journal of Clinical Epidemiology*, Vol. 44, pp. 81s-89s.

Clarke, Angus (1991), 'Is Non-directive Genetic Counseling Possible?', *Lancet*, Vol. 338, pp. 998–1001.

Cook-Deegan, R. (1994), '*The Gene Wars: Science, Politics, and the Human Genome Project*' W. W. Norton, New York.

Dahm, Lisa L. (1997), 'Using the DNA Profile as the Unique Patient Identifier in the Community Health Information Network: Legal Implications', *The John Marshall Journal of Computer & Information Law*, Vol. 15, No. 2, Winter, pp. 227–275.

Decruyenaere, Marleen; Evers-Kiebooms, Gerry; Van den Berghe, Herman (1993), 'Perception of Predictive Testing for Huntington's Disease by Young Women: Preferring Uncertainty to Certainty?', *Journal of Medical Genetics*, Vol. 30, pp. 557–561.

Drmanac, Radoje; Labat, Ivan; Brukner, Ivan; Crkvenjakov, Radomir (1989), 'Sequencing of Megabase Plus DNA by Hybridization: Theory of the Method', *Genomics*, Vol. 4, pp. 114–128.

Edelstein, Ludwig (1943), 'The Hippocratic Oath: Text, Translation and Interpretation', *Supplement to the Bulletin of the History of Medicine*, Vol. 1, p. 3.

Fersko, R.S.; Connolly, M.J. (1992), 'Intellectual Property Issues in a Clinical Trial: A Corporate Perspective', *Quality Assurance*, Vol. 1, No. 3, June, pp. 237–248.

Fisher, Lawrence M. (1996), 'Cancer-Research Venture Announced', *New York Times*, 21 August, p. C6.

Fodor, Stephen P.A.; Read, J.Leighton; Pirrung, Michael C.; Stryer, Lubert; Lu, Amy Tsai; Solas, Dennis (1991), 'Light-Directed, Spatially Addressable Parallel Chemical Synthesis', *Science*, Vol. 251, pp. 767–773.

Giardiello, Francis M.; Brensinger, Jill D.; Brensinger, Jill D.; Peterson, Gloria M.; et al. (1997), 'The Use and Interpretation of Commercial APC Gene Testing for Familial Adenomatous Polyposis', *The New England Journal of Medicine*, Vol. 336, No. 12, 20 March, pp. 823–827.

Gibbs, Richard A. (1995), 'Pressing Ahead with Human Genome Sequencing', *Nature Genetics*, Vol. 11, No. 2, October, pp. 121–125.

Gilbert, Walter (1992), 'A Vision of the Grail' Kevles; in Hood, Daniel J. Leroy (ed.), *The Code of Codes: Scientific and Social Issues in the Human Genome Project* Harvard University Press, Cambridge, pp. 83–97.

Greely, Henry T. (1995), 'Conflicts in the Biotechnology Industry', *Journal of Law, Medicine and Ethics*, Vol. 23, No. 4, Winter, pp. 354–359.

Green, Mark (1996b), *Compromising Your Drug of Choice: How HMOs are Dictating Your Next Prescription* Office of the Public Advocate for the City of New York, New York.

Green, Mark (1996a), *What Ails HMOs* Office of the Public Advocate for the City of New York.

Hilts, Philip J. (1997), 'Researcher Profited After Study by Investing in Cold Treatment', *New York Times*, 1 February, p. A6.

Hubbard, Ruth; Wald, Elijah (1993), *Exploding the Gene Myth* Beacon Press, Boston.

Hudson, Kathy L.; Rothenberg, Karen H.; Andrews, Lori B.; Kahn, Mary Jo Ellis; Collins, Francis S. (1995), 'Genetic Discrimination and Health Insurance: An Urgent Need for Reform', *Science*, Vol. 270, pp. 391–393.

Huggins, Marlene; Bloch, Maurice; Kanani, Shelin; Quarrell, Oliver W.J.; Theilman, Jane; Hedrick, Amy; Dickens, Bernard; Lynch, Abbyan; Hayden, Michael (1990), 'Ethical and Legal Dilemmas Arising During Predictive Testing for Adult-Onset Disease: The Experience of Huntington Disease', *American Journal of Human Genetics*, Vol. 47, pp. 4–12.

Kolata, Gina (1995), 'When Patients' Records are Commodities for Sale', *New York Times*, 15 November, pp. A1, C14.

Kolata, Gina (1996), 'Breaking Ranks, Lab Offers Test Assessing Cancer Risk', *New York Times*, 1 April, p. A1.

Koshland, Daniel (1989), 'Sequences and Consequences of the Human Genome', *Science*, Vol. 246, p. 189.

Kozal, Michael J.; Shah, Nila; Shen, Naiping; Yang, Robert; Fucini, Raymond; Merigan, Thomas C.; Richman, Douglas D.; Morris, Don; Hubbell, Earl; Chee, Mark; Gingeras, Thomas R. (1996), 'Extensive Polymorphisms Observed in HIV-1 Clade B Protease Gene Using High-Density Oligonucleotide Arrays', *Nature Medicine*, Vol. 2, No. 7, pp. 753–759.

Lewontin, R.C. (1992), 'The Dream of the Human Genome', *New York Review of Books*, 19 November, pp. 31–40.

Lockhart, D.J.; Dong, H.; Byrne, M.C.; Follettie, M.T.; Gallo, M.V.; Chee, M.S.; Mittmann, M.; Wang, C.; Kobayashi, M.; Horton, H.; Brown, E.L. (1996), 'Expression Monitoring by Hybridization to High-Density Oligonucleotide Arrays', *Nature Biotechnology*, Vol. 14, pp. 1675–1680.

McNally, Ruth; Wheale, Peter (1996), 'Biopatenting and Biodiversity', *The Ecologist*, Vol. 26, No. 5, pp. 222–228.

Nelkin, D.; Lindee, M.S. (1995) *The DNA mystique: The gene as a cultural icon* W. H. Freeman and Company, New York.

OMIM (1996), *Online Mendelian Inheritance in Man* Center for Medical Genetics, Johns Hopkins University (Baltimore, MD) and National Center for Biotechnology Information, National Library of Medicine (Bethesda, MD); World Wide Web URL: http://www3.ncbi.nlm.nih.gov/omim.

Pease, A.C.; Solas, D.; Sullivan, E.J.; Cronin, M.T.; Holmes, C.P.; Fodor, S.P.A. (1994), 'Light Generated Oligonucleotide Arrays for Rapid DNA Sequence Analysis', *Proceedings of the National Academy of Sciences*, Vol. 91, pp. 5022–5026.

Pokorski, R. (1994), 'Use of Genetic Information by Private Insurers' in Murphy, T.; Lappe, M. (eds.), *Justice and the Human Genome Project* University of California Press, Berkeley, pp. 91–109.

Raab, Selwyn (1996), 'N.J. Officials Say Mob Ring Infiltrated Health Care', *New York Times*, 21 August, pp. A1, C19.

Silverman, Paul H. (1995), 'Commerce and Genetic Diagnostics', *Hastings Center Report*, Vol. 25 No. 3 Supplement, pp. S15-S18.

Stone, Richard (1993), 'Biotech Sails into Heavy Financial Seas', *Science*, Vol. 260, No. 5110, 14 May, pp. 908–910.

Task Force on Genetic Information and Insurance (1993), *Genetic Information and Health Insurance: Report of the Task Force on Genetic Information and Health Insurance* National Institutes of Health, Washington, D.C.

Wade, Nicholas (1997), 'Where Computers and Biology Meet: Making a DNA Chip', *New York Times*, 8 April, p. C1.

Wadman, M. (1996), 'Drug Company 'Suppressed' Publication of Resaerch [news]', *Nature*, Vol. 381, No. 6577, 2 May, p. 4.

Watson, James D. (1990), 'The Human Genome Project—Past, Present, and Future', *Science*, Vol. 248, pp. 44–49.

Watson, James D. (1992), 'A Personal View of the Project' Kevles, in Daniel J.; Hood, Leroy (eds.), *The Code of Codes: Scientific and Social Issues in the Human Genome Project* Harvard University Press, Cambridge, pp. 164–176.

11

ETHICAL QUESTIONS IN THE PURSUIT OF GENETIC INFORMATION

Geneticization and BRCA1

Susan Sherwin and Christy Simpson

Department of Philosophy, Dalhousie University
Halifax, Nova Scotia, Canada

Breast cancer is a disease that will directly affect more than 10% of all women (and a small percentage of men) in North America. All who realize they have this disease will have their lives disrupted; most will undergo surgery to remove the offending cells and surrounding tissue. Many will also undergo radiation, chemotherapy, and/or long term prophylactic drug regimens; all will be advised to seek ongoing medical surveillance of their breasts and related tissues for the rest of their lives. An unacceptably high number of these people will die from this disease or from complications associated with it. The lives of all those who are close to people who experience breast cancer will be dramatically changed and their own sense of safety will be noticeably reduced. And most women, including many of the more than 85% of the North American population which will never develop breast cancer, will live their lives in fear of developing this disease; many will come to depend upon regular medical reassurances that they are (for now) cancer free.

These are some of the facts about breast cancer which form the starting point of our investigation of the geneticization of a disease which threatens the lives and well-being of millions of women worldwide. Improvements in screening and detection account for some of this increase but do not explain it all (Ursin, Bernstein, and Pike, 1994). The first place most of us are inclined to look for answers in the face of this frightening threat is to biomedicine. Biomedicine is the dominant approach to health matters in the West and in most other parts of the globe. It incorporates a world view that encourages us to see disease (and also, by implication, health) as an internal property of a distinct, isolated human organism. Under the biomedical model, we understand disease as something that affects the body when some internal mechanism becomes disrupted either because of an inborn weakness or flaw or because of some sort of invading alien organism or hostile environment (or a combination of these factors). Because the individual body is the site of health and illness, it is also considered the appropriate site of all interventions aimed at promoting or restoring health (Sherwin, 1998). Biomedical practice relies on science and technol-

Genetic Information, edited by Thompson and Chadwick.
Kluwer Academic / Plenum Publishing, New York, 1999.

ogy to provide accurate surveillance of each body in order to allow for early detection of some breakdown in one of that body's internal systems. Once problems are detected and measured, biomedicine draws on a variety of invasive strategies to attack the offending cells, system, or organism. In the case of cancer, the tools to date have largely consisted of efforts to "slash, burn, or poison" the problem cells into submission.

Effective technological approaches are, surely, welcome. Breast cancer, like other diseases, affects particular people, and, once it does, it must be addressed by, at least, treating the bodies in which it is found. Because cancer does reside and develop within the discrete bodies of individual women and men and because individuals experience illness and, sometimes, death from this disease, those affected must indeed seek out the best available treatment for their conditions: that involves subjecting their bodies to the most effective interventions science and technology have produced. At present, there is no meaningful alternative for those with breast cancer, or suspected breast cancer, than that of exploring biomedical solutions to their disease. We are grateful for the successes biomedicine has achieved in increasing the life span and quality of life for many who develop breast cancer and we join others in eagerly awaiting further scientific understanding of this frightening disease and the development of new, more effective, biomedical treatments.

We are, however, concerned about the ways in which the public as well as the private focus on biomedical solutions tends to interfere with other types of approaches to this (and other) serious public health problem(s). As a society, we have directed almost all of our attention and resources towards biomedical strategies for diagnosing or treating breast cancer. By focusing on the ways breast cancer occurs within individual bodies, we tend to lose sight of the relations that exist among those bodies. Although breast cancer is a disease of epidemic proportions, society has yet to respond with appropriate sorts of public health measures. In particular, most governments have not provided adequate resources to explore possible environmental, dietary, or social links, nor do they display any real willingness to counter environmental threats as they are identified. Biomedicine keeps the public focus on the micro-level of individual illness; that strategy interferes with our ability to think or act on the macro-level. As such, it provides an inadequate response to the disturbing incidence of breast cancer.

Consider, for example, the ways in which biomedicine directs attention primarily to the level of the individual by encouraging each woman to develop high levels of fear about her own risk of contracting breast cancer. In the name of health promotion, women are repeatedly warned of the risk each faces of developing this disease. Many women are easily frightened into a mind-set where they are preoccupied with monitoring their bodies for threatening signs of change (or where they guiltily avoid checking their breasts for fear of discovering some suspicious lump); if a lump is detected, they are primed to quickly seek expert advice and biomedical interventions. This focus on individual responsibility and surveillance occurs as part of a much larger public health campaign aimed at self-regulation: not only are women asked to pay worried and regular attention to their breasts, they are also advised to change lifelong eating habits, develop healthy exercise programs, learn stress management techniques, figure out the appropriate "moderate" level of drinking to engage in, and break free of any addictions they may have succumbed to (be it to alcohol, drugs, or even cigarettes, pills, or chocolate). And, since most women do not live alone, most find they are also expected to try to cajole or otherwise affect the various lifestyle patterns of family members. Fulfilling these multiple health responsibilities can come close to a full time job in itself, but, of course, most women must also earn a living, manage their households, and spend quality time with family and friends. The main preventive strategies regarding breast cancer, like those for many other major diseases, involve stern warnings to individuals to engage in regular suspicious surveillance of their bodies—a practice that invites women to treat their bodies as alien, threatening objects—

along with demands that they change lifestyle patterns and live up to moralistic demands of "healthy living"; this demand is sometimes so strong that some have taken to calling this new morality "healthism" (Lock, 1998).

The problem is not that this is bad advice; it is that it is so unrelentingly directed at placing the full responsibility for health on individuals when there is abundant evidence that many other factors come into play. It is widely known, for example, that socio-economic conditions play a major role in health status; life expectancy and general levels of health are well correlated with income brackets and also with vulnerability to oppression. Even though income and social status provide no guaranteed protection from any form of cancer, disadvantaged populations seem to suffer disproportionately from many cancers: aboriginal women in Canada suffer from some forms of cancer (e.g., cervical cancer) at many times the rate of other women and, in the U.S., African-American women die much more frequently from breast and other forms of cancer than do white women (National Cancer Institute, 1994). Moreover, environmental factors have a significant impact on health. Depletion of the ozone layer, production of greenhouse gases, contamination of water supplies, acceptance of lead in paint, and pollutants in the air—all dramatic planetary changes associated with ever-increasing industrial activity—are directly correlated with many illnesses (e.g., skin cancer, asthma, and a new disease of general "environmental illness") (Last, 1993). At the same time, full-scale industrialization of agriculture through the use of pesticides, herbicides, growth hormones (and feeding practices that spread "mad cow" disease) may have devastating impacts on health. Yet, research funds, both private and public, are primarily directed towards developing further technologies that will bring increased profits to industry. Even so-called "pure" research funds, available to academics through publicly-funded research agencies, often encourage applicants to find a "private partner" to co-sponsor the research, ensuring that research efforts will be concentrated in areas that promise economic benefits for some despite the potential of serious health costs.

The pursuit of the "breast cancer genes," BRCA1 and BRCA2, is a clear example of how these two trends—biomedicine's focus on the individual and support of profit-oriented research—come together. These two genetic mutations are associated with some inherited forms of breast cancer. Research in support of this discovery is part of the massive global public-private research project known as the Human Genome Project, aimed at "mapping" the full human genome with the expectation that once we better understand where each gene is located, we might be able to figure out what each does, and, it is hoped, we will then be in a position to manipulate it. This is the lead project in a general move towards what Abby Lippman has deemed "geneticization" (Lippman, 1998).

Geneticization is the attitude that the differences among people can be reduced to differences in their genetic makeup; it assumes that most disorders are largely attributable to genetics. The ideological and practical dimensions of geneticization encourage us to look to the genetics of the individual to find both the source of each disease and the (single) site at which to respond to it. This perspective carries significant political and practical implications; hence, there are important ethical questions to be asked about this approach.

Consider how geneticization operates in the case of the breast cancer genes. There has been a major push to discover the genes "responsible" for breast cancer. It has been estimated that 5–10% of all breast cancers have a heritable element: i.e., when a person inherits a mutated gene, such as BRCA1 or BRCA2, it predisposes her to developing breast (or ovarian) cancer (Burke, Daly, Garber, et al., 1997). Those with these mutated genes are told that they are at "high risk" of developing breast cancer. But what does this designation mean? What should an individual do about this warning?

Let us start with the first question about the meaning of "high risk" and break it down even further to ask, what does the term "risk" mean? It represents a statistical difference, but not just any statistical variation. "Risk" is a heavily value-laden term; typically, it connotes fear or anxiety. We speak of "risks" of being struck by lightning but of "chances" of winning the lottery.

As K.S. Shrader-Frechette has pointed out, values can enter into the determination of and discussions about risk or risk assessment at each of three stages of analysis: choice of topics, methods/procedural considerations, and evaluation (Shrader-Frechette, 1991). In the first, decisions must be made as to what should be researched. What do we need to know the risk(s) of? In this case, what sorts of causal links will we explore regarding the likelihood of someone developing breast cancer? Decisions such as these ultimately rest on one's view of such factors as society, disease, and culture and the relative importance attached to these different areas. Second, once a topic is chosen for investigation, how the research is done is also affected by methodological value judgments—what data are deemed relevant, what statistical tools are used, and what assumptions are made can all affect what the risk is estimated or computed to be. Finally, once data are collected, it is necessary to move to the evaluation stage in which decisions are made as to whether the risk is minimal, above acceptable standards, or is even something about which we should be concerned. These decisions will all be based on the values of those setting the research agenda as to what is important and what is not; these values will also affect the measures that are developed or used to minimize or avoid certain risks. So, what researchers want to know the risk of, how they determine what the risk is, and finally, how they interpret the data are all affected by values. In the case of breast cancer, the values of the biomedical model (such as the importance of intervention at the micro-level of the individual body) are dominant and tend to shape both understandings of the risk of having the "breast cancer genes" and the options available to counteract this risk. These values also limit the exploration of other risk factors that may be equally or more important.

Consider, then, what being at "high risk" means. Someone has decided that the risk of carrying the BRCA1 and BRCA2 genes is high. We must ask who, why, and how. By what methods and criteria? Different studies have estimated that the cumulative risk of developing breast cancer for persons carrying these genes can range from 40–85%; these levels are significantly higher than the estimated 12% cumulative risk of developing breast cancer for the general population (Schrag, Kuntz, Garber, and Weeks, 1997). It is seldom made clear, however, that these figures are for *lifetime* risk, not an immediate indication of onset of disease; indeed, 63% of women with these genes will not develop breast cancer until after the age of 40 (Weijer, 1995). Moreover, having the breast cancer gene is not equivalent to having the disease, though it is commonly interpreted as such, especially when the 85% cumulative risk figure is the one which is quoted. Even though from 15–60% of women who carry these genes will not in fact develop breast cancer, there is little discussion of the role of the environment within which the person having these genes lives and works. The geneticization move is an effort to focus on what is wrong internally in the individual that she should develop such a disease. Yet women with these genes need access to knowledge about matters such as what sorts of environments are dangerous and what sort are safe for individuals with these "high risk" genes.

Breast cancer is not unique in this respect. The correspondence between the having of certain genes and of developing a certain disease is rarely as simple and as straight-forward as it is commonly presented (Ottman, 1996). In many cases, so-called genetic disease may have at least as much to do with the environment in which these genes are found and expressed than it does with the having of the genes themselves. One clear example which illustrates these points is the case of phenylketonuria (PKU). People who have this "genetic disease" lack the ability to make (enough of) the enzyme, phenylanalanine hydroxylase, due to a mutation in this gene. Without this enzyme, phenylalanine is not converted into tyrosine and accumulates in the body. High levels of phenylalanine can cause severe mental retardation, hyperactive reflexes, and an early death. But, if PKU is discovered soon enough, these effects can be avoided by putting the individual on a low phenylalanine diet. Thus, if a person happened to live in an environment where diets were as a matter of course low in phenylalanine, this genetic mutation and corresponding "disease" would not be detected and would not create any problem for the individual. The importance of the environment in which genes are found cannot be ignored. There remains much to learn about the breast cancer genes and their interaction with different environments. While we know that the "solution" for reducing the incidence of breast cancer in those with these predisposing genetic mutation(s) is unlikely to be as simple as for PKU, this example does help to illustrate the importance of pursuing a macro-level approach to disease.

If we are to test for "the breast cancer genes" we must reflect on the implications of test results. For example, what follows for those individuals who do test positive for the "breast cancer genes"? We currently have very little understanding of what the implications should be for these people. We do know that such individuals will probably have a much more difficult time obtaining health and life insurance and employment than those who test negative or who do not know their genetic status, that they will be encouraged to undergo untested and highly invasive prophylactic measures that have serious health consequences of their own and that they will likely experience increased levels of anxiety for the remainder of their lives. Nonetheless, the company that has developed the test for these genes is already marketing it to a frightened population as a "preventive" test.

Does this test really qualify as a preventive measure? Can the knowledge it gives really help prevent breast cancer? At present, the only recommended medical options are those of a prophylactic mastectomy and/or increased surveillance (i.e., regular mammography and clinical/self breast exam); neither of these strategies is preventive in the ordinary sense of the term, since neither can ensure that breast cancer will not develop. At best, increased surveillance may help to prevent a woman's premature death from breast cancer (a valuable but very different form of prevention) because the cancer is caught early; but having a mammogram or performing breast self-examination will have no effect upon whether or not breast cancer develops. What about prophylactic mastectomy? If breast tissue is removed, the risk of developing breast cancer will be reduced, but the data are not yet available to determine by how much.

A theoretical model was constructed to attempt to give a preliminary answer to this question: it determined that a prophylactic mastectomy on a 30-year-old woman who carries either a BRCA1 or BRCA2 mutation would, on average, increase her life expectancy by 3–5 years, but if the woman is over 60, having this procedure would (on average) result in only one extra year (Schrag, Kuntz, Garber, and Weeks, 1997). A 10 year study to determine whether a double prophylactic mastectomy is the best way to reduce the risk of breast cancer in women who carry the mutated genes has just begun (Kirkey, 1997). In the meantime, no one really knows what the effect of a prophylactic mastectomy will be for an individual woman and whether it is really the best approach. Bernadine Healy, in a recent editorial on the breast cancer genes and the option of prophylactic mastectomy, cautions against "letting statistical prophecies lead to irreversible decisions" (Healy, 1997). Thus, it seems that this genetic testing and the options available to those who test positive do not at this point qualify as satisfactory "preventive" measures.

The genetic testing approach feeds off widespread existing fears about breast cancer—studies show that most women have an exaggerated view of their risk of developing breast cancer (Fallowfield and Clark, 1991). It also helps to feed those fears by insisting that such levels of surveillance are worth the risks to avoid the alternative of (advanced) breast cancer even though the vast majority of breast cancers are not associated with either gene. It seems worth noting that these tests are developed and marketed by companies that profit by the existence of these fears. The higher the levels of fear, the greater the use of the tests. Genetic tests are marketed by the same industries (and sometimes even the same companies) that produce the pesticides, emissions, and greenhouse gases that contribute to increased risks of disease. Companies profit by producing and selling hazardous chemicals and by engaging in harmful production practices, then by offering tests to determine the effects of our exposure to those chemicals, and, if we do indeed become ill, by selling technologies and pharmaceuticals to relieve our illnesses. Illness production, illness detection, and illness care are all profitable. There is far less promise of profits available in true prevention than in screening and treating.

What about the women who test negative for BRCA1 or BRCA2? What does the negative result mean with respect to their risk of developing breast cancer? It seems that there are two main possibilities for these women: (1) they may still be at high risk according to the other known risk factors for breast cancer or (2) they do not fall into any of the identified high risk categories. The other risk factors for breast cancer include (among others): young age at the beginning of menstruation, late age at menopause, not having a child before the age of thirty, use of hormone replacement therapy, and previous breast, ovarian, or endometrial cancer. For women who have these risk factors, increased and continual surveillance are the only options which the biomedical approach offers other than pharmacological manipulations based on the presumed role of estrogen in the development of breast cancer, e.g., the Breast Cancer Prevention trial with tamoxifen currently underway (National Cancer Institute, 1995). It is important to note, however, that all of these risk factors combined, including the presence of BRCA1 or BRCA2, account only for approximately 30% of all breast cancer cases (Henderson, 1993).Thus, we must ask, what causes breast cancer in the other 70% of cases? Are we sure that the above-given risk factors are the most critical ones? Such information should affect how we define being at high risk for breast cancer. Women who neither test positive for the breast cancer genes nor fall into any of the other risk categories are not identified as being at high risk, yet they make up the majority of breast cancer cases. Hence, we need to re-evaluate the direction in which the biomedical model has been taking us in investigating this disease and in telling us who is at high risk.

There are still other features of this genetic testing which need to be addressed. Consider, for example, the information which is (potentially) gained about other family members as a result of this testing and what the effects of either a negative or positive test may be within the family setting. Should information be disclosed to other family members? How will one person's genetic test affect the ability of a relation to get health insurance? As well, very little attention is paid to the psychological impact of finding out the status of one's genes. Responses to a positive test may include coming to view oneself as already sick and doomed to die—even though no cancer has been detected—or it may lead to a double mastectomy followed by a mistaken sense of security. Responses to a negative test may also induce a false sense of security about one's risk of developing breast cancer and lead to a failure to monitor one's breasts.

There are conflicting interests at stake in our current patterns. A biomedical approach to illness that is focused on treating individuals after they become ill serves not only the interests of patients who will inevitably fall ill, but also the interests of many private businesses and many health care providers. But it does not serve the interests of those whose disease might have been prevented if we truly understood its causes. A different approach might yield genuine preventive strategies which could dramatically reduce the incidence of the disease. The difficulty is breaking out of the mindset that sees illness (in this case breast cancer) as a private problem best addressed through primarily personal strategies. It should also be seen as a political issue that must be addressed through political challenges of the interests that profit from ill health. Somehow we must find ways of getting past our private fears in order to demand political change, a change whereby research money goes into exploring ways of limiting the disease, of actually preventing it, and not just into finding profitable ways of detecting and treating it. This will, for example, require public funding of public interest projects which investigate ways of cleaning up and protecting the environment in order to create healthier spaces in which we can live and work.

If we are to encourage such alternative research projects, we need to find ways to transform some of our private fears into anger. We must challenge breast cancer strategies that emerge from geneticization assumptions and ignore the role of social and environmental factors. We should object to cancer prevention policies that consist solely in informing individuals that they can reduce their risk by accepting personal responsibility for their health-related behaviors or by establishing their risk status. While such private strategies have a role to play, the messages must be carefully framed lest they convey other, dangerous messages as well, such as "if we do as directed, we will be safe." Interpreted that way, these private responsibility approaches suggest that those who do contract cancer were somehow responsible for their fate—they must have failed to take proper care of themselves. The temptation is for those without cancer to reassure ourselves that it is within our own power to avoid such diseases if we are only conscientious enough to pursue all of the options the biomedical model provides. Thus, while fearful, we can try to lure ourselves into the complacency of believing that full compliance with expert advice will protect us from developing cancer.

Despite the good intentions and (mostly) good sense behind most health warnings, there is, then, something deeply disturbing about the pattern of advice. They are problematic in ways that are reminiscent of the effect of the many messages women receive about the physical dangers that sexism poses for them. By reporting stories of the violence women experience when not in the company of a male protector, the culture encourages women to be fearful of attack and to believe that they can avoid danger by personal strategies of complying with social norms for proper femininity. But such strategies do not keep women safe anymore than breast self-examination or determination of one's genetic status will keep a woman from contracting breast cancer. In the case of violence against women, it was only when women learned to see violence against women as a political issue, as a matter of dominance and power rather than of personal failure, that we were able to get beyond our private struggles to try to avoid it and start working on finding ways to end it (Sherwin, 1996).

For similar reasons, then, it seems likely that the action required to develop effective campaigns for cancer prevention will require widespread recognition of the fact that cancer is a political as well as a personal issue. Social and political action is necessary to change the basic assumptions of society that encourage and often subsidize carcinogenic-producing industries and that support the funneling of health research money almost entirely towards expensive, technological interventions on individuals. An important step in this campaign is that we each understand that acting solely as individuals we are least able to prevent cancer or even protect ourselves.

Geneticization, represented by the concentration of resources directed towards developing and marketing the tests for BRCA1 and BRCA2, involves an effort to further privatize anxieties about breast cancer. These tests are designed to encourage women to trust science to diagnose their risk of developing breast cancer and to provide interventions that will lessen their chances of dying from the disease. They distract from awareness that genes operate differently in different environments; we have no idea what genetic composition makes anyone safe from developing breast cancer in an environment that is hazardous. If we do continue to pursue genetic strategies to detect those at high genetic risk of developing breast cancer, it is essential that we also develop strategies of reducing the impact of other factors on breast cancer rates.

Ethics is about values: in this case the values at issue are those that underlie public policies towards breast cancer. As long as health research is limited to the individualized focus of the biomedical model and as long as governments pursue uncritical pro-business research policies the dominant approaches to breast cancer are unlikely to serve women's interests well. Women need health-promoting policies which go well beyond advising them on how to live their personal lives. We need to ensure that research is conducted into the environmental, socio-economic, and psycho-social dimensions of breast cancer. Governments, industry, and the health research community must work jointly with community based health activists to find ways of preventing illness rather than merely profiting from it.

REFERENCES

Burke, Wylie; Daly, Mary; Garber, Judy; et al. (1997), 'Recommendations for Follow-up Care of Individuals with an Inherited Predisposition to Cancer', *Journal of the American Medical Association*, Vol. 277, No. 12, pp. 997–1003.

Fallowfield, Lesley and Clark, Andrew. (1991), *Breast Cancer*, Routlege: London, Britain.

Healy, Bernadine. (1997), 'BRCA Genes—Bookmaking, Fortunetelling, and Medical Care', *New England Journal of Medicine*, Vol. 336, No. 20, pp. 1448–49.

Henderson, Craig. (1993), 'Risk Factors for Breast Cancer Development', *American Cancer Society*, Professional Education Publication: United States, March 15.

Kirkey, Sharon. (1997), 'Canadian to Assess Preventive Mastectomy', *Ottawa Citizen*, pp. A1, Wednesday, July 9.

Last, J.M. (1993), 'Global change: ozone depletion, greenhouse warming, and public health', *Annual Review of Public Health*, Vol. 14, pp. 115–136.

Lippman, Abby. (1998), 'The Politics of Health: Geneticization vs Health Promotion', in Sherwin, Susan, et al. *The Politics of Women's Health: Exploring Agency and Autonomy*, Temple University Press: Philadelphia.

Lock, Margaret. (1998), 'Situating Women in the Politics of Health', in Sherwin, Susan, et al. *The Politics of Women's Health: Exploring Agency and Autonomy*, Temple University Press: Philadelphia.

National Cancer Institute. (1995), *Questions and Answers: The Breast Cancer Prevention Trial*, NCI, CancerFacts: United States, June.

National Cancer Institute. (1994), *Racial Differences in Breast Cancer Survival*. NCI, CancerFacts: United States.

Ottman, Ruth. (1996), 'Gene-Environment Interaction: Definitions and Study Designs', *Preventive Medicine*, Vol. 25, pp. 764–770.

Schrag, D., Kuntz, K.M., Garber, J.E., and Weeks, J.C. (1997), 'Decision Analysis—effects of prophylactic mastectomy and oophorectomy on life expectancy among women with BRCA1 or BRCA2 gene mutations', *New England Journal of Medicine*, Vol. 336, No. 20, pp. 1465–71.

Sherwin, Susan. (1996), 'Cancer and Women: Some Feminist Ethics Concerns', in Sargent, Carolyn F. and Brettell, Caroline B. (eds.), *Gender and Health: An International Perspective*, Prentice Hall: Upper Saddle River NJ.

Sherwin, Susan (Coordinator), Feminist Health Care Ethics Network. (1998), *The Politics of Women's Health: Exploring Agency and Autonomy*, Temple University Press: Philadelphia.

Shrader-Frechette, K.S. (1991), *Risk and Rationality*, University of California Press: Berkeley.

Ursin, G., Bernstein, L. and Pike, M.C. (1994), 'Breast Cancer', *Cancer Surveys: Trends in Cancer Incidence and Mortality*, Vol. 19/20, Cold Spring Harbor Laboratory Press: United States.

Weijer, Charles. (1995), 'Our Bodies, Our Science', *The Sciences* , pp. 41–45, May-June.

12

THE ETHICS OF GENE PATENTING

Ruth Macklin

Department of Epidemiology and Social Medicine
Albert Einstein College of Medicine
Bronx, New York 10461

Is it ethical to patent human genes and gene sequences? Like many questions about what is ethically permissible or impermissible, this query has layers of complexity. These include fundamental questions about what should and should not be patentable from an ethical perspective, as well as what is or is not patentable under existing patent laws. A second layer of legal questions asks whether existing patent laws are adequate to address the somewhat unique ethical concerns raised by patenting human genes. A related inquiry looks into differences in these laws in different countries. Even if current patent laws can be interpreted to allow gene patenting (Looney, 1994, pp. 231 272, Resnik, 1997, pp. 43–61) we can always pose an ethical challenge to an existing law. The United States and European countries have so far taken divergent approaches to gene patenting. The biotechnology industry in the U.S. is aggressively seeking patents for its "discoveries" or "inventions" whereas countries in the European Union have steadfastly resisted the patenting of human genes (Looney, 1994). The European Parliament has voted several times against gene patenting in the past few years.

An ethical analysis begins by asking: Are human genes somehow unique, and should they therefore be treated differently from other things for which the biotechnology industry has obtained patents? Do genes and gene sequences occupy a zone of personal privacy that would be violated by assigning patent rights to others? (Looney, 1994, p. 238). Are genes and gene sequences parts of our common heritage, so that information about them belongs in the public domain? Theologians have weighed in on the side of clear opposition to gene patenting, contending that all aspects of life are gifts of God and neither individuals nor companies should be granted "property rights" to such items. Also opposed are groups in the United States such as the Council for Responsible Genetics (CRG), a non-governmental organization of scientists, public health advocates, trade unionists, environmentalists, feminists, disability activists, and other concerned citizens. CRG opposes gene patenting for several related reasons, which can be grouped under the heading, "rejection of the commercialization of life" (Teitel, 1996, pp. 1–3).

Genetic Information, edited by Thompson and Chadwick.
Kluwer Academic / Plenum Publishing, New York, 1999.

How can we sort out the multiple and complex issues surrounding the debates over gene patenting? In this, as in many ethical and policy debates, reasons and arguments comprise a mix of consequentialist and nonconsequentialist elements. Sometimes these elements are pitted directly against one another, as illustrated by one debate about the likely consequences of gene patenting. One side contends that granting patent rights to an individual or organization results in limiting the application of knowledge from scientific work; the opposing position maintains that pursuit of patents leads to an increase in the general knowledge base and in the creation of useful products (Looney, Resnik).

At other times, a principled opposition to gene patenting holds fast despite consequentialist arguments promising great benefits that will accrue from allowing patents (Resnik, 1997). An example is the view that it is simply wrong to patent life—whatever financial gains or scientific advances may result—since patenting human elements of any sort involves a commodification or commercialization that should be rejected. Still another debate pits one conception of justice against another. Before turning to these opposing views, let me address the "threshold" question: What can and cannot be patented?

1. WHAT IS PATENTABLE?

The argument over whether genes and gene sequences should be considered patentable from a legal standpoint persists. The legal line distinguishes between a *discovery* of something that exists in nature, which is not patentable, and a true *invention*, which requires that human beings contribute something of significance (Looney, 1994, Resnik, 1997). One prevalent opinion holds that there is a legal basis for patenting invented, non-naturally occurring genes and DNA sequences, parts or combinations of chromosomes, as well as the uncontroversial processes for manufacturing, analyzing, sequencing, or recombining human genes. What is not patentable are naturally occurring human genes or their combinations (Resnik, 1997). To show that human genes and sequences meet the threshold test, it is necessary to demonstrate significant human intervention. Courts in the United States have already ruled that human innovation in the biotechnology realm enables living things to be patented. Nevertheless, a dispute remains at the conceptual level over whether gene sequences are a discovery or an invention (Looney, 1994).

In 1991 the U.S. Patent and Trade Office (PTO) refused an attempt by the National Institutes of Health (NIH) to patent gene sequences discovered as part of the Human Genome Project. Critics at the time argued that human genes are not inventions and should not, therefore, be patented (Resnik, 1997). However, the PTO's rejection of the NIH patent application rested on technical, not moral considerations. U.S. patent law requires that patentable items must demonstrate utility and novelty. The absence of known functions associated with the gene sequences in question resulted in failure to pass the utility test; and a finding of several of the claimed sequences in existing genetic databases led to failure to pass the novelty test (Looney, 1994, p. 252, n. 90). The then-director of the NIH defended the patent bid based on the need of the U.S. to protect its global market position, given the uncertainties in the international arena regarding patent laws (Looney, 1994). Although Harold Varmus, the subsequent NIH director, chose not to appeal the decision by the PTO, this has not deterred the private biotechnology industry from continuing its aggressive pursuit of patents.

One problem in the international arena is that the legal standards for patenting are not precisely the same in the United States and in Europe. U.S. patent law allows for a broader range of patentable items. Laws in Europe pertaining to biotechnology appear to be still evolving, but a comparison with European and US laws revealed that US law makes no provision for many exceptions to be found in European law, which make certain products and processes involving living matter unpatentable (GAEIB 1993, p. 29). The Group of Advisers on the Ethical Implications of Biotechnology (GAEIB) to the European Commission issued a recent opinion that included the following items:

> 2.2 The traditional distinction between discovery (not patentable) and invention (patentable) involves, in the field of biotechnology, a particular ethical dimension. It follows from this distinction that the knowledge related to the human body or its elements is relevant to scientific discovery and cannot be patented. It has to be clearly specified that the simple knowledge of the complete or partial structure of a gene cannot be patented.
>
> 2.3 The human body...as well as its elements, do not constitute patentable inventions. Such an exclusion does not come only from the usual conditions of patentability, but it is also inspired by the ethical principle of non-commercialisation of the human body (GAEIB 1996).

These statements would appear to preclude patenting genes or gene sequences. However, the opinion includes the following subsequent clause:

> 2.5 Concerning the inventions issued from the knowledge of a human gene or a partial human gene sequence, the granting of a patent is acceptable only if, on the one hand, the identification of the function attached to a human gene, or a partial human gene sequence allows new possibilities (for instance the production of new drugs) and, on the other hand, if the intended use of the patent is sufficiently specific and identified (GAEIB 1996).

This latter statement in the opinion of the Group of Advisers looks very much like the requirement of "novelty" in U.S. patent law. Moreover, it appears to take a step in the direction of U.S. law and away from a previous opinion by this same group. An opinion less than four years earlier included the following provisions:

> Genes and partial gene sequences whose functions are unknown should be made expressly unpatentable to end the international debate on the matter...Furthermore, the Community should take a stand against the commercial exploitation of the human body (GAEIB 1993, p. 35).

If the present legal uncertainty relies on a resolution of the conceptual controversy over whether genes and gene sequences should be considered discoveries or inventions, a conceptual decision must be taken. As in any conceptual dispute, the answer cannot be reached by "discovering" the right answer, but only by providing persuasive arguments for adopting one interpretation rather than the other.

This brings us directly to the ethical basis for the dispute over patenting genes and gene sequences. Items that are patented are considered "intellectual property," and a patent grants to the holder certain rights over that property (Resnik, 1997). Opponents of gene patents contend that genes are not a subject matter for which individual property rights should be granted because they are part of our common human heritage. Therefore, they should not belong to individuals or to corporations but rather, should remain in the public domain. Defenders of patenting genes in the U.S. argue that ownership of artificial human genes or artificial combinations of genes are no different, in principle, from many other items relating to human beings for which patents have been granted. But even if U.S. patent law could properly allow for the patenting of genes and gene sequences, the deeper ethical question remains: should gene patenting be permitted?

2. NONCONSEQUENTIALIST ARGUMENTS

2.1. The Theological Position

In 1995 a group of 186 religious leaders from all major faiths called for a moratorium on patents of human and animal life based on the premise that genes are creations of God and not human inventions. Theologians contended that patents on animal as well as human genes are a violation of the sanctity of life. Although this religious viewpoint may appeal to people who readily accept the theistic first premise, it is unlikely to be compelling in the secular realm of science. Moreover, if the patentability of genes and gene sequences requires a showing of human invention or innovation, then it is not the genes themselves, as found in nature, that are patentable but rather, the processes invented by humans to identify and sequence them. The theological argument can get off the ground only if the conclusion is reached that genes and gene sequences are "discoveries" rather than "inventions." But in that case, they would not be patentable even under the broad provisions of U.S. patent law. To the extent that the religious position rests on theological distinctions, it raises issues separate from the philosophical inquiry. I shall not discuss the religious objection further.

Not all nonconsequentialist arguments begin with a religious premise, however. Others start with a premise of human dignity or a conception of our humanness that is thought to be undermined by patenting. The following two nonconsequentialist arguments rely not on theological underpinnings but rather, on these secular notions.

2.2. Human Gene Patents and Human Dignity

The first nontheological argument is described by a philosopher, David Resnik, as taking a Kantian perspective. The argument contends that gene patenting is wrong because it treats persons as things that can be bought and sold, traded, or modified. In three simple steps, the argument proceeds as follows: 1) the practice of patenting human genes treats persons as property; 2) it is morally wrong to treat persons as property; 3) the practice of patenting human genes is morally wrong (Resnik, 1997, p. 54).

Resnik contends that gene patenting does not treat persons as property if it allows only for ownership of inventions for analyzing, sequencing, manipulating, or manufacturing human genes. This is analogous to patents for making other kinds of artificial human body parts, such as hair, bones, or hearts, and so the Kantian perspective does not preclude patenting of genes or gene sequences. However, the Kantian perspective would not allow for patents on genetically engineered humans to extend to the whole human animal, as patents have already done for genetically engineered mice. Resnik concludes that a prohibition should exist against patents on processes for making entire human beings, but that ownership of a process for making or manipulating a part of a human body need not constitute ownership of a person (Resnik, 1997, pp. 54–55). This conclusion is seconded by Caplan and Merz, who write: "...while strong theological reservations exist, it is hard to equate assigning a patent to a DNA strip with ownership of a human body. Selling bodies into slavery is exploitative, because our personal identity is so intimately tied to our bodies. It is not so obviously a violation of the human spirit to assign rights to exclusive use and development over a segment of chromosome 13 to a government agency or a biotechnology concern" (Caplan and Merz, 1996).

2.3. Patenting Human Genes and Our Humanness

The next nonconsequentialist argument, like the preceding one, appeals to the dignity of human beings. It relies on a conception of humanness as somehow morally "sacred". Patenting human genes is thus dehumanizing because it alters our view of humans from beings with dignity and respect into objects that can be bought, sold, or modified. Resnik rejects this view first, by noting that there is no good reason to think that the practice of patenting human genes will be any more "dehumanizing" than our present and past uses of the human body (p. 56). Second, Resnik observes that there is a variety of different subjective conceptions of what constitutes our "humanness," and no one conception can serve as a basis for a public policy banning gene patenting (Resnik, 1997, p. 57).

2.4. The "Common Property" Argument

Still another nonconsequentialist argument is the view that human genes should be treated as common property, not belonging to a single individual or corporation (Resnik, 1997, p. 57). This view, Resnik argues, rests on a mistaken understanding of gene patents. This takes us back to the threshold question, "what is patentable?" since patenting does not allow ownership of naturally occurring genes, only inventions relating to genes and gene sequences. Since ownership of the processes of copying, sequencing, modifying, and analyzing human genes does not constitute ownership of our naturally occurring, common genes, this argument is flawed.

But there is a different sort of "common property" argument, one that does not rely on a determination of what is patentable. This position begins by noting that in the U.S., at least, the resources for mapping and sequencing the human genome have come largely from the publicly funded National Institutes of Health. According to one commentary: "If government funds have been used to map and sequence the human genome, why should the fruits of that effort be turned over to a single owner? Permitting patents of simple segments of the genome, rather than for products and inventions, would seem to be contrary to the public interest" (Caplan and Merz, 1996).

2.5. The Commodification Argument

A supporter of the Council for Responsible Genetics puts forward a paradigm of the commodification position opposed to gene patenting. There is some overlap with the two arguments pertaining to human dignity and conceptions of our humanness, but the commodification view centers more on the aspect of commercialization. Martin Teitel refers to those seeking to patent genetic information as "the new commodifiers": "Just as the new commodifiers abruptly lay claim to the cultural heritage of generations of traditional societies, they also assert their ownership of the fantastically intricate genetic code that represents the current end point of millions of years of biological evolution. It is difficult to imagine a greater presumption" (Teitel, 1996, p. 2). This position does not mount an argument, but relies on a value conception that rejects the idea that everything may be subjected to the forces of the marketplace. The position is summed up in statements like this:

> As we recognize the nature and scope of the new commodification, we need to begin investigation, analysis, education and action. In spite of the power and reach of the huge corporations, universities and governments promoting the commercialization of life, we are in truth faced only with the actions of ordinary greedy people, who can and should be stopped (Teitel, 1996, p. 3).

Since patents serve a primarily economic function (Looney, 1994, p. 233), it becomes necessary to decide whether progress in the science of human genetics and its applications should be governed solely by economic considerations.

2.6. Competing Conceptions of Justice

A final nonconsequentialist argument appeals to one conception of justice. A view of gene patenting derived from the concept of distributive justice argues that the proper distribution of benefits and burdens in society (in this case, the world) requires that no group be deprived of the benefits of genomic research. Since less developed countries lack the resources of wealthier countries, it would be unjust for the benefits of genomic research to be located only in those richer nations that sponsor the research and obtain the patents (Looney, 1994). However, distributive justice is only one of several different conceptions of justice. A competing conception relies on a marketplace conception of justice as fairness. This view grants to researchers and financial investors a fair return on their efforts and expenditures. Since they have spent time and money on the genetic research yielding information about the human genome, they are entitled to the just reward of patent protection. Both of these opposing views rely on a conception of justice, but whereas the former claims that justice in distribution is the relevant conception, the latter maintains that providing just deserts should be the ruling view (Looney, 1994, pp. 240–42).

3. CONSEQUENTIALIST ARGUMENTS

At least four separate arguments on both sides of the issue appeal to the consequences of gene patenting. They revolve around the question: Does gene patenting result in limiting or increasing knowledge and its applications? These four arguments are as follows.

3.1. Delays in Disseminating Information

One view holds that researchers awaiting a patent they have applied for are likely to withhold information until the patent is granted, therefore delaying dissemination of important information (Looney, 1994, p. 244). An opposing view contends that patents provide protection and thus permit disclosure of scientific knowledge before an actual product is ready for the market. The consequence of not being allowed to patent would lead to the more detrimental consequence of maintaining secrecy in research.

A leading U.S. geneticist, David Botstein, who shares in a number of genetic technology patents, argues that not only do patents protect a scientist's self-interest, but they also promote the dissemination of research findings. Botstein agrees with those who say that the one of the primary purposes of patents is to reduce the economic motivation for excessive trade secrecy, which has a negative effect on progress (Hoke, 1995, p. 1). Jonathan King, a molecular biologist, proposes a directly opposite view:

> Contrary to the claims of the biotech industry, gene patents retard progress in the biomedical arena, introduce secrecy where openness is essential, and slow the publication and sharing of important results. This follows from the fact that once a result is reported publicly, it cannot be patented. Thus researchers drawn into the web of the patent process do not report their results, even informally, until they have passed through the expensive patent application and granting process (King, 1996, p. 11).

3.2. Development of New Drugs

An analogous pair of opposing consequences are envisaged with respect to development of new drugs. One position argues that "the next generation of modern medicines will never get out of research labs if efforts to halt the patenting of genes are successful" (Feldbaum, 1996, p. 10). This view is supported with evidence taken from other areas of biomedical technology, where patents have been granted for blood clotting agents for hemophiliacs, products for breaking up blood clots, a vaccine for Hepatitis B, along with numerous other drugs and diagnostic products (Feldbaum, 1996). The opposing position contends that the extraordinary advances in biomedical knowledge and technology in the past 40 years have been largely a result of public funding of biomedical research (King, 1996).

3.3. Promoting versus Stifling International Collaboration

Still another pair of opposing consequences appears in the discussion of international collaboration. In the absence of international agreements that would prohibit gene patenting, countries simultaneously doing research may feel compelled to seek to obtain patents in order to avoid losing a competitive advantage. After the U.S. NIH initially sought a patent application for gene sequences in 1991 (later abandoned), the UK Medical Research Council (MRC) defensively sought patents of its own (Looney, 1994, p. 245). These actions suggest that gene patenting would stifle international collaboration in genetic research. In contrast is the view that patenting promotes, rather than stifles international research. This latter view relies on the notion that an inability to patent leads to fewer inventions, since sponsors will not invest in research that does not promise a good return that would be guaranteed by having exclusive commercial rights.

3.4. Assigning Different Weights to Different Consequences

A different sort of opposition arises out of assigning higher or lower values to the different potential consequences of gene patenting. One position contends that patent protection promotes efficiency, reducing duplicative research and wasteful funding that would otherwise occur when independent efforts simultaneously go forward. On this view of consequences, more knowledge and its applications can be obtained more quickly by allowing patenting. In contrast is the consequence that some genetic research unlikely to be profitable will fall by the wayside, therefore resulting in loss of benefits to classes of persons afflicted by rare disorders.

3.5. Restrictions on What Is Patentable Related to the Consequences

Even among those who would allow patenting of some genetic information, concerns about stifling research lead to calls for restrictions. For example, in a letter to the Commissioner of Patents and Trademarks in the U.S. Patent and Trademark Office, the President of the Council of the National Academy of Sciences (NAS), Bruce Alberts, wrote: "I write to encourage you to make every effort to insure that any future patents granted for DNA sequences do not unfairly impede research and innovation in biotechnology" (Alberts, 1997). The specific concern Alberts identified was that EST (expressed sequence tags) patents will become impediments to research, slowing progress in biomedical research. On behalf of the Council of the NAS, Alberts suggested that "DNA sequences per se should not be patentable unless the patent clearly discloses specific 'real world' utilities for the particular DNA sequences in question that can be implemented without substantial further developmental research."

The concern over patenting ESTs goes back to the aborted attempt by the NIH to obtain a patent. Scientists criticized that attempt because no basic biological knowledge accompanied the partial sequences. Some researchers claim that such patents would foreclose much future research because the patents would be held by a private biotechnology company, The Institute for Genomic Research (TIGR). The Vice President of TIGR, a scientist who had been an intramural researcher at NIH at the time it sought the patent, makes the opposite prediction: to put the partial gene sequences in the public domain would result in the full genes being unpatentable, and that would remove the incentive for companies to develop them into useful drugs (Hoke, 1995). It is apparent from the concerns about EST patents that it is not simply the permissibility or impermissibility of patenting that has consequences for scientific progress, but also what specific items may or may not be patentable even if patenting in general is allowed.

4. CONCLUSION

At the time of this writing, no patents for human genes or gene sequences have yet been granted, despite the many applications made by U.S. biotechnology companies. Mounting concerns in Europe and the prospect that European countries will bend to the pressures of scientific and economic competition suggest the need for a thorough and unbiased examination of the probable consequences of gene patenting in the areas outlined in the preceding section.

The central ethical question about patenting human genes turns on the consequences: whether patenting results in more benefits than harms. An answer to the conceptual question—whether the methods to identify genes and gene sequences constitute an invention or a discovery—is partly legal and partly philosophical. It requires further discussion and debate. But even assuming that the threshold question yields the answer that genes are, indeed, patentable, only an examination of the likely consequences, for all who stand to be affected, will be persuasive for making public policy.

From a global perspective, as well as within nations, it is important to determine whether patenting of genes and gene sequences is in the public interest. Who stands to gain, and who stands to lose by the patenting of genetic material? Is national or regional regulation warranted? Should a high priority be placed on international harmonization of patent laws? A major impediment to determining what are the likely consequences of gene patenting is the lack of clear evidence that would support one side or the other. A multidisciplinary, systematic study should be conducted; it could begin by examining the effect of patents granted in other areas of biotechnology. Also needed are the views and experiences of biomedical researchers who do not have ties to the increasing number of biotechnology companies that are among the most aggressive pursuers of patents on genes and gene sequences. Although a systematic inquiry is needed in order to provide relevant evidence that could shed light on the controversy, the more gene patents that are aggressively pursued in the United States and the more time that elapses, the harder it will become to turn back the clock in the event that the predicted negative consequences of gene patenting begin to emerge.

On the nonconsequentialist side, my own view is that commodification of human material for any purpose is unsavory, and ought to be avoided whenever possible. Yet "unsavoriness" may not be a category of moral disvalue strong enough to warrant prohibition. If commodification of genetic material does not involve a violation of a moral principle or the rights of any person or group, then a stronger basis for a ban on gene patents would have to be found. Although it seems correct to say that commodification does not violate any ethical principle, it can nevertheless be viewed as an unsavory feature of modern society, at least in those countries where almost everything is subject to market forces. This involves a judgment about what kind of society we value: one in which almost everything can be subjected to commercial exchanges or one in which some things should be treated as social goods for the common benefit of all. Although my own values strongly favor the latter view, it is hard to find a principled moral argument that would exempt genes and gene sequences from the commercial forces that govern other aspects of modern biomedical technology.

REFERENCES

Albert, B. (1997), letter to Bruce A. Lehman, Commissioner of Patents and Trademarks (published June 19 on the World Wide Web).

Caplan, A.L. and Merz, Jon. (1996), 'Patenting Gene Sequences,' *BMJ* Editorial, Vol. 312, 13 www.bmj.com/bjm/archive/7036e.htm.

Feldbaum, C. (1996), 'Gene Patents Deemed Essential to Next Generation of Cures,' *GeneWatch*, Vol. 10.

GAEIB. (1993), Opinions Submitted by *The Group of Advisers on the Ethical Implications of Biotechnology* to the European Commission, Brussels; No. 3 Opinion on Ethical Questions Arising from the Commission Proposal for a Council Directive on Legal Protection for Biotechnological Inventions, 30 September.

GAEIB. (1996), 'Opinion of the Group of Advisers on the Ethical Implications of Biotechnology of the European Commission,' Brussels; No. 8 Ethical Aspects of Patenting Inventions Involving Elements of Human Origin, 25 September.

Hoke, F. (1995), *The Scientist*, Vol. 9, April 17.

King, J. (1996), 'Gene Patents Retard the Protection of Human Health,' *GeneWatch*, Vol. 10.

Looney, B. (1994), 'Should Genes Be Patented? The Gene Patenting Controversy: Legal, Ethical, and Policy Foundations of an International Agreement,' *Law and Policy in International Business*, Vol. 26.

Resnik, D. B. (1997), 'The Morality of Human Gene Patents,' *Kennedy Institute of Ethics Journal*, Vol. 7.

Teitel. (1996), 'The Commercialization of Life,' *GeneWatch*, Vol. 10.

13

PARADIGMS OF AUTHOR/CREATOR PROPERTY RIGHTS IN INTELLECTUAL PROPERTY LAW

Ethical Implications for the Acquisition, Access, and Control of Genetic Information

Robin Mackenzie

School of Law, University of Kent
Canterbury, United Kingdom

This paper seeks to examine the theoretical models which underlie intellectual property law today and the ethical implications of these insofaras they impact on the acquisition, access and control of genetic information. I shall analyse some present attempts to combine legal regulation of biotechnology with ownership in order to argue that intellectual property law must be revisioned or reimagined in order to overcome the ethical shortcomings which its present paradigms embody.

1. THEORETICAL MODELS IN INTELLECTUAL PROPERTY LAW AND THEIR SHORTCOMINGS

The problem which intellectual property law seeks to solve is how to allot ownership and property rights in objects which are seen as having resulted from some sort of creative process and also as bringing benefit to society as a whole as well as having commercial value. Typically, artistic productions attract rights under copyright, inventions under patents, and there is specific protection for other products of intellectual labour such as trademarks, databases and character merchandising. The focus here will be on patent law in relation to intellectual property rights in genetic information.

Intellectual property law's theoretical basis for assigning property rights is typically based on a liberal conception of society as composed of rational autonomous individuals engaged in civic debate in public, commercial transactions in the marketplace and actions which concern only themselves in private (Drahos, 1996, Fisher, 1988). The conception of

Genetic Information, edited by Thompson and Chadwick.
Kluwer Academic / Plenum Publishing, New York, 1999.

property which accompanies this model is Lockean: we should have rights in the products of our labour. Typically here labour is seen as creating value by manipulating nature, which is both freely available and without value in itself as value is created through human effort. There is also a strong thread of utilitarian justification for allotting intellectual property rights; here that it benefits society as a whole to assign finite property rights to creators provided that their creations bring beneficial innovations to society.

These conceptions raise significant public goods problems. If the products of our intellectual endeavours are to be owned, what becomes of the intellectual commons? What creative work is possible if its constituent elements are already owned by others rather than available to all, or where they are available only to those who can afford to purchase them? These questions involve intellectual property law in line drawing exercises. [1] In patent law an invention attracts property rights provided it exhibits novelty, is non-obvious in that it involves an inventive step and has utility or is capable of industrial application (2),[2] whereas a discovery is unable to be protected by intellectual property rights, on the assumption that it should form part of the intellectual commons, where its presence is likely to give rise to inventions which will benefit society. Natural laws and mathematical formulae are thus unable to be patented.

The rhetoric here is that of commercial exploitation and national interest. Limited monopolies over information such as a process for manufacturing better widgets are granted in exchange for this information being shared with others: the limited monopoly equates with state support to exclude others from using this information for a specific period of time, typically twenty years,[3] without authorisation of the owners of the property rights. National interests in the commercial exploitation of intellectual property rights come particularly to the fore now that we can analyse today's world in terms of its being an information society. This label translates into the economic fact that information, whether genetic, electronic or artistic, is now one of the main sources and forms of wealth.

2. ENTITLEMENT AND ETHICS: ECONOMICS AND INCOMMENSURABILITY

The language of entitlement we choose to use in this context has enormous ethical consequences. The range of options we have where this choice is concerned, and how far ethical issues can enter the conversation here, and on what terms, are all strongly influenced by recent technological advances which render copying cheap, easy and difficult to police. According to the president of the Pharmaceutical Manufacturers Association, Gerald J. Mossinghoff, in 1986 it took a company about ten years, spending over US$125,000,000, to discover, test and secure approval for a new drug in the United States (Mossinghoff, 1987). But another pharmaceutical company in a country where patent protection differs from that in the United States may be able to lawfully copy and sell this drug for the equivalent of a few United States cents. Pfizer, the multinational pharmaceutical company, reported in 1986 that whereas it had sold US$47,000,000 worth of specific pharmaceuticals in one year, 'pirates' who had copied the drugs concerned without authorisation had sold US$42,000,000 worth of the same products (Comment, 1986). What are viewed as similar acts of piracy also result from the ease of copying software, music CDs, videos and many other products of modern technology.

Hence in the Uruguay Round of the GATT, the General Agreement on Tariffs and Trade, intellectual property rights were brought into international trade negotiations for the first time, largely at the instigation of the United States. After strenuous lobbying from industrial interests, particularly those of the multinational pharmaceutical companies, the United States succeeded in imposing its criteria for patent protection over resistance from developing countries whose view of appropriate intellectual property rights and entitlements differed markedly. The GATT negotiations were completed in December 1993 and ratified in Morocco in April 1994. They resulted in TRIPs: Trade Related Intellectual Property rights whereby in order to participate in world trade networks, countries had to agree to United States style intellectual property legal structures (Byrne, 1995, Leaffer, 1991, Reichman, 1995).

I wish to focus on these here as they seem to exemplify the problem of incommensurability which I believe haunts discussions about the ethics of acquiring and controlling access to genetic information. Many developing countries excluded pharmaceuticals from patent protection as on public policy grounds they favoured cheap medicine (4).[4] Pfizer, for example, has for thirty years charged around US$30 per bottle of a product which costs around US$3 to manufacture (Sherer, 1980, at 450). India estimated that royalty payments to foreign pharmaceutical companies could cost around US$10 per year per head of population; clearly an unaffordable amount (Desai, 1989). The language of life and death versus corporate profit is one which does not admit of easy solutions because of the incommensurability of these languages. There is no obvious common ground, nor a clear means by which compromises can be reached. Although the TRIPs agreement and patent acts in general preclude patents being granted if this is seen as being 'against the public interest' or 'encouraging offensive, immoral or antisocial behaviour' there is no agreement on where the demarcating lines here should be drawn. Indeed, there is a strong body of opinion which holds that patent law is a technical body of rules only, with ethical questions more appropriately decided elsewhere (Crespi, 1995, Bently and Sherman, 1995, Nuffield, 1995).

3. LEGAL ATTEMPTS TO REGULATE BIOTECHNOLOGY AND ASSIGN OWNERSHIP

Legal documents concerning biotechnology and property rights over genetic information exemplify these difficulties. The Recitals to the Proposal for the EC Directive on the Legal Protection of Biotechnological Inventions give a clear picture of patent law as outside ethical considerations (5) and concerned with attracting biotechnological industries to remain in the EC, harmonising regulatory control within the EC and rewarding the inventor.[5] The salient provisions here are as follows:

> Whereas the investments required in research and development, particularly of genetic engineering, are especially high and especially risky and the possibility of recouping that investment can only effectively be guaranteed through adequate legal protection; [Recital 2]
>
> Whereas without effective and harmonized protection throughout the Member States of the Community such investments might well not be made; [Recital 3]
>
> Whereas difficulties exist in the legal protection of biotechnological inventions offered by the laws and practices of the Member States; whereas such differences could create barriers to trade and to the creation and proper functioning of the internal market; [Recital 5]
>
> Whereas the uncoordinated development of national laws on the legal protection of biotechnological inventions in the Community could result in the creation of new disincentives to

> trade to the detriment of the industrial development of such inventions and of the smooth operation of the internal market; [Recital 7]
>
> Whereas a patent for invention does not authorize the holder to implement that invention, but merely entitles him to prohibit third parties from exploiting it for industrial and commercial purposes and whereas patent law is not capable of calling into question national and Community law on the monitoring of the applications of research and of the use or commercialization of its results, notably from the point of view of the requirements of public health, safety, environmental protection, animal welfare, the preservation of genetic diversity and compliance with certain ethical standards; [Recital 11]
>
> Whereas such moral considerations must be given greater weight in the examination of biotechnological inventions because this branch of technology is concerned with living matter and because of the often enormous implications of the inventions to be examined; whereas these considerations do not, however, change the nature of patent law as a primarily technical body of law and are no substitute for the other legal checks which biotechnological inventions are required to undergo from the start of their development or at the marketing stage, particularly with regard to safety; [Recital 23]
>
> Whereas, in view of the fact that the function of a patent is to reward the inventor with an exclusive but time-bound right for his creative efforts and thereby encourage inventive activities, the holder of the patent should be entitled to prohibit the use of patented self-reproducible products, namely in respect of the patented product itself. [Recital 26]

Recital 22 also states that the Directive should provide a general guide to interpreting references to concerns of public policy and morality which potentially curtail patentability: in fact it can be argued that it fails to do so.[6]

How has patent law managed to be seen under western law as neutral and somehow removed from cultural or ethical considerations? Historically the subject matter of patents, inventions, have been associated with science as positive, objective and progressive. Clearly this has coloured the view that patent law is merely the body of apolitical technical rules to which Recital 23 refers. The question of patenting genetic information, particularly that of humans, has brought ethical concerns into the debate, but salient matters have still to be resolved. For instance, while Article 53(a) of the European Patent Convention states that patents are not to be granted for inventions, the publication of which would be contrary to ordre public or morality, the European Patent Office traditionally decides issues under this section in a utilitarian risk/benefit balancing act, an approach which is explicitly endorsed in Recital 21. In practice, cases involving genetic modification tend to be decided here on how serious the decisionmaking body considers the problem concerned to be. The case of the Harvard Oncomouse, which had been genetically engineered to be particularly prone to cancer, was decided on the basis that the use to humanity would outweigh the mouse's suffering or any environmental dangers.[7] Conversely, a mouse which had been genetically engineered to be prone to hair loss would not attract patent protection.

The European Patent Office has explicitly eschewed an ethically based orientation in its decisions on biotechnological patent applications. In a case where Greenpeace objected to a patent application for a genetically engineered plant, the Opposition Division stated that it was not required to consider the Article 53(a) exclusion unless the invention was universally regarded as outrageous; hence in the vast majority of cases it would not be necessary to consider ethical concerns.[8] Following this, where the Opposition Division was required to pronounce on Relaxin, a patent application for DNA sequences of a naturally occurring substance which relaxes the uterus in childbirth, it stated that '[t]he European Patent Office is not the right institution to decide on fundamental ethical questions'.[9] Faced with a choice between the scientific understanding of DNA as chemicals and the social understanding of DNA as life, the Opposition Division preferred the former (Straus, 1995).

This approach, however, may not be sustainable indefinitely. Recently, arguments have been made that the European Convention on Human Rights is binding on biotechnological decisions under the European Patent Convention, on the proposed Directive on the Legal Protection of Biotechnological Inventions and on legislation embodying the TRIPs Agreement. Decisions on patentability of human, animal and plant tissue which raised questions of morality would thus fall to be decided under the guarantees of human rights embodied in the Convention (Beyleveld and Brownsword, 1993, pp. 68–71, Ford, 1997). Inconsistencies between obligations under the morality exceptions in Article 53(a) of the European Patent Convention, Article 9 of the Proposed Directive on the Legal Protection of Biotechnological Inventions and Article 27.2 of the TRIPs Agreement also suggest that it may become increasingly difficult for patent decisions to continue to be regarded as primarily simply technical. Article 53(a) provides that patents shall not be granted for inventions the publication or exploitation of which would be contrary to ordre public or morality. Article 9 states that inventions will be considered unpatentable where exploitation would be contrary to public policy or morality. Article 27.2 permits signatories of the TRIPs Agreement to exclude from patentability inventions, the prevention within their teritory of the commercial exploitation of which is necessary to protect public order or morality, including to protect human, animal or plant life or health or to avoid serious prejudice to the environment. Salient legal issues here are the differing interpretations of ordre public and public order, the use of permissive rather than mandatory language in Article 27.2 and the conflicting obligations which logically arise for member states bound by these agreements. Legal uncertainties here are likely to give rise to challenges to decisions on biotechnological patents under these morality exceptions. Bioethical concerns will thus become increasingly unable to be sidelined or marginalised where biotechnological patents are concerned.

4. THE ROMANTIC AUTHOR/CREATOR/ENTREPRENEUR PARADIGM IN INTELLECTUAL PROPERTY LAW

Patent law is thus confronted with the need to translate considerations such as risk to the environment, sacredness of life, or harm and suffering caused to animals into a language which can incorporate these concerns, as well as find an acceptable conceptual, procedural and institutional framework to do so. If bioethical concerns are to be taken seriously, patent law must be revisioned and reimagined, so that guidelines may be provided in order to safeguard and regulate the permissive boundaries of the commercialization of genetic information. My argument here is that an essential component of such an enterprise would be a revision of the present system of allocation of intellectual property rights as based on the stereotype of the Romantic author/creator.

If we return to intellectual property law's theoretical basis, this is the liberal picture of society as divided into the polity, where citizens should have free access to information in order to engage in informed debate, the marketplace, where information has commercial value, and the private, a mysterious arena where noone knows what takes place which serves as a location for the sacred. This liberal picture presents obvious contradictions where managing information in terms of allocating property rights in it is concerned. The solution enshrined in presentday intellectual property law is the valorization of the Romantic author/creator figure. Where artists, inventors and entrepreneurs are identified with the Romantic author figure who creates in the private sphere from sacred sources of inspiration, the conflict between the public polity argument that information should be free and the marketplace argument that information should be able to be bought and sold is concealed. Thus, granting artists, inventors and entrepreneurs intellectual property rights which constitute limited monopolies allows information into the polity to be debated while also assisting the marketplace. The contradiction between the need for freedom of information in the civic realm, as against its commercial value in the commercial arena, is disguised by this privatisation of information production.

Recent commentators have pointed out that this model suppresses the claims of the sources and audiences of such creations: the intellectual commons shrinks, and what was previously free and available to all becomes commodified (Boyle, 1996). Aoki labels as 'author reasoning' the characteristic legal decisionmaking process whereby things which were once considered unownable, such as DNA or words in the English language, become viewed as unauthored, uncreated sources or facts which may then be translated as susceptible to ownership according to the degree of intervention by an author/creator figure who is seen as having transformed them into an original creation (Aoki, 1993/94). Others have charted the historical connection between presentday intellectual property rights and the Romantic conception of the author in order to point out that such notions of authorship are clearly historically contingent (Boyle, 1988, Jasri, 1991, Woodmansee, 1984). Before the Romantic conception of the author took hold, those who created books were considered ordinary craftspeople, such as the medieval scribe or the corporate team member in mid-eighteenth century Germany. Notions of artistic excellence then typically centred upon what we would today consider as plagiarism, but at the time was seen as respectful imitation. Forms of authorship which are serial, collective or collaborative, once commonplace, have thus become marginalised as deviant.

Applied to patent law, this paradigm renders both the scientist/inventor and the entrepreneur as Romantic creator figures who bring innovation and progress for the good of society as a whole. A typical example here would be the United States Patent Code, which was written with the eighteenth century lone inventor in mind, a 'hero-inventor' whose main resource was his own ingenuity (Cherensky, 1993). This figure became hybridised with the entrepreneur as hero in the form of the heroic corporate research and development team (Reich, 1987). Other forms of scientific knowledge or innovation which fall outside this model, such as the folk medicine of indigenous peoples, fail to attract the protection of intellectual property law. Those that do so, however, are handsomely rewarded. Commercial law is based on the legal fiction that a company may be considered as a legal person; hence multinational pharmaceutical companies funding research teams with the equivalent of a small nation's GNP, cloaked with the mantle of the Romantic author/creator, are able to receive intellectual property protection for a period of twenty years for pharmaceuticals which may well have been based on genetic material or medicinal knowledge gained through contact with indigenous people.[10] Even the Convention on Biological Diversity, an international agreement purporting to address conservation and sustainable use of biodiversity which entered into force in December 1993, supports the Romantic author/creator paradigm by supporting bilateral deals rather than multilateral mechanisms which would provide effective guidelines for 'recognising and rewarding the contributions of indigenous peoples and other informal innovators who are responsible for nurturing, using and developing biodiversity worldwide' (Rural Advancement Foundation International, 1994, p. 2).

The paradigm of the Romantic author/creator/entrepreneur thus logically joins with notions of nationality and territoriality: the power to exclude others from using one's property becomes, in times of globalization, a desire to impose western style intellectual property models of entitlement on the rest of the world, as in the TRIPs Agreement (Aoki, 1996, Brush and Stabinsky, 1996). Alternative modes of creation and ownership are thus marginalised. The implications of this conceptual nexus for the access, acquisition and control of genetic material are difficult to overestimate. Genetic material from the biodiversity of the South is notoriously responsible for enriching the North. 80% of biodiversity is in the South, plant genetic resources form the basis of most pharmaceuticals and pharmaceutical companies routinely patent medicines derived from plant genes located via the shamanistic knowledge of indigenous peoples[11]. Without mechanisms to monitor the bilateral contracts which exist between pharmaceutical companies and indigenous peoples, injustices are all too likely.[12]

The lack of these mechanisms has been one of the factors provoking support for a revisioning of intellectual property law's territorial ambitions. Indigenous peoples gathered in the Mataatua region of Aotearoa/New Zealand in 1993 to consider cultural and intellectual property issues in relation to indigenous knowledge and claims on natural resources.[13] The resulting Mataatua Declaration called for a global moratorium on any further commercialization of traditional plants, medicines and human genetic materials until appropriate mechanisms had been developed, asserting that existing western intellectual property law needed modifying in order to provide adequate protection here via standardised codes of ethics and research agreements. The first international agreement formulated by indigenous peoples on this subject, it was presented to the UN Working Group on Indigenous Peoples in 1993 and has since been signed by over 800 indigenous nations and organisations as well as non-indigenous organisations. The 1995 Treaty for a Lifeforms Patent-Free Pacific and Related Protocols, which argues for recognition of collective rights, communal heritage and common property where indigenous peoples' resources are concerned, has also attracted much support throughout the Pacific and beyond (Te Pareake Mead, 1997).

Such arguments over what ought to be able to be patented are not new. They have haunted the history of intellectual property law. Only some of the actors have changed parts. The negotiations during the attempts to arrive at the first International Convention for the Protection of Industrial Property rehearsed the same incommensurabilities: the industrialised countries favouring higher intellectual property protection while the developing countries fought for compulsory licensing or exemptions from patentability on ethical grounds. 'Battailles tres chauds' took place over whether chemical products, pharmaceutical preparations and foodstuffs should be patentable (Penrose, 1951 at 51–4). Contemporary conflicts over the access, acquisition and control of genetic information in relation to intellectual property rights are directly comparable. Historically, since the first patent systems of the 1400s, a contested issue has been whether a patent system is desirable and, if so, how far it can assist in national development. Nation states' attitudes have varied with their degree of industrial development. Take Britain. Before its industrial revolution, the British definition of an inventor was not someone who created something new but someone who imported something new from abroad. The compulsory licensing of foreign inventions to promote domestic industries was also legalised. Once it had achieved economic dominance, a volte face in favour of stronger patent protection took place (Penrose, 1951 at 84). The United States has also notoriously moved from piracy to piety where intellectual property laws are concerned. Patent law is thus perhaps more plastic than it might at first appear. Hence any attempt to create a transgenic patent system which combines ethical and economic considerations must look beneath historically contingent rhetoric.[14] It is my contention here that the Romantic author/creator/entrepreneur as embodied in existing western intellectual property law forms part of this historically and culturally specific attempt to extend national sovereignty and that this figuration must needs vanish, or at least allow in other possible visions, before ethical mechanisms which regulate the access, acquisition and control of genetic information can be put in place.

5. CONCLUSION

Any reimagining of patent law would thus need to take into account notions of creation and ownership outside the Romantic author/creator/entrepreneur paradigm, as only then could the ethical issues around the access, acquisition and control of genetic information be addressed at all adequately. Opportunities for this to arise include public debate such as that envisaged over the proposed Directive on the Legal Protection of Biotechnology, the assessment of Article 27.2 of the TRIPs Agreement scheduled to take place in 1999 and the challenges to patentability of biotechnological claims rendered inevitable by the inconsistencies in the morality exceptions analysed above. My point here is that these opportunities must include a reimagining of intellectual property law as appropriate in an information age, with a consequent disempowering of the Romantic author/creator/entrepreneur paradigm.

NOTES

1. Judge Learned Hand commented here that '[n]obody has been able to fix the boundary, and noone ever can', Nichols v Universal Pictures, 45 F. 2d 119, 121 (2d Cir. 1930), cert denied, 282 US 902 (1931).
2. Under the TRIPs (Trade Related Intellectual Property Rights) agreement, article 27(1).
3. Under the TRIPs agreement, articles 27,28 and 33, domestic patent laws must provide a uniform term of twenty years protection from the filing date.
4. Mossinghoff (1987 at 315) cites Brazil's Law No 5772 of 21 December 1971, article 9(c) from its Code of Industrial Property, which states that patents are not available for 'food and chemical-pharmaceutical substances, mátter, admixtures or products and medicaments of any kind, as well as for the respective processes for obtaining them or modifying them', along with India's Patents Act 1970 No 39 of 1970, where Article 3(i) states that 'any process for the medicinal, surgical, curative, prophylactic or other treatment of human beings or any process for a similar treatment of animals or plants to render them free of disease or to increase their economic value or that of their products' is not to be considered an invention and hence is unpatentable
5. Under Amendment 78 in the version agreed to by the European Parliament on July 16 1997, an Ethics Committee is to be set up 'to assess all ethical aspects of biotechnology and its utilisation, particularly with regard to patents'.
6. A strong argument that many of the problems of the proposed Directive would disappear if the existing provisions of the European Patent Convention were reformulated to correspond with those in the EU Community Plant Variety Rights Regulation, a mechanism which provides an accepted form of protection for living material, is made in Llewellyn (1995).
7. HARVARD/Oncomouse T19/90 [1990] EPOR 501.
8. PLANT GENETIC SYSTEMS/Glutamine synthetase inhibitors ('PGS') T356/93 [1995] EPOR 357.
9. Hormone Relaxin Opposition OJ EPO 1995, 388.
10. The Rural Advancement Foundation International, a non-governmental organization, reports that 'one in 10,000 chemicals derived from mass screening of plants, animals and microbes eventually results in a potentially profitable drug. However, Shaman Pharmaceuticals Inc, the US based company that collects

plants by talking to indigenous healers and watching them work, claims a success rate of 50% ... the filter of indigenous knowledge [is] 5,000 times more effective than random collection' (Rural Advancement Foundation International, 1994, p. 2).

11. RAFI estimates that medicinal plants and microbes from the South contribute more than US$30 billion annually to the North (Rural Advancement Foundation International, 1994, p. 1).
12. Under agreements negotiated via the International Cooperative Biodiversity Group, a collaboration between three United States government agencies (National Institute of Health, National Science Foundation and Agency for International Development), discoveries are ostensibly shared so that benefits accrue equitably to local communities and indigenous populations involved in the discovery of the natural product. The terms of the benefit sharing, however, are strictly confidential. RAFI has revealed the terms of one such leaked contract, an agreement to allow Searle, a subsidiary of Monsanto, to collect Peruvian medicinal plants. The licence option agreement states that where Searle determines that a valuable plant extract was 'known or otherwise available' to Searle, Searle has no obligation to compensate the source country or community but retains the right to make, use or sell worldwide the active agent or any derivative of it. The royalty payments which Searle committed itself to in the agreement constitute US$15,000 annually for four years, to be distributed by Searle's agent, Washington University, 'for the benefit of the local inhabitants of the collection area as compensation for the collection and use of plant material'. Monsanto's after tax profits in 1993 amounted to US$494 million. (Rural Advancement Foundation International, 1994, p.6).
13. A claim that the government of Aotearoa/New Zealand had no right to sign the GATT Uruguay Round Agreement without definitive legal advice on the ownership of the country's natural resources under the Treaty of Waitangi (claim WAI262) put forward on behalf of the Maori tribes is also awaiting a hearing.
14. The phrase transgenic patent system is not mine, see Bentley and Sherman (1995).

REFERENCES

Aoki, K. (1996) '(Intellectual) Property and Sovereignty: Notes towards a Cultural Geography of Authorship', *Stanford Law Review*, Vol. 48, p. 1293. Regrettably, a detailed consideration of how far this model might be capable of offering appropriate protection to the rights and interests of nonwestern people and their knowledge is outside the scope of this article. A helpful blend of theoretical and pragmatic viewpoints here may be found in

Aoki, K. (1993/4), 'Authors, Inventors and Trademark Owners: Private Intellectual Property and the Public Domain', *Columbia-VLA Journal of Law and the Arts*, Vol, 18 Part I, p. 1; Part II , p. 191.

Bently, L. and Sherman, B. (1995),'The Ethics of Patenting: Towards a Transgenic Patent System, *Medical Law Review*, Vol. 3, p. 275.

Beyleveld, D. and Brownsword, R. (1993) *Mice, Morality and Patents*, Common Law Institute of Intellectual Property: London.

Boyle, J. (1996) *Shamans, Software and Spleens*, Harvard University Press: Cambridge, Mass.

Boyle, J. (1988), 'The Search for an Author: Shakespeare and the Framers', American University Law Review, Vol. 37, p. 625.

Brush, S. and Stabinsky, D.(eds.) (1996) *Valuing local knowledge: indigenous people and intellectual property rights* Island Press: Washington, D.C.

Byrne, J. (1995) 'Changes at the frontier of intellectual property law: an overview of the changes required by GATT', *Duquesne Law Review*, Vol. 34, p. 121.

Cherensky, S. (1993) A penny for their thoughts: employee-inventors, preinvention assignment agreements, property and personhood. *California Law Review*, Vol. 81, p. 597.

Comment (1986), 'Intellectual piracy captures the attention of the President and Congress', *National Journal* 22 February, p. 443.

Crespi, S. [1995] 'Biotechnology Patenting: the Wicked Animal Must Defend Itself., p. 431.

Desai, A. (1989) 'India and the Uruguay Round', *Journal of World Trade Law*, Vol. 39, p. 47.

Drahos, P. (1996), *A philosophy of intellectual property*, Dartmouth: Aldershot.

Fisher, W. (1988) 'Reconstructing the fair use doctrine', *Harvard Law Review* , Vol. 101, p.1659.

Ford, R. [1997] 'The Morality of Biotech Patents: Differing Legal Obligations in Europe', *European Intellectual Property Review*, p. 315.

Jaszi, P. (1991) 'Toward a Theory of Copyright: the Metamorphoses of Authorship', *Duke Law Journal*, Vol. 47, p. 455.

M.A. Leaffer, M. (1991) 'Protecting United States intellectual property abroad: toward a new multilateralism', *Iowa Law Review*, Vol. 76, p. 273.

Llewellyn, M. [1997] 'The Legal Protection of Biotechnological Inventions', p.115.

Mossinghoff, G. (1987) 'Research based pharmaceutical companies: the need for increased patent protection worldwide', *Journal of Law and Technology*, Vol. 2, p. 307.

Nuffield Council on Bioethics. (1995), *Human Tissue: Ethical and Legal Issues*, Nuffield: London.

Penrose, E. (1951) *The Economics of the International Patent System* John Hopkins: Baltimore.

Reich, R. [1987] 'Entrepreneurship Reconsidered: the Team as Hero', *Harvard Business Review* p. 77.

Reichman, J. (1995) 'Universal minimum standards of intellectual property protection under the TRIPs component of the WTO agreement', *International Lawyer*, Vol. 29, p. 345.

Rural Advancement Federation International. (1994) 'Bioprospecting/biopiracy and Indigenous Peoples', *RAFI Communique* Nov/Dec p. 1.

Sherer, F. (1980) *Industrial Market Structure and Economic Performance*. 2d ed., Rand McNally: Chicago.

Straus, J. (1995), 'Patenting Human Genes in Europe: Past Developments and Prospects for the Future', *IIC*, Vol. 26, p. 920.

A. Te Pareake Mead, A. (1997) Resisting the Gene Raiders, *New Internationalist*, Vol 297, p. 26.

Woodmansee, M. (1984) 'The Genius and the Copyright: Economic and Legal Conditions of the Emergence of the 'Author'', *Eighteenth Century Studies*, Vol. 17, p. 425.

14

REGULATING THE COMMERCIALIZATION OF HUMAN GENETICS

Can We Address the Big Concerns?

Timothy Caulfield

Health Law Institute, Faculty of Law
University of Alberta, Canada

1. INTRODUCTION

Given the amount of media attention and academic pontification on the subject, one would think that the "genetic revolution" had already delivered on its much publicized promise of medical and scientific breakthroughs (Time, 17.1.95, BusinessWeek, 10.3.97). In fact, there are currently very few genetic services which are readily available to the citizens of the "genome nations." There have undoubtedly been incredible advances in our knowledge of human genetics. Yet, outside of the diagnostic, carrier and presymtomatic tests for monogenic disorders there are only a handful of commercially available genetic services. There has, however, been an extraordinary amount of financial investment, both from public and private sectors, in the emerging "genetech" industry. Indeed, if this financial anticipation is any indication, we seem to be just around the corner from the next phase of the genetic revolution—the phase of practical application. And as we leave the current stage of initial discovery, it seems almost certain that the new era will not be dominated by the conventional laboratory scientist, but by the commercial entities that fund the research and that will ultimately disseminate the genetic services.

Biotechnology has become a dynamic element of many countries' economies (Novarro, 55; Cohen, 767). In Canada, for example, it is one of the fastest growing industries. Total revenue generated by biotechnology has increased 45% since 1994 and is now valued at approximately $1.2 billion; the number of companies identified as biotech in orientation has risen from 121 in 1994 to 224 in 1997; and biotech's share of Canada's GDP, while still relatively small, has more than doubled since 1994 (Ernst and Young, Heller). Much of this growth, in Canada and elsewhere, can be attributed to the enthusiasm which is associated with the recent advances in genetics. Genetics has become big business (Thomas, 387, Damsell, 24, Day, Dodet, 182, Culotta, 914, Anderson, 300).

Genetic Information, edited by Thompson and Chadwick.
Kluwer Academic / Plenum Publishing, New York, 1999.

With the growth of the genetech industry has come a plethora of social, ethical and legal issues. This paper will briefly discuss a number of these concerns with the goal of exploring whether they can be meaningfully addressed in this current complex, "pro-commercialization", environment. In the end we will see that the concerns which are associated with the commercialization process can, roughly, be divided into three categories: 1) those concerns around which there is a degree of consensus and which we can (and should) deal with through thoughtful regulatory policy (e.g., the fear of premature implementation of genetic services); 2) those concerns which are likely valid but where the advisability of a regulation involves an initial weighing of laudable social values and goals (e.g., the argument that industry/academia collaborations skew the direction of university based research); and 3) the controversial concerns which are difficult both to verify and regulate.

This paper will primarily focus on the last of these categories (see section III). Specifically, I will explore how our consumer culture, market forces, and our reverence for the notion of autonomy may be relevant to the nature, scope and (probable) resiliency of the most serious, yet speculative, of the commercialization concerns—the fear of eugenic pressures.[1] I will begin, however, by examining several issues which can be placed in the first two of the above mentioned categories. These are concerns which, given enough political motivation, could be addressed through a comprehensive regulatory policy—one that sensibly integrates available and novel regulatory mechanisms.

Readers should be alert to two important limitations of this paper. First, it is introductory. I seek merely to broach a number of important issues and concepts relevant to the regulation of the commercialization process. Second, much of the background material used in this paper flows from the United States and Canada. Accordingly, many of the conclusions would seem to have more immediate relevance to North America. Regardless, the economic dynamics which drive the biotech industry are shared throughout the world and this suggests that the North American experience may be instructive.

2. THE CURRENT CLIMATE: BALANCING LEGITIMATE CONCERN AGAINST THE NEEDS OF THE PRIVATE SECTOR

The commercialization process has begun to attract a significant amount of attention in the genetics literature (Malinowski and Blatt, 1211, Erramouspe, 961, Malinowski, 8). And while not all commentators agree on many of the more controversial issues (e.g., the "eugenic" potential of market forces), there are emerging areas of apparent consensus—such as concern about the impact of commercialization on the data sharing practice of academics, the potential skewing of university based research, and the pressure to introduce genetic services before they have been appropriately evaluated. However, even these agreed upon areas of concern are sometimes balanced by countervailing, largely economic, forces emanating from the biotechnology sector. Compounding the weight of the biotech industry's position is the fact that it is largely mirrored by governments who desire to enhance local economies by facilitating the growth of biotechnology (Cervelli; Alberta Science and Research Authority). This tension between (arguably) well founded concerns and an understandable wish to support a burgeoning sector of the economy will, obviously, lead to conflicts in the development of commercialization policy.

2.1. A "Category 1" Concern: The Premature Implementation of Genetic Services

Few in the ELSI community would argue with the need for safeguards to ensure that commercial forces do not lead to the premature or improper implementation of a genetic service (Malinowski and O'Rourke, 163, Motulsky, 603, Kolata, 1). The availability of appropriate counselling services, a clear understanding of the clinical benefits and potential risks, and a high degree of quality control, are among the many professional norms associated with the provision of genetic services (Hoedemaekers, ten Have and Chadwick, 135, Knoppers and Chadwick, 2033, Caulfield, Danish Council of Ethics, Institute of Medicine). Despite the long existence of such norms, it has been argued that the few diagnostic tests that are available commercially have been implemented prior to an appropriate assessment of, for example, the psycho-social implications of the test (Malinowski and Blatt, 1211). The recent commentary questioning the efficacy of BRCA1/2 testing (Couch et al, 1409, Krainer, et al., 1401, Schrag, et al, 1465)—a test first introduced commercially by Myriad Genetic Laboratories, Inc.—has underscored the need for an appropriate evaluation process—one that examines both the clinical benefits and the potential legal, ethical and social ramifications (Healy, 4487, Giardiello, et al., 823). Likewise, the early introduction and subsequent withdrawal of a commercially available predisposition test for APOE4, an allele considered a risk factor for Alzheimer's disease, stands as another example of the need for caution (Relkin, et al., 149, Post, et al., 832).

The concern of premature implementation serves as an example of an issue which warrants an immediate regulatory response—a concern which likely cannot be overcome even with persuasive economic arguments. While policy makers may need to ensure that the regulatory measures are not overly burdensome, as some would argue is the current situation with the regulation of pharmaceuticals (Novarra, 55), protections are clearly needed. Let us turn next to a concern which, though undoubtedly valid, may meet more legitimate opposition. That is, while the concerns may be real (or even empirically demonstrated), there are those who may feel that regulation would impair the attainment of a valid social goal.

2.2. A "Category 2" Concern: The Impact of Commercialization on the Nature and Direction of University Research

The impact of commercialization on University based research, particularly the effect of emphasizing the patenting process, has been the subject of a large degree of commentary (Heagerty, 588, Rosenberg, 392, Ducor, 13, Kurland, 761). For example, there is evidence that "involvement with commercialization and participation in academic-industry relationships are significantly associated with the tendency to withhold the results of research" (Blumenthal, et al., 1224, 1227, Blumenthal, 368, Blumenthal, 1291, Frankel, 1297). What impact does this climate of secrecy have on the advancement of science? Are clinically relevant data being withheld for the purposes of securing patent protection (Percy, 1)? Moreover, on a broader level there is concern that the emphasis on university/industry partnerships is, *inter alia*, affecting the objectivity of the researchers, eroding the quality of teaching, and distorting the priorities of universities.

While most of these concerns (to some extent) are undoubtedly well founded, they must be weighed against the reality that private investment, which often hinges on the potential for intellectual property protection, is viewed as a key ingredient of contemporary biomedical research. "In a society in which private industry is the only mechanism by which we can bring new knowledge to bear on practical application, relationships between industries and academics and health care institutions are essential" (Patterson and Emanuel, 316). A balance must be struck.

2.3. Responding to These "Addressable" Concerns

Despite the few direct refutations of the legitimacy of the concerns associated with the premature implementation of genetic services and with the potential impact that commercialization will have on the nature and climate of university of research, to date the dominant tone in many of the government policies in the area of biotechnology is one not of caution but of encouragement (Alberta Science and Research Authority, British House of Commons Science and Technology Committee, Ernst and Young). This position is understandable. Genetic research is extremely expensive and a government which is perceived as "anti-biotech" may discourage much needed private sector investment or drive research funds to another region or country (Novarro, 55, Burk and Boczar, 791). Similarly, if we wish to reap the benefits of the advances in human genetics, we must recognize that the vast majority of genetic services will ultimately be refined and delivered by industry. A regulatory environment hostile to biotech may impede this process.

Also, and perhaps most importantly, governments have probably had little reason to consider the concerns associated with the mix of genetics and the biotechnology industry, as the concerns are seldom articulated in connection with the broader commercialization policy. In other words, although there exists a great deal of commentary on legal, ethical and social issues, this commentary has not penetrated much of the industry's policy analysis. Indeed, it is possible for a government to have, on the one hand, a policy which addresses the concerns associated with the commodification of human genetics (e.g., Canada's failed Bill C-47, *The Human Reproductive and Genetic Technology Act*) (Caulfield, Hirtle and Le Bris, 3) and, on the other hand, blindly encourage the expansion of genetech and the collaboration of academia and industry (Prouty, 955, Alberta Science and Research Authority, Science and Technology Committee). However, given the increase in the commentary focusing on these issues, and given the fact that we are just now entering the era of practical application, it seems unlikely that this policy paradox can be sustained. A comprehensive commercialization policy, one that integrates the ethical and legal concerns, is essential.

In fact, despite the presence of numerous conflicting positions, I believe that many—but not all—of the commercialization issues, such as the fear of premature implementation, can be resolved in a manner which is advantageous to both industry and the general public. Indeed, resolving the legal and ethical concerns surrounding the implementation of genetic technologies by establishing sufficient safeguards will enhance consumer confidence which, in turn, may increase the uptake of genetic services (Marshall, 782, Bekker, et al., 1584). Some issues, such as the impact of commercialization on university research, will require a careful weighing of potential benefits (e.g., the attraction of research capital) and adverse side-effects (e.g., restricting the academic community's ability to collaborate). Nevertheless, if there exists the necessary political will, there is no reason to believe that these issues cannot be addressed.

There is a broad array of existing regulatory mechanisms which need to be considered and integrated in any comprehensive commercialization policy, including: research ethics boards; the common and civil law (e.g., negligence and fiduciary law); professional practice guidelines and accreditation mechanisms; university conflict of interest provisions; and drug and device legislation (e.g., Canada's *Food and Drug Act*). In addition, as has been done in some jurisdictions, a specific regulatory framework could be developed to meet the issues which arise in connection to genetic technologies (Caulfield, Hirtle and Le Bris, 3).

3. THE BIG ISSUES (CATEGORY 3): SPECULATIVE FEARS AND THE CONSUMER CULTURE

Though many of the issues which are associated with the commercialization process can probably be (and should be) addressed through thoughtful regulatory policy, it seems doubtful that the biggest issues, those issues which I believe are driving the broader fears which surround the "genetic revolution", can be confronted through traditional regulatory mechanisms. Specifically, I am speaking of the speculation that the commercialization process will facilitate a new "laissez faire eugenics" (Kitcher), create a market driven and defined view of human normalcy (Testart, 304), and further stigmatize of those with disabilities. Justifiably or not, many of these concerns are rooted in the eugenic history of genetic technology (King,1,6, Tibbetts, A1, Caulfield and Robertson, 59), in an inaccurate notion of genetic determinism (McGee) and in an almost instinctive fear of "genetic engineering" technologies—as evidenced by the reaction to the announcement of Dolly the lamb (Time, Moysa, A3). However, I will argue that the forces which are perpetuating many of these concerns are also intertwined in the fabric of a broader socio-political trend—that is, our belief in the market economy and the social values that necessarily accompany it.[4]

Obviously, this is not a new idea. Many commentators have expressed similar concerns about the prospect of combining human genetics and capitalist forces (Cuttle, 531, Testart, 304, Roy, Williams and Dickens, Leopold, 1993, Burstyn, 1993). In order for there to be a demand for genetic technologies there must be a perceived need for the service—be it for an individual genetic test, prenatal diagnosis or a screening program. And the creation of this need by the biotech industry may, at least theoretically, cause individuals to re-evaluate the notion of disease, disability and normalcy. The wider the definition of disease and the narrower the view of normalcy the bigger the "genetics market". However, what has not been considered in any depth, and what I hope to explore briefly below, is whether we can realistically do anything about this speculative concern.

3.1. The Complex Commercialization Environment: Consumerism, Autonomy, and the Regulation of Genetic Technologies

Commercially available genetic technologies are emerging at a time when and, to a large degree, in a place where (i.e., the United States) the market economy has gained a position of unprecedented paramountcy. And, as argued by numerous commentators, the individual's "right" to participate in this economy, that is, to select and consume the products produced by it, has become *the* emblem of individual freedom (Gitlin, Hacker). "In a market paradigm, citizens are seen as consumers and rights are understood as rights to consume" (Schneideman, 165). While this is undoubtedly a phenomenon which is most clearly seen in the United States, a growing reverence for the commercial culture can be seen throughout the world (Barber). How will this climate of consumer empowerment effect the implementation and utilization of genetic services?

Minimally, it seems likely that this general pro-commercialization environment is a significant ingredient in the hesitancy displayed by governments in the realm of biotech regulation. Indeed, as noted above, policy statements from numerous governmental organizations explicitly reflect this "anti-regulation" stance. It is more difficult, however, to predict how this consumer culture will impact the uptake of genetic services on the level of the individual patient/geneticists. Will we see a strengthening of the already strongly held, but largely legally inaccurate, belief in a right of access to health care services (Chapman, Canadian Bar Association, Jackman, 3, Jackman, 54, Windwick, 20)?

The connection between consumerism and the notion of patient autonomy, the ethical principle which fuels the right of access ethos (Scheiderman and Jecker, Veatch, 3, Emanuel and Nevelloff, 323), is hinted at in the work Dorothy Wertz. For example, in a survey of 409 American and Canadian parents, which was done in the pediatric genetics context, Wertz found that 59% believed that "patients are entitled to any service they can pay for out-of-pocket" (Wertz, 30). While the views of Canadian genetic professionals and US primary care physicians differed markedly from the more autonomy-oriented parents (only 11% of the Canadian and 26% of the American health care providers agreed with the statement), Wertz believes that "[t]he parents' views may represent the wave of the future."

> What consumers want is every service they ask for, without limit. They believe that nothing should be withheld (this would be a "denial of patients' rights") and that patients are "entitled" to whatever service they request, as long as they can pay for it out-of-pocket (Wertz, 30).[5]

Moreover, genetic professionals also seem to place a high value on the patient's right of autonomy. For example, Wertz's survey of Canadian genetic professionals found that 53% percent of those surveyed thought they should refer a patient outside of Canada if domestic law forbids a requested genetic service—a statistic that has clear implications for the potential effectiveness of future regulation (Wertz, 59).

The principle of autonomy also plays a dominant role in health law jurisprudence. In Canada, for example, it is consistently used by our courts to support: an extensive obligation to obtain informed consent prior to treatment (Ciararliello v. Schacter); the right to refuse treatment (B.(R) v. Children's Aid Society of Metropolitan Toronto, Fleming v. Reid, Re Baby R., Walker v. Region 2 Hospital Corp., Malette v. Shulman); and, as argued by some authors and courts (Shanner, 823, Jackman, Ryan, 6), a notion of reproductive freedom. While it is as yet uncertain how this body of law will be interpreted in the context of clinical genetics, it clearly sets a tone which supports the sway of individual choice—particularly in the realm of reproduction.[6] As recently argued by the American legal scholar John Robertson in relation to the regulation of prenatal genetic tests:

> If we take procreative liberty seriously, however, there may be no way to avoid recognizing the prebirth liberty of parents to exercise control over offspring characteristics. Only a greatly changed view of the importance of reproductive choice or of responsibilities toward unborn persons would bring about a change in the prebirth right to select or shape offspring characteristics (Robertson, 482).

Taken together, the apparent legal paramountcy of the notion of autonomy and the pervasiveness of the consumer culture creates an environment which both encourages the commercialization process and which may make it difficult to impose legal limitations on individual decisions—unless those limitations can be clearly justified.[7] This is particularly so if the social goal we wish to achieve by limiting individual choice, such as avoiding eugenic pressures, remains speculative and ill-defined.

3.2. Shaping Decisions

While the individual's "right to choose" has never been stronger nor more universally supported, we cannot ignore the fact that "choices" are not shaped in a vacuum. The decisions individuals make, even in the realm of health care and reproduction, are fashioned by numerous forces. Subtle pressure to use genetic services in order to contain costs (Clarke, 1145), the way in which information about genetics is conveyed by health care providers, the popular media's spin on a given issue—all of these forces could arguably have an impact on individual utilization decisions. A dominant non-scientific factor will undoubtedly be market pressure. Indeed, one of the primary goals of the market, of the "invisible hand", is to influence what individuals (including health care providers) decide to consume. Many commentators, such as John Kenneth Galbraith, have gone so far as to argue that the needs of the market, as manifested in marketing strategies, have subsumed real choice. "[T]he consumer is very substantially in the service of the business firm. It is to this end that advertising and merchandising in all their cost and diversity are directed; consumer wants are shaped to the purposes and notably to the financial interests of the firm" (Galbraith, 134, Seabrook)[8] This is no less true in the sphere of health care (Annas, Chren and Landefeld, 684, Postman).

To date, market pressure in this context has arguably had negligible *direct* impact on the patient/consumer. That is, there has been very little, if any, marketing about genetics which has been aimed directly at the Canadian health care consumer. However, marketing occurs in a variety of ways which may, in turn, *indirectly* impact consumers' perceptions of need and, as a result, their views of genetics and its importance to "who they are." For example, a group of scientists may promote a genetic discovery in the hope of attracting investors, or a biotech company may speculate (or even exaggerate) on the implication of a new genetic discovery in order to hype the company before a public offering of shares. Likewise, there is a great deal of advertising which is aimed at (often under-informed) health care professionals (Milunsky, 627). Though these marketing strategies are usually not meant for public consumption, they are probably the most explicit and pervasive form of genetic "literature". In fact, this "inter-industry hyping" of genetic services is often picked up by the media and becomes part of the popular culture's view of genetics (The Globe and Mail, Waldholz, 1997a, Waldholz, 1997b).

It is also important to note, however, that there are a number of emerging sources of relevant marketing strategies which the general public can access directly. The Internet and toll-free telephone numbers, for example, have become an important, and potentially worrisome, sources of consumer information.[10] In addition, there seems to be growing acceptance, particularly in the United States and Canada, of direct-to-consumer advertising of biotech products (e.g., pharmaceuticals) (Ingersoll, B1, Wysong, 32, News, 1329).

While it is far from certain whether marketing strategies in the sphere of human genetics will have an overall "eugenic" effect, there is no doubt that marketing can have an impact on public perceptions of disease and on the utilization of a given health care service (e.g., the impact of the marketing of recombinant growth hormone on the definition of short stature as a disease) (Leopold, 215, Cuttle, 531). Because of this fact, and because of the great potential for exploitation, the emerging consensus is that we should carefully regulate advertising in the area of genetics. As noted by the Working Group of the Stanford Program in Genomics, Ethics, and Society with respect to testing for breast cancer susceptibility:

> Many women are terrified of breast cancer. Marketing campaigns might easily prey on this fear, by first increasing women's anxiety and then offering testing as a solution to this heightened concern. Genetic testing for BRCA1/2 mutations holds too many dangers to allow unrestricted advertising (Draft Report).

It remains to be seen whether such advice will be heeded, however, given the current pro-commercialization climate, any form of regulation will undoubtedly meet a degree of resistance.[11] In any event, due to the numerous marketing avenues available to the biotechnology industry (e.g., inter-industry hyping, press releases, etc.), a complete ban on direct-to-consumer advertising would have only a marginal impact on the broader social issues. To have commercial entities involved in human genetics *requires* some degree of marketing—be it to health care professionals, health authorities, venture capitalists or otherwise. You can't have it both ways; either the private sector is involved, with all its trappings, or it's not.

3.3. Facing the Consequences

If the private sector is involved—which, of course, it will be—then the systemic effects of marketing in the realm of genetics need to be more thoroughly explored. In the end, however, it seems doubtful that patchwork regulatory measures can meaningfully reverse the detrimental impact of the commercialization process. Individual autonomy, private enterprise, consumerism—the social forces which would carry this (still theoretical) laissez faire eugenics movement are imbedded in the fabric of our current culture. Unlike the first eugenics movement, where one could point to a particular social policy as the vehicle of eugenic ideology, it will be difficult to pinpoint one mechanism or policy to which we can attribute a "eugenic" effect. Do we stop the involvement of commercial entities? Do we significantly curtail individual autonomy? Can we check the broadly based impact of the emerging consumer culture? Even if we chose to act on one of these factors (the easiest being limiting the availability of genetic services), it would likely not be sufficient to stop the broader phenomena.

4. CONCLUSION

> Scientific research was not unlimited and free, if only because it required resources which were in limited supply. The question was not whether anyone should tell researchers what to do or not to do, but who imposed such limits and directions, and by what criteria (Hobsbawm, 555–56).

I started this paper by suggesting that the concerns associated with the commercialization of human genetics can be divided into three categories. Action can and should be taken to manage the issues which are "addressable" (categories 1 and 2). This will not be easy. There are numerous conflicting social forces at play and the interests of a broad number of stakeholders need to be considered. Nevertheless, the issues are identifiable, the relevant social values apparent and the regulatory tools available. With regard to the third category, that of the fear of eugenic pressures, I have painted a decidedly pessimistic picture. Minimally, the current consumer culture, the dominance of individual autonomy, and the increasing involvement of commerce, warrant real concern about the re-emergence of a eugenic ethos. In addition, these are powerful forces capable of effectively resisting, if not fending off, regulatory measures. Can we do anything? Clearly, more research is needed to explore the efficacy of many of the above mentioned concerns. In addition, we should seek to promote the notion of genetic equality and a broad definition of human normalcy.

Am I a "genetic Luddite"? On the contrary, I believe that the current genetic revolution will eventually deliver on the promise of medical and scientific breakthroughs. Likewise, commercial entities inject much needed funding into the genetic research environment and serve as effective and efficient conduits of dissemination. However, some form of regulation is required if only to counter-balance the emerging paramountcy of market forces. As noted above by Hobsbawm, science has always been limited (or regulated) in some fashion. We need to be cognizant of how those limitations and regulations shape the direction of scientific inquiry and, concomitantly, the impact which science has on society as a whole.

NOTES

1. I use the term "eugenics" with some hesitancy as it is a term which is infused with a large degree of historical and ideological baggage. Nevertheless, it is the term often used by commentators to summarize the broad concerns relating to commercialization pressures.
2. See H. Healy who argues: "It is too early to use BRCA gene testing in everyday clinical practice, because it violates a common-sense rule of medicine: don't order a test if you lack the facts to know how to interpret the result" (Healy, 1997). See also F. Giardiello, et al., where a study evaluating the clinical use of commercial adenomatous polyposis coli gene testing is reported. The authors conclude that: "[N]early 20 percent of tests were ordered for indications considered unconventional according to current knowledge." And later: "Offering genetic counselling before the test and obtaining informed consent for testing are considered essential, but neither was done in over 80 percent of the cases" (Giardiello, 1997).
3. The preamble to Bill C-47, *An Act respecting human reproduction technologies and commercial transactions relating to human reproduction*, 2nd Session, 35th Parliament, 45 Elizabeth II, 1996, reads as follows: "[T]he Parliament of Canada acknowledges the health and ethical dangers inherent in the commercialization of human reproduction..." The Bill never made it through the enactment process (Caulfield, Hirtle and Le Bris, 1997).
4. It is worth noting, but beyond the scope of this paper to discuss in depth, that at times of economic inequality, such as today and in the 1920s, justification for inequality based in biology thrive. "The economic boom of the [1920s]] benefited the affluent, especially stock speculators, so much that inequality widened considerably. And in that decade eugenics flourished and was applied against "darker" European immigrants. We are now in a similar era" (Fischer, et al., 1996 at 206; and Frank and Cook, 1996). Of course, the late 1800s was an era of little government intervention—a time when market forces were re-shaping the western world. Interestingly, the rising science of genetics also played well during this period of history, providing a biological "explanation" for why there was such a divergence between the economic winners and losers (Hobsbawm, 1975 at 248).
5. Wertz has done a variety surveys of providers and patients and has generated similar findings (e.g., Wertz, 1995). For example, in a 1994 Fletcher and Wertz concluded thus:"Most [patients] thought withholding any service was a denial of patient rights (80%), providers unwilling to do some procedures for moral reasons should offer referrals (79%), prenatal tests should be available on request (75%), and consumers were entitled to whatever services they can pay for out of pocket (61%)" (Fletcher and Wertz, 1994).
6. Interestingly, even Ruth Hubbard, a well known critic of the Human Genome Project, has noted this tension between individual autonomy and concerns about eugenic pressures. "Despite the eugenic implications of prenatal testing, if tests are available then women must have the option to take or refuse them" (Hubbard, 1993, p. 30).
7. It should be noted that I am not arguing that we need to erode the justifiably firmly imbedded notion of patient autonomy. On the contrary, ensuring that the individual (or family) remains the focal point of health care decisions is one sure way to guard against the atrocities associated with the first genetic revolution. However, I do believe that growing reverence for individual autonomy is more a miss-placed by-product of the consumer culture than a laudable extension of individual rights.
8. As noted by J. Seabrook, this relationship between marketing strategies and individual wants is itself a tension which ultimately resolves itself by, of course, emphasizing the paramountcy of the market. "Market research seems to be creating a kind of neo-Darwinian sociology, a hybrid of science and business, that is on its way to replacing class, race, gender, and cultural identity with patterns of consumption behaviour. If things keep going this way, there will soon be no difference between what the market wants to hear and what the individual is allowed to say. The only morality will be the morality of the demo, the first commandment of which is *The market is always right*" (Seabrook, 1997, p. 185).

9. And, as noted by Neil Postman, the informational value of marketing may be minimal as it often has little to do with the conveying of facts. With the rise of modern advertising techniques "a fundamental principle of capitalist ideology was rejected; namely, that the producer and consumer were engaged in a rational enterprise in which consumers made choices on the basis of a careful consideration of the quality of a product and their own self-interest" (Postman, 1993, 169). Given the already tenuis nature of health care markets, and given the complexity of the issues involved in the context of genetics, this seems particularly worrisome.
10. "The Internet's international scope and immediate consumer access to such information sources as medical data and informal chat boxes offer unprecedented opportunities for obtaining highly reliable or, on the other hand, distinctly unreliable information" (Pines, 1997, p. 65).
11. As noted by C. Dykes, "As more disease susceptibility genes, and their causative mutations, are identified, the pressure from some concerned individuals to have such tests performed will increase. Advertisements offering genetic testing directly to private individuals have already appeared in the national press in the UK and USA."

REFERENCES

"The Biotech Century" (10 March 1997) *BusinessWeek.*

"Will There Ever Be Another You?: A Special Report on Cloning" (10 March 1997) *Time.*

"Genetics: The Future is Now" (17 January 1995) *Time.*

"Ethics Issues in Academic-Industry Relationships in the Life Sciences: The Continuing Debate" (1996) 71 Academic Medicine 1291.

"Clinic to Sift Out Bad Genes" (24 September 1997) *The Globe and Mail* A1.

"Sector 6: Biotechnologies" (1994–1995) in *Canada's International Trade Business Plan: An Integrated Plan for Trade, Investment and Technology Development;* (Ottawa; Industry Canada).

Alberta Science and Research Authority, (November, 1996) *The Commercialization of Biotechnology in Alberta.*

Anderson, A. (1996) "Commercialization of Genetic Screening" 5 Euroscreen 1.

Anderson, C. (1993) "Genome Project Goes Commercial" 259 Science 300.

Annas, G. (1993) *Standard of Care: The Law of American Bioethics* (New York: Oxford University Press).

B.(R) v. Children's Aid Society of Metropolitan Toronto (1995) 1 S.C.R. 315.

Barber, B. (1995) *Jihad vs. McWorld* (New York: Times Books).

Bekker, H. Modell, M. Denniss, G. et al., (1993) "Uptake of Cystic Fibrosis Testing in Primary Care: Supply Push or Demand Pull" 306 *BMJ* 1584.

Blumenthal D. et al, (1996a) "Relationships Between Academic Institutions and Industry in the Life Sciences—An Industry Survey" 334 *New Eng. J. Med.* 368.

Blumenthal, D. (1996b) "Ethics Issues in Academic—Industry Relationships in the Life Sciences: The Continuing Debate" 71 *Academic Medicine* 1291.

Blumenthal, D. et al., (1997) "Withholding Research Results in Academic Life Science: Evidence From a National Survey of Faculty" 277 *JAMA* 1224.

British House of Commons, Science and Technology Committee, (July, 1995) *Third Report: Human Genetics: The Science and Its Consequences.*

Burk, D. and Boczar, B. (1994) "Biotechnology and Tort Liability: A Strategic Industry at Risk" 55 *University of Pittsburgh Law Review* 791.

Burstyn, V. (June, 1993) "Breeding Discontent" *Saturday Night* 15,

Canadian Bar Association Task Force on Health Care, (1994) *What's Law Got To Do With It?: Health Care Reform in Canada*, (Ottawa: Canadian Bar Association).

Caulfield, T. (1995) ed. "Professional Norms in the Practice of Human Genetics: Special Edition" 3 *Health Law Journal.*

Caulfield, T. and Robertson, G. (1996) "Eugenic Policies in Alberta: From the Systematic to the Systemic?" 35 *Alberta Law Review* 59.

Caulfield, T. Hirtle, M. and LeBris, S. (1997) "NRTs: Is Criminalization the Solution for Canada?" 18 *(Health Law in Canada)* 3

Cervelli, R. (1996) "Product Commercialization in Small Biotech Start-Ups" in *Canadian Biotechnology.*

Chapman, A. (1994) *Health Care Reform: A Human Rights Approach* (Georgetown University Press).

Chren, M. and Landefeld, C.S. (1994) "Physicians' Behaviour and Their Interactioons With Drug Companies" 271 *JAMA* 684.

Ciararliello v. Schacter, (1993) 2 S.C.R. 119 at 135.

Clarke, A. (1990) "Genetics, Ethics and Audit" 335 *Lancet* 1145.

Cohen, J. (1997) "The Genomics Gamble" 275 *Science* 767.

Couch, F. et al., (1997) "BRCA1 mutations in women attending clinics that evaluate the risk of breast cancer" 336 *New England Journal of Medicine* 1409.

Culotta, E. (1993) "New Startups Move in As Gene Therapy Goes Commercial" 260 *Science* 914.

Cuttle, L. (1996) "Short Stature and Growth Hormone Therapy: A National Study of Physician Recommendation Patterns" 276 *JAMA* 531.

Damsell, K. (16 August 1996) "Young Biotechs Mingle in UBC's Research Halls" *Financial Post* at 24.

Day, K. (22 March 1995) "The Rights to a Medical Dream: Local Firm to Commercialize Gene Therapy" *The Washington Post* A1.

Dodet, B. (1993) "Commercial Prospects for Gene Therapy—A Company Survey" 11 *Biotech* 182.

Ducor, P. (1997) "Are Patents and Research Compatible?" 387 *Nature* 13.

Dykes, C. (1996) "Genes, Disease and Medicine" 42 *British Journal of Clinical Pharmacology* 683.

Editorial, "Capitalizing on the Genome" (1996) 13 *Nature Genetics* 1.

Emanuel E. and Nevelloff Dubler, N. (1995) "Preserving the Physician-Patient Relationship in the Era of Managed Care" 273 *JAMA* 323.

Emson, H.E. (1995) "Rights, Duties, and Limits of Autonomy" 4 *Cambridge Quarterly of Healthcare* 7.

Ernst and Young, (1997) "Ernst and Young's Fourth Report on the Canadian Biotechnology Industry", *Canadian Biotech '97: Coming of Age.*

Erramouspe, M. (1996) "Staking Patent Claims on the Human Blueprint: Rewards and Rent-Dissipating Races" 43 *UCLA Law Review* 961.

Fischer, C. Hout, M. Sanchez, M. et al., (1996) *Inequality by Design* (New Jersey: Princeton University Press).

Fleming v. Reid (1991), 4 O.R. (3d) 74 (C.A.).

Fletcher, J. and Wertz, D. (October 21 1994) "Patients' ethical views on autonomy, new reproductive technology and prenatal diagnosis: a pilot study" (Poster presentation at the 44th Annual Meeting of the American Society of Human Genetics, Montreal)[abstract #1712].

Frank R. and Cook, J. (1996) *The Winner-Take-All Society* (New York: Penguin).

Frankel, M. (1996) "Perception, Reality, and the Political Context of Conflict of Interest in University—Industry Relationships" 71 *Academic Medicine* 1297.

Galbraith, J. (1992) *The Culture of the Contentment* (New York: Houghton Mifflin).

Giardiello, F. Brensinger, J. Petersen, M. et al. (1997 "The Use and Interpretation of Commercial *APC* Gene Testing for Familial Adenomatous Polyposis") 336 *New England Journal of Medicine* 823.

Gitlin, T. (1995) *The Twilight of Common Dreams* (New York: Henry Holt and Company).

Hacker, A. (1997) *Money: Who Has How Much and Why* (New York: Scribner).

Heagerty, A. (1997) "Industry-Sponsored Research" 349 *Science* 588.

Healy, H. (1997) "BRCA Genes—Bookmaking, Fortunetelling, and Medical Care" 336 *New England Journal of Medicine* 1448.

Heller, J. (1996) "Highlights from the Recent Background Economic Study of the Canadian Biotechnology Industry" in *Canadian Biotechnology.*

Hobsbawm, E. (1975) *The Age of Capital* (New York: Vintage Books).

Hobsbawm, E. (1994) *Age of Extremes: The Short Twentieth Century (1914–1991)* (London: Abacus). 555–56

Hoedemaekers, R. ten Have, H. and Chadwick, R. (1997) "Genetic Screening: A Comparative Analysis of Three Reports" 23 *Journal of Medical Ethics* 135.

Hubbard, R. (1993) *Exploding the Gene Myth* (Boston: Beacon Press)

Ingersoll, B. (1997 August 8) "FDA to Clear the Way for Blitz of TC Drug Ads" *The Wall Street Journal* B1.

Institute of Medicine, (1993) *Assessing Genetic Risks: Implications for Health and Social Policy.*

Jackman, M. (1993) "The Constitution and the Regulation of New Reproductive Technologies" in Research Studies of the Royal Commission on New Reproductive Technologies *Overview of Legal Issues in New Reproductive Technologies*, vol. 3 (Ottawa).

Jackman, M. (1994) "The Regulation of Private Health Care Under the *Canada Health Act* and the Canadian *Charter*" 6 *Constitutional Forum* 54.

Jackman, M. (1995) "The Right to Participate in Health Care and Health Resource Allocation Decisions under Section 7 of the *Canadian Charter*" 4 *Health Law Review* 3.

King, D. (1997) "Eugenics Scandals Shock Europe" 19 *GenEthics News* 1.

King, D. (1997) "Genetic Testing by Post: How Should it be Controlled?" 15 *GenEthics News* 6.

Kitcher, P. (1996) *The Lives to Come: The Genetic Revolution and Human Possibilities.*

Knoppers, B.M. and Chadwick, R. (1994) "The Human Genome Project: Under an International Ethical Microscope" 263 *Science* 2033.

Kolata, G. (27 March 1995) "Tests to Assess Risks for Cancer Raising Questions: Wide Marketing Debated" *The New York Times* 1.

Krainer, M. Silva-Arrieta, S. FitzGerald, M. et al., (1997) "Differential Contributions of BRCA1 and BRCA2 to Early Onset Breast Cancer" 336 *New England Journal of Medicine* 1401.

Kurland, C. (1997) "Beating Scientists into Plowshares" 276 *Science* 761.

Langgreth, R. and Waldholz, M. (26 February 1997) "Drug Firm Dreaming of Designer Critters" *The Globe and Mail* A6.

Law Reform Commission on Canada, (1992) *Medically Assisted Procreation* (Ottawa).

Lehrman, S. (1997) "Investment Analysts Paint Rosy Prospect for Biotech" 385 *Nature* 189.

Leopold, M. (December 1993) "The Commercialization of Biotechnology" *Annals of New York Academy of Sciences* 215.

Malette v. Shulman (1990), 67 D.L.R. (4th) 321 (Ont. C.A.).

Malinowski, M. (1995) "Bringing Human Genome-Related Biotechnology to Market: A Matter of Survival" *The Health Lawyer* 8.

Malinowski, M. and Blatt, R. (1997) "Commercialization of Genetic Testing Services: The FDA, Market Forces, and Biological Tarot Cards" 71 *Tulane Law Review* 1211.

Malinowski, M. and O'Rourke, M. (1996) "A False Start? The Impact of Federal Policy on the Genotechnology Industry" 13 *Yale Journal on Regulation*. 163.

Marshall, E. (1997) "Gene Tests Get Tested" 275 *Science*. 782.

McGee, G. (1997) *The Perfect Baby: A Pragmatic Approach to Genetics*. Lannam: Rowman and Littlefield.

Milunsky, A. (1993) "Commercialization of Clinical Genetic Laboratory Services: In Whose Best Interest?" 81 *Obstetrics and Gynecology*. 627.

Minister of Health, (June 1996) *New Reproductive and Genetic Technologies: Setting Boundaries, Enhancing Health*.

Motulsky, A. (1994) "Predictive Genetic Diagnosis" 55 *Am. J. Hum. Gen.* 603.

Moysa, M. (1 March 1997) "Next... Designer Children?: Risk in Cloning is Being Able to Redesign the Human" *Edmonton Journal* A3.

News, (1993) "Physicians Say Direct-to-Consumer Advertising Affects Patient Behaviour" 50 *Am. J. Hosp. Pharm.* 1329.

Novarra, L. (5 August 1997) "Drugs and Money" Hospital and Health Networks 55.

Patterson, W. and Emanuel, E. (1996) "Physician-Drug Company Conflict of Interest" 14 *Journal of Clinical Oncology* 316.

Percy, S. (21 January 1997) "Research Conflict: Doctor May be in Legal Battle After Reporting Negative Findings of Drug Company Study" *Medical Post* 1.

Pines, W. (1997) "New Challenges for Medical Product Promotion and Its Regulation" 52 *Food and Drug Law Journal*. 62

Post, S. Whitehouse, P. Binstock, R. et al., (1997) "The Clinical Introduction of Genetic Testing for Alzheimer Disease" 277 *JAMA* 832.

Postman, N. (1993) *Technolopoly: The Surrender of Culture to Technology* (New York: Vintage).

Prouty, A. F. (1996) "Quebec Biotechnology: The Quiet Revolution" 14 *Nature Biotechnology* 955.

Re Baby R. (1988), 53 D.L.R. (4th) 69 (B.C.S.C.), *rev'g* (1987), 9 R.F.L.(2d) 415 (Prov. Ct.).

Relkin, N. Kwon, Y. Tsai, J. and Gandy, S. (1996) "The National Institute on Aging/Alzheimer's Association Recommendations on the Application of Apolipoprotein E Genotyping to Alzheimer Disease" *Annals New York Academy of Science* 149.

Robertson, J. (1996) "Genetic Selection of Offspring Characteristics" 3, 421 *Boston University Law Review* 482.

Rosenberg, S. (1996) "Secrecy in Medical Research" 334 *New England Journal of Medicine* 392.

Roy, D. Williams, J. and Dickens, B. (1994) *Bioethics in Canada* (Prentice Hall: Scarborough).

Ryan, M. (July/August, 1990) "The Argument for Unlimited Procreative Liberty: A Feminist Critique" *The Hastings Center Report* 6.

Schneiderman, D. (1996) "Consumer Interests and Commercial Speech: A Comment on *RJR-MacDonald v. Canada (A.G.)*" 30 U.B.C. Law Review 165.

Schneiderman, L. and Jecker, N. (1995) *Wrong Medicine* (Baltimore: The Johns Hopkins University Press).

Schrag, D. Kuntz, K. Garber, J. and Weeks, J. (1997) "Decision Analysis—Effects of Prophylactic Mastectomy and Oophorectomy on Life Expectance Among Women with BRCA1 and BRCA2 Gene Mutations" 336 *New England Journal of Medicine* 1465.

Seabrook, J. (20 October 1997) "The Big Sellout" *The New Yorker*.

Shanner, L. (1995) "The Right to Procreate: When Rights Claims Have Gone Wrong" 40 *McGill Law Journal* 823.

Stanford University Center for Biomedical Ethics. (November 1996) *Draft Report of the Working Group of the Stanford Program in Genomics, Ethics, and Society*.

Testart, J. (1995) "The New Eugenics and Medicalized Reproduction" 4 *Cambridge Quarterly of Healthcare Ethics* 304.

The Danish Council of Ethics, (1993) *Ethics and Mapping of the Human Genome.*

Thomas, S.M. et al., (1996) "Ownership of the Human Genome" 380 *Nature* 387.

Tibbetts, J. (30 June 1996) "Gov't Fights Sterilization Claims" *Edmonton Journal* A1.

Veatch, R. (1992) "Allocating Health Resources Ethically: New Roles for Administrators and Clinicians" 1 Frontiers of Health Services Management 3.

Waldholz, M. (2 May 1997) "Team Closes In On Identifying Alzheimer Gene" *Wall Street Journal* B1.

Walker v. Region 2 Hospital Corp. (1994), 116 D.L.R. (4th) 477 (N.B.C.A.).

Wertz, D. (1995) "Professional Perspectives: A Survey of Canadian Providers" 3 *Health Law Journal* 59.

Wertz, D. (1997) "Ethical Issues in Pediatric Genetics: Views of Geneticists, Parents, and Primary Care Physicians" (draft chapter for upcoming publication) 30 (manuscript).

Windwick, B.F. (1994) "Health-Care and Section 7 of the Canadian Charter of Rights and Freedoms" 3 *Health Law Review* 20.

Wysong, P. (6 August 1996) "Direct To Consumer Drug Ads Spark Debate" *The Medical Post* 32.

ETHICAL IMPACTS OF HUMAN HEALTH-RELATED BIOTECHNOLOGY IN BRAZIL

Marilia Bernardes Marques*

Center for Scientific and Technological Information
Oswaldo Cruz Foundation
Av. Brasil 4365, 21045-900 Rio de Janeiro Brasil

1. INTRODUCTION

The expression "modern biotechnology" alludes to a very diffuse and expansive arena of diverse interest groups that has been rising internationally in the last three decades. In fact, since the 1970s, modern biotechnology has been surrounded by remarkable controversies and public concern, with numerous debates polarised between economic arguments and ethical dilemmas. This situation was stimulated by the sharp flow of genetic engineering methods, which has increased the potential of biotechnology to modify and interfere with life.[1] It was also during the 70s and the 80s that the word bioethics became progressively popular, contemporaneously to this outburst of genetic engineering techniques.[2]

The growth of this polemical area of scientific research has been associated with a corresponding expansion of the notions of intellectual property of living things (Churchill, 1994). Nowadays, the main international bioethics debate focuses on patent applications for human DNA sequences, with the expression human health biotechnology making visible this modern trend of extending intellectual property rights to constituents of the human body. [3] The race to obtain patents on gene sequences is creating generalised ethical concerns over the fact that a group or nation could gain sole control on fundamental life information (Kiley, 1992, Adler, 1992). The regulation of human health-related biotechnology has become a challenge that merges ethics with economic issues, in areas where the limits between science and industrial interests are blurred (Blumenthal et al, 1986, Mudurg, 1995, Mervis, 1995).

* Phone: 55 21 2700295Fax: 55 21 2702668, E mail: bmarques@dcc001.cict.fiocruz.br

Genetic Information, edited by Thompson and Chadwick.
Kluwer Academic / Plenum Publishing, New York, 1999.

This article describes the most relevant and controversial bioethics aspects presently focused on by the regulation process of human health-related biotechnology in Brazil. The two next sections present a brief description of modern biotechnology and an introduction to the genome's new archetype for understanding diseases. These steps are necessary to comprehend the reasons why, for Brazil, a South American developing country, bioethics has also become a fundamental public policy issue in the ending years of the present century.

2. MILESTONES OF MODERN BIOTECHNOLOGY

In the last few decades, pharmacology and drug research has been boosted by an authentic biological research revolution, driven largely by the advances observed in the domains of genetics, molecular and cell biology—disciplines mostly dedicated to the exploration of the chemical basis of heredity—and information sciences.

Since the fantastic breakthrough represented by the identification, in 1944, by Avery, McLeod & McCarty of the DNA as the genetic material, i.e., the molecule that can store and pass on hereditary characteristics, humankind has worked towards gaining the ability to specify, recombine, and express at will the genes of virtually all living organisms (Healy, 1992).

Other important breakthroughs of this biologic revolution were: in 1953, the discovery of the double helix structure of DNA by Watson & Crick; in 1973, the first cloning of a gene and development of rDNA technology by Boyer & Cohen; in 1975, the development of the hybridomas technique and the production of monoclonal antibodies by Kohler & Milstein and the outlining of the first guidelines for rDNA research (Asilomar Conference); in 1976, the foundation of the first firm to explore rDNA technology (Genentech).[4]

In 1980, patenting became a capital issue in the rDNA debate. That year, following the now famous *Diamond vs Chakrabarty* affair, the USA Supreme Court ruled that microorganisms could be patented and a patent was issued to Cohen & Boyer, on the rDNA technique. This is considered a critical moment in the development of biotechnology industry in the USA, which started to be built afterwards, amid an explosion of patent applications (Bureau of National Affairs, 1989, USA Congress, Office of Technology Assessment, 1989).

In 1988, the USA Patent and Trademark Office (PTO) issued its first patent on a living animal: the Harvard mouse or Onco-Mouse (USA, Congress, Office of Technology Assessment, 1989, p.99). Since then, ethical disputes about patenting higher life forms persist.[5 6] According to Churchill (1994, p.270), "Over the last decade, United States patents have been issued for human-derived hybridomas and cell-lines, and for non-human multicellular living organisms (transgenic mice and genetically "altered" pigs and cattle)".

The most ambitious project in the short history of the advancements of health-related biotechnology started at the end of the 80s. In 1989, in the USA, the Human Genome Project (HGP) was formally established, as a jointly sponsored effort of the Department of Energy (DOE) and the Department of Health and Human Services (DHHS), which created the National Center for Human Genome Research at the National Institutes of Health (NIH). Special genome efforts have also been implemented in the UK, the European Community (EC), Japan, France, Italy, Canada and former USSR (whose genome project survived as a component of the Russian science program). In Latin America, scientists instituted a regional network with the aim of stimulating collaboration on genome research with laboratories in North America and Europe and at the regional level (Cook-Deegan, 1994).

The Human Genome Project rapidly evolved into an international effort supported by many governments and the EC. Research on social, legal and ethical implications of genome research became an important goal, with the potential threat of eugenics, biological weapons, and of many other moral abuses playing an important role.

3. THE GENOME: A NEW ARCHETYPE FOR UNDERSTANDING DISEASE

The word genome broadly defines the totality of genes and DNA sequences existing in the cell nucleus. Health-related biotechnology research is gradually converging upon the restricted belief that the genome determines the form, development, chemical composition and all functions in an organism, be it a micro-organism or a higher organism.

This reductionist approach is embodied in much biomedical research and in a large number of health-related industrial biotechnology R&D projects, and science is ultimately broadening the knowledge of the genetic bases of human health and disease and of basic life functions, including development.

Nowadays these ideas are motivating the expanding employment of genomic and complementary DNA (cDNA) sequencing techniques. Powerful computers and advanced robotics have improved these procedures, which are now rapidly yielding considerable amounts of information on large segments of expressed genes, which are valuable for genes' complete identification. Thanks to these research efforts an authentic genomics industry has been fostered, including the manufacturing of automated devices for DNA sequencing and cDNA libraries. At the present time, databases containing mainly expressed tagged sequences (ETS)—short cDNA pieces—as well as the DNA sequence of entire genes have become priceless for pharmaceutical research. It can be said that the core process of drug discovery is being redirected by this new archetype for understanding disease.

Nevertheless, determining biological function for a particular sequence is by far the most important and difficult step. This process requires straightforward, increasingly mechanical, analysis of private or public database information and specific experiments that are individually tailored to the particular gene. It is a major enterprise with an unknown probability of success. Understanding of biological function and the use of genes and gene products in the diagnostics and treatment of human disease remains a challenge needing great creativity (Caskey et al, 1995).

This huge undertaking has introduced a new information-intensive environment for molecular medical research, with scientists scanning the gene digital libraries and selectively tackling, by molecular pharmacological means, biological targets, relevant for a given patho-physiological process. This new framework requires the interpretation of the disease phenomenon as an information disturbance—a deficit, a defect, a redundancy or a regulatory disorder—occurring at the genetic level (Drews, 1996).

Appropriate for this new archetype for understanding disease, the corresponding central purpose of medical diagnostics becomes to depict particular information states. Following the new canon, the therapeutic act experiments with changes, its task progressively becoming to repair an information disturbance (Drews, 1996, p.23–24). The potential therapeutic applications of diverse human cells transformed into tissue cultures are crucial for the biotechnology industry.

Still, it remains to be seen how effective this information-strategy will be at generating new clinically active agents (Weinsteim et al, 1997). Nevertheless, it already became invaluable to health-related biotechnology industry.

4. HUMAN HEALTH RELATED BIOTECHNOLOGY IN BRAZIL

Brazil has an area of 8,547,403 square kilometres and a population of 152 million inhabitants, 80% of them living in urban areas. Brazilian GNP, in 1996, reached US$ 750 billion, 12% corresponding to agriculture and cattle. The country invests in science and technology roughly 0.8% of the GNP, corresponding to US$ 6 billion. Of this total, 70% to 80% is governmental funding.

Presently the country experiments with fundamental structural and political reforms. A process of privatising some of the biggest state-owned companies and opening up commercial frontiers to international trade is ongoing. After 1994, a successful economic plan has reduced the huge previous levels of inflation of 2,000% a year to less than 10% a year. It is expected that these changes will have a positive impact on endogenous science and technology development.

During the 1980s, Latin American and the Caribbean countries contributed scientific papers published in international journals that were estimated by several authors to comprise 1–2% of all scientific publications in the world (Martínez-Palomo and Sepúlveda, 1989, Polanco, 1990, Pellegrini Filho, 1993). Notwithstanding the small budgets allocated to science and technology, efforts made since the 1980s have intensified scientific activity in the health field at the regional level. In the group of countries formed by Argentina, Brazil, Chile, Mexico and Venezuela, which produce approximately 90% of the scientific papers originated in the Latin America region, Brazil contributes more than 30% of this total.

It is estimated that Brazil has 4 researchers for every 10 thousands inhabitants, a very low proportion compared to the situation observed in the developed countries of 40 researchers for 10 thousands inhabitants.

Nevertheless, Brazil shows a concentration of research in biological areas and a considerable capability in biomedical science. Nearly 26% of its researchers operate in biological and health areas. This statement is relevant, taking in to account that a large proportion of the Brazilian health necessities constitutes strategic targets for health-related modern biotechnology. These goals include vaccines, sera, anti-toxins, biological reagents, drugs, medicinal plants and pesticides.

In a previous paper this author has already argued that the extent to which Brazil has been contributing to the international pool of advances in biomedical field during the last hundred years is underestimated (Marques, 1996a). Among the nearly thirty new infectious diseases described world-wide, in the last two decades, three of them were identified by or with the strong participation of Brazilian biomedical scientists: Brazilian purpuric fever caused by the *Haemophilus aegyptius,* first described in 1984; Sabiá virus hemorrhagic fever, first described in 1990; Rocio virus encephalitis, first described in 1975 (Lederberg et al, 1992). Still, the industrial production of some of the main Brazilian public institutes like the Oswaldo Cruz Foundation, created in 1900, and the Butantan Institute, created in 1889, currently provides the internal market requirements for vaccines against measles, serum-type A/C meningococcal meningitis, yellow fever, and other biologic substances applicable to human and animal health.

The utilisation of modern biotechnology methods is now a reality in the country and vaccines against hepatitis B and yellow fever via recombinant DNA are foreseen in both the aforementioned institutions. Regarding diagnostic products, it is estimated that in the medium term, the Brazilian biotechnology research institutes will be manufacturing reagents for Chagas disease using DNA probe techniques and for Schistosomiasis, Leishmaniasis and Malaria, using monoclonal antibodies.

We know that in developed countries most health-related industrial biotechnology occurs in the private sector, being conducted by highly specialised small to medium-sized firms in combination with large pharmaceutical corporations. This contrasts with the framework currently observed in Brazil, where in the three last decades, investments in biotechnology R&D and industrial production have been largely the responsibility of federal government.

Brazil contributes to approximately 40% of patents issued in Latin America, most of this total pertaining to biological areas. Nevertheless, the great majority of patent rights granted in Brazil belong to foreign entrepreneurial groups, especially from the USA. Brazilian governmental institutions (universities and research institutes) are allowed to obtain patents; nonetheless, patents are not well understood by the local scientific community, which operates on a principle of communal possession of research results rather than private ownership. Recent developments in the regulatory framework for biotechnology have at least partly changed this situation. Brazilian scientists are becoming less indifferent to intellectual property rights and the main universities and research institutes are introducing internal rules regarding the ownership of research results.

Transfer of technologies applicable to the field of health from developed countries to the country has been a prime concern. Strategies for achieving such a transfer, involving the integrated actions of international agencies and national government, have included incentives to promote co-operation between Brazilian universities and research centres and groups of the developed countries.

Since the 1980s Brazil has been an active participant in a number of biomedical programs and initiatives of multilateral agencies. These include the World Health Organization (WHO) Special Program for Research and Training in Tropical Diseases (Tropical Disease Research—TDR); the United Nations Development Program (UNDP) / UNESCO / UNIDO Regional Biotechnology Program for Latin America and the Caribbean; the Pan American Health Organization (PAHO) Program for the Regional Development of Biotechnology as Applied to Health; the PAHO / WHO Expanded Program on Immunization (EPI); and PAHO's Regional System of Vaccines (SIREVA).

Brazil was an active participant of the two main multilateral agreements for the debate on patenting of human genetic materials: the Agreement on Trade-Related Aspects of Intellectual Property Rights (TRIPS) and the Convention on Biological Diversity.

According to the Biodiversity Convention, the expression biological resources comprises genetic resources, organisms or parts of them, populations and any other biotic-component of ecosystems, with real or potential utility or value for humankind. The capacity of modern biotechnology to identify and incorporate biological resources into commercial products has led to the growing importance placed by Brazil on its own rich biological resources. Brazil is the most important Latin America country with respect to the number of mammal species, reptiles and amphibians. At the world level, in terms of biodiversity Brazil is ranked the first in amphibians, the third in birds and the fourth in mammals and reptiles.

One important issue raised at the United Nations Conference on Environment and Development, held, in 1992, in Rio de Janeiro, was the debate over the risk of private and foreign appropriation of Brazilian biological resources, particularly the rain forest genome, which would impede local R&D and manufacturing in the field of medicinal plants. In the Brazilian Amazon, there are 55 thousands species of plants. Regarding the status of the Earth's living resources, Brazilian ecosystems comprise more then 10% of the nearly 1.4 million life species described by universal science. At the present time, patents are not issued for higher life forms in Brazil; nevertheless, patents are issued for engineered micro-organisms.

Since the 80s, the main concern in Brazil about patenting biotechnology has been the possible impact on national efforts to research and develop new drugs for prevention, diagnosis and therapy of diseases such as Malaria and Chagas. Brazil, like most developing countries, considers that the advanced technologies are a critical tool for its development strategies, considering its position in the international market.

5. TRENDS IN THE CONTEMPORARY BIOETHICS REGULATORY APPROACH

In contemporary medicine, research with human subjects gives rise to diverse bioethical dilemmas. Most of them are—directly or indirectly—related to the development of intellectual property rights in the human body: to name a few, use of cell and tissues culture technology, together with genetic engineering, to develop human cell lines and ultimately commercial products human-derived; gene therapy; modern genetics associated with medically assisted procreation; and the use of tissue from aborted foetuses and human cloning.

Currently, the regulation of modern biotechnology is a challenge in many countries, simultaneously raising technical, philosophical, ethical, legal, cultural, environmental, military, commercial and international trade policy issues (USA Congress, Office of Technology Assessment, 1991). It is usually recognised that the ethical questions associated with human health biotechnology-research need to be approached in a meaningful way: besides the views of health professionals and the theological viewpoint, bioethics should also contemplate the distinctive attitudes of societies and groups (Rabino, 1994).

The limits of regulatory approaches and the role of government in troublesome bioethics dilemmas vary from one country to another. Thus, while the French government regulates bioethical issues at the federal level, in other European countries the model of professional self-regulation predominates (Shapiro, 1994, Lenoir, 1991, Edelman, 1991, Champagne and Marchetti, 1994). The United States regulatory model, although envisioning public funded research, is based on arbitration of controversies and in norms broadly respected both in the public and private sectors (Hanna et al, 1993).

Presently, most developed countries recognise the need for a national bioethics commission (Gillam, 1994, Keats, 1995). Nevertheless, the question is whether a government should legislate on bioethical issues, including applicable punishments or, on the contrary, whether its main role should be to interpret and to implement mediation to the myriad of disputes involving biomedical science (Miller, 1994).

Many assert that legislation on bioethics risks the minimisation of some complex problems, threatens research freedom in medicine and will have negative impacts on the current vogue of paradigmatic changes in the biomedical field. Lastly, to submit bioethics to Law will necessarily discharge scientific facts to the domain of Justice, which often is not prepared to deal with Science (Swinnerton-Dyer, 1995).

One controversial aspect is the process of choosing the members of a national bioethics commission. This point sustains the polemic on elitism and legitimacy. Albeit qualified to assess the complex network of ethical issues, the expert members of a commission will not invariably represent the values and the interests of all the social sectors, so as to accomplish moral judgements on behalf of society (Marques, 1996b). This implies that morality is subject to democratic principles.

In the last few years, the idea of intellectual property has been expanded to include human living materials. The marketing for human tissue and its products, viewed as commodity items, is causing disputes over who should profit from this business and how. The contemporary reality is an international human tissue trade, where human beings give blood or tissue samples for clinical or research purposes. Thousands of genetic profile databases derived from the analysis of these cells are available.

The American Type Culture Collection (ATCC) is the world's largest depository of cultured biological material, including animal and human cell lines. Around the world there are 25 smaller collections.[7] The amounts of biological material of human origin deposited in these collections will expand with the Human Genome Diversity Project (HGDP), associated with the NIH's Human Genome Project. The aim of the HGDP is to collect genetic materials from approximately 700 indigenous communities and store these materials in the ATCC. [8] The likelihood of commercial utilisation of these human materials reinforces the ethical dilemmas related to the patenting of human materials *vis à vis* social inequalities in the obtaining of informed consent. The central point is how to define equitable arrangements for sharing the financial gains of such commercialisation.

The unrestricted flow of human genetic resources among military and civilian researchers across international borders is barely monitored. Human tissue sampling from citizens and indigenous people of South America, Pacific, and possibly African countries is extensive among European and North American civilian and military scientists (Crucible Group, 1994). The potential application of biotechnology and genetic research to the making of biological weapons is another ethical issue currently causing international concern (Rosenberg, 1993).

Samples in the human tissue trade are collected through a variety of means, including surgery, volunteer participation in research, transplant/transfusion tissue donations and from cadavers. In many cases, research subjects are not completely informed about the potential commercial uses of their cells. Recently, patent claims made on human T-lymphotropic viruses (HTLV-1) derived from the immortalised cell lines of indigenous peoples in Panama, Papua New Guinea, and the Solomon Islands have caused apprehension and animosity (Rosenberg, 1993).[9] Brazilian anthropology researchers expressed concerns regarding the trade of genetic information obtained from two indigenous groups living in the Amazon Region—the Karitiana and the Suruí, from Rondonia Province—whose cells has been offered through the Internet by an USA non-profit research organisation and tissue bank (Santos and Coimbra, 1996). This organisation distributes cell lines from diverse peoples and places.

In the mid-1990s, two multilateral agreements addressed the most important debates on regulation of biotechnology research: the TRIPS and the Convention on Biological Diversity. The TRIPS requires that signatory states adopt intellectual property laws covering both microbial materials and plant varieties. It is important to note that human genetic material is not specifically excluded from the TRIPS.

The Convention on Biological Diversity also requires signatory states to recognise the ownership of genetic materials by countries or companies. Cell lines collected in a country prior to the Biodiversity Convention coming into force must be regarded as the property of the country that now stores the material. Thus, the human cell lines of people in Panama, Papua New Guinea, and the Solomon Islands stored and under patent claim by the USA government become America's legal property. According to corporate interpretations of the Convention, the donors—peoples and countries—will have to pay for future access to their donated human materials and any medical products derived from them.

These examples show that the bioethical regulation of research involving human beings does not focus on a disinterested field, where scientific information circulates freely between scientists with the sole intention of benefiting humankind. Unfortunately, the rhetoric of ethics frequently only masks vested interests existing in the international human health market (Jasanoff, 1995).

In this field filled with controversies and economic interests, citizen/human subjects will be morally authorised to submit their bodies and minds to the advancement of science, provided they give their informed consent (Nowak, 1994). Ultimately, we can consider that, under the guard of civilian rights, the currently predominant bioethical regulatory approach has, inherently, a liberal viewpoint of the human body submitted to scientific research.

6. LAW ENCOUNTERS MODERN BIOTECHNOLOGY IN BRAZIL

After a long period of controversy between Brazil and the USA on patents in pharmaceuticals and biotechnology, a new and more strict Brazilian intellectual property law favouring patents was finally approved in 1996 (Brasil, Law No. 9279, 1996). In a previous paper (Marques, 1994), I discuss how the USA and other developed countries have tended to standardise for developing nations, such as Brazil, the same international property rights rules and commercial policies designed for stronger competing countries, like Japan. I concluded that current legal standardisation trends do not discriminate developed from developing competitors and thus may jeopardise the development of local research in some newly industrialised countries like Brazil.

In fact, considering the growing information-intensive environment of molecular medical research, newly industrialised countries will have increasing difficulties in accessing the bulk of scientific and technological information, which could endanger the autonomous control of the potential effects—positive and negative—of modern biotechnology in the health sector.

Stimulated by this new information-based scientific environment, the contemporary dominant bioethical approach, focusing upon the edge of science and technology and given precedence to individual autonomy, has been receiving much attention in Brazil in the last few years. A remarkably meaningful approach to bioethics has developed, incorporating the issue of the country's development rights in the international scenario.

Under this new regulatory climate, a new set of laws and guidelines has been formulated, focusing upon broad problems such as local biodiversity, environmental and human safety, human dignity and research integrity in Brazil.

A new bill was introduced in the Brazilian Senate in 1995 including the essential elements of the Biodiversity Convention (Brasil, Senado Federal, 1995). So far, there is no consensus on whether or not the human genome issue should also be included in this bill. However, this author feels that it is only a matter of time until the two houses of Congress, the Federal Supreme Court, and diverse debaters in the country find themselves addressing the issue of whether or not to grant property rights for living, multicellular organisms and human-derived cells.

In order to regulate the introduction of genetically engineered organisms into the environment, Brazil approved, in January 1995, the *Biosafety Law* (Brasil, Lei No. 8974, 1995). Under this law were introduced a National Commission of Biosafety and institutional biosafety review boards. Despite these advances the present version of this law introduced rigorous legal punishments concerning genetic engineering techniques in a vague way. This risk of misinterpretation puts in jeopardy further developments in this area. Furthermore, the scope of this law goes far beyond environmental biosafety, including some crucial human bioethical topics: genetic manipulation of human germ-line cells is banned; *in vivo* interventions on human genetic material are also banned (except when used to treat a genetic deficiency); to produce, store or manipulate human embryos is also prohibited. Ethical questions raised by the cloned sheep called Dolly, in March 1997, echoed the internal flaws of this Law. The scientific proof that a mature tissue from an adult animal can still redirect the development of a complete organism, demonstrated the fragility of the present version of the Law regarding the ethical differences between germ and somatic cells. As a consequence, a review process of some definitions found in the Brazilian Biosafety Law is now expected.

As the welfare of people is at stake, another ethical dilemma related to human biotechnology gave rise to a new Law in 1996: the gathering of organs for human transplants. Among many other important measures to make more organs available for transplants in the country, the recently approved law introduced the presumed consent rule: now, Brazilian hospitals may gather human organs unless the deceased has explicitly forbidden them to do it. This decision provoked apprehension and anxiety in many people in the country, in light of a suspected illegal trade of organs and human tissues.

Considering all the current bioethical complexities, defining appropriate ethical guidelines for research involving humans in a country like Brazil, should be taken as a challenge by the entire society, in order for the process to be a success. In Brazil, the bioethical challenge is monumental because the broad participation of society in such an undertaking will entail confronting the barriers imposed by cultural diversity, deep social inequalities and wide-spread poverty.

Pressed by diverse allegations of moral abuses in unethical experiments involving humans, the National Health Council decided to launch, in 1995, an open-minded inquiry with the aim of defining new ethical guidelines for clinical trials in Brazil. This assignment introduced diverse interest groups—physicians and pharmaceuticals; lawyers; religious people; women's rights activists; scientists and HIV/AIDS activists, among others—into the discussion on the limits of public policy and the role of government on thorny ethical questions surrounding human health-related biotechnology research all over the country.

Finally, in October 1996, the new guidelines for research on humans were published (Brasil, Conselho Nacional de Saúde, 1996). These are not compulsory rules, but they rely on voluntary compliance, i.e., they take stock of individual and institutional responsibilities. The new Brazilian model is based in a National Ethics Commission for Human Research and in a network of Institutional Ethics Review Boards distributed all over the country. All these are independent and multidisciplinary committees and their common interest is to protect the moral integrity of research involving human beings in Brazil. Astonishingly, 75 local committees for checking research on humans were installed all over the country in the six month period that followed the new guidelines' approval. This number is growing fast, accompanied by increasing popularity of the subject of research involving human subjects.

Federal regulations apply to research on humans done at universities and research institutes financed by both public and private money. Regulations are quite tight in the following research areas: human genetics; assisted reproductive technologies; research on drugs and devices; research on Brazilian indigenous peoples; research carried out in the country financed with foreign money.

To assure the autonomous decision making of vulnerable people—poor or defenceless—and culturally diverse groups, including indigenous communities, is an item deserving special attention in the new Brazilian guidelines: informed consent should not be denied to these individuals. Regarding the collection of samples, a declaration is required, containing information about the possible uses and final destination of human materials and data, with the promise that they will be used exclusively for the purposes previously established in the research project.

According to the Declaration of Helsinki, Brazilian researchers should characterise the benefits for the participants and for the country that are expected to result from the research conducted inside the national boundaries and involving foreign collaboration.

The new bioethical guidelines also suggest that researches receiving financial support from abroad should consider the potential for clinical and research training, aiming to increase Brazilian scientific capability and technology transfer. Full description of financial transactions and of intellectual property rights agreements are recommended, demonstrating that restrictive clauses disclosing research results are not included.

Animal welfare is another important concern for the scientific community. This relevant bioethical issue recently was addressed by a law project in the Congress (Brazil, Congresso Nacional, 1995b), making animal experimentation in Brazil heavily regulated in the near future.

7. CONCLUDING REMARKS

In this article it has been shown that the development of human health-related biotechnology research, in the last few decades, have been associated with complex bioethical dilemmas, resulting from strong international economic interests and alliances within science and the technology transfer among nations. As we saw here, new bioethical challenges have important implications for the development of local capability to carry out work in modern biotechnology in developing countries like Brazil. In the same way that the ethical disputes about patenting micro-organisms represented, in the 1970s, a critical moment in the development of the biotechnology industry in the USA, bioethical questions should be seen and understood as important factors by developing countries in the current global negotiation and legal standardisation process concerning modern biotechnology.

In this article was presented a succinct description of the Brazilian attempt, observed in the last few years, of addressing crucial bioethical issues. This process suggests that in the future Brazil will continue restrictive regulation focusing on patenting and commercialisation of human biotechnology research. Presently, the general arguments favouring a more restrictive regulatory framework, incorporate many bioethical issues, like human and animal welfare, into the economic end of developing a burgeoning biotechnology industry in the country. Factors contributing to these ideas are the fragility of Brazil's position in an increasingly open and interdependent economic world and the urgent necessities of increasing the national income, the employment possibilities and the general well-being of people.

For this author, the present growth-spurt of a legal and bioethical framework in Brazil, encompassing patenting in biotechnology, biodiversity, biosafety, animal-based and human-based research, is rooted in the pressing issue of sustainable development. Finally, the author also considers the current regulatory process an essential strategy for achieving the societal benefits of science and, as we saw here, there is strong evidence that this trend will affect positively the coherence and consistency of national science and health policies in Brazil.

NOTES

1. Recently, Axel Kahn defined genetic engineering as the entire range of techniques which, benefiting from the universality of the genetic code, allow us to instruct living organisms to execute the genetic program contained in one or more genes coming from another organism. Kahn, A. 1995 Transgenic plants. In: *Genethics* Edited by: Bernhard, H. & Cookson, C. Published by: Ciba-Geigy Limited, Ciba Communications, Basel, Switzerland.
2. The word bioethics was used initially in English, including subjects of environmental ethics and medical ethics and subsequently, the term was embraced by many other idioms. Potter V.R. *Bioethics: bridge to the future.* Englewood Cliffs: Prentice-Hall, 1971, quoted in Macer D.R.J. *Bioethics for the people by the people.* Christchurch: Eubios Ethics Institute, 1994.
3. The designation human health-related biotechnology usually includes scientific research, technologic development, and industrial production of both recombinant and other biologic substances applicable to medicine; i.e., products of therapeutic, diagnostic, or immunologic value
4. Monoclonal antibodies and recombinant DNA (rDNA) techniques allows to identify the genetics basis of cellular functioning, to confirm disease diagnosis in a very precise way, to produce vaccines with higher levels of efficacy and safety. Monoclonal antibodies contribute to identify with precision different microorganisms strains and help to obtain information about genetic composition of parasites. So far, the DNA probe techniques and the polymerase chain reaction (PCR) are the two main innovative tools constructed under this new approach.
5. See the discussion paper *Ethical issues associated with the patenting of higher life forms* Westminster Institute for Ethics and Human Values, McGill Centre for Medicine, Ethics and Law, Canada, 12 December 1994 Study Team: Schrecker, T; Elliott, C.; Hoffmaster, C.B.; Keyserlingk, E.W.; Somerville, M.A.
6. Ibid, page 62
7. Ibid, page 82
8. See also http://www.rafi.ca Rafi Communique / Rural Advancement Foundation International, May, 1993, Jan./Fe. 1994, Mar/Apr 1996, Jan/Feb 1997

REFERENCES

Adler, R.G. (1992), 'Genome Research: fulfilling the public's expectations for knowledge and commercialization', *Science* Vol. 257, pp. 908–914.

Blumenthal, D.; Gluck, M.; Louis, KS.; Wise, D. (1986), 'Industrial support of university research in biotechnology', *Science* Vol. 231, pp. 242–246.

Brasil, Senado Federal (1995), Projeto de Lei No 306, Autoria: Senadora Marina Silva.

Brasil, Congresso Nacional (1995), Lei No 8974, Janeiro.

Brasil, Congresso Nacional (1995b), Projeto de Lei No 1153, Autoria: Deputado Sergio Arouca.

Brasil, Congresso Nacional (1996) Lei No 9279, 14 de Maio.

Brasil, Conselho Nacional de Saúde (1996), *Resolução No. 196* , 10 de Outubro.

Bureau of National Affairs, Inc., (1989) *Biotechnology Law for the 1990s: Analysis and Perspective* Special Report, N 4.

Caskey, T.; Eisenberg, R.S.; Lander, E.S.; Straus, J., (1995) *HUGO Statement on the Patenting of DNA Sequences*, Document, The Human Genome Organization, USA.

Champagne, P. and Marchetti, D. (1994), 'L'information médicale sous contrainte. A propos du scandale du sang contaminé', *Actes de La Recherche en Sciences Sociales* , Mars, pp. 40–62.

Churchill, I.J. (1994), 'Patenting humanity: the development of property rights in the human body and the subsequent evolution of patentability of living things', *Intellectual Property Journal* Vol. 8, pp.249–284.

Cook-Deegan, R.M., (1994), 'Origins of the Human Genome Project' *Risk: Health, Safety & Environment,* Vol.5 N.2, pp.97–118.

Crucible Group (1994), *People, Plants and Patents: the Impact of Intellectual Property on Biodiversity, Conservation, Trade, and Rural Society,* The International Development Research Centre, Ottawa, Canada.

Drews, J., (1996) *Intent and Coincidence in Drug Research: The Impact of Biotechnology* Editiones Roche F.Hoffmann-La Roche Ltd, Basel, Switzerland .

Edelman, B. (1991), 'Expérimentation sur l'homme: une loi sacrificielle', *La Recherche* Vol. 22, pp. 1056–1065.

Gillam L. (1994), 'Bioethics and public policy in Australia', *Politics and the Life Sciences* Vol. 3, pp. 87–88.

Hanna, K.E.; Cook-Degan, R.M. and Nishimi, R. (1993) 'Finding a forum for bioethics in U.S. Public Policy', *Politics and the Life Sciences* Vol. 12, pp. 205–219.

Healy, B., (1992) 'Special Report on Gene Patenting', *The New England Journal of Medicine* Vol. 327 N. 9, pp. 664–668.

Jasanoff, S. (1995), 'Procedural choices in regulatory science', *Technology in Society* Vol. 17, pp. 279–293.

Keats, T. (1995), 'Bioethics in Italy', *The Lancet* Vol. 345, pp. 182.

Kiley, T.D. (1992), 'Patents on random complementary DNA fragments?', *Science* Vol. 257, pp. 915–918.

Lederberg, J.; Shope, R.E.; Oaks, S.C. (1992), *Emerging Infectious: Microbial Threats to Health in the United States,* Institute of Medicine, Washington, DC: National Academy Press.

Lenoir, N. (1991), *Aux Frontières de la Vie: Une Éthique Biomédicale à la Française* , Paris: La Documentation Française.

Marques, M.B. (1994), 'Brazil-US controversy on the impact of patenting in biotechnology: some relevant questions for pharmaceuticals', *Science and Public Policy* Vol. 21 N. 3, pp. 165–172.

Marques, M.B. (1996a), 'An overview of health-related industrial biotechnology in Latin America and the Caribbean', *Bulletin of PAHO* Vol. 30 N.2, pp. 95–105.

Marques, M. B. (1996b), 'In search of a national forum for bioethics in Brazilian public policy', *Cadernos de Saúde Pública*, Rio de Janeiro, Vol. 12 N. 4, pp. 443–454.

Martínez-Palomo, A. and Sepúlveda, J. (1989), 'Biomedical research in Latin America: old and new challenges', in Bloom B.R. and Cerami A. (ed.) *Biomedical Science and the Third World: under the volcano*, New York: Annals of The New York Academy of Science Vol. 569, pp. 36–44.

Mervis, J. (1995), 'Conflict of interest: final rules put universities in charge', *Science* Vol.269, pp.294.

Miller, J. (1994), 'What to do until the philosopher kings come: bioethics and public policy in Canada', *Politics and the Life Sciences* Vol. 3, pp. 93–95.

Mudurg, G. (1995), 'New rules push researchers closer to biotech industry', *Science* Vol. 269, pp.297–298.

Nowak, R. (1994), 'Problems in clinical trials go far beyond misconduct', *Science* Vol. 264, pp. 1538–1541.

Pellegrini Filho, A. (1993), 'La investigación en salud en cinco países de América Latina', *Bol Oficina Sanit Panam* Vol. 114 N. 2, pp. 142–157.

Polanco, X. (1990), *Naissance et développement de la science-monde: production et reproduction des communautés scientifiques en Europe et en Amérique Latine.* Paris: UNESCO, Conseil de l'Europe, Èditions La Découverte.

Rabino, I. (1994), 'How European and US genetic engineering scientists view the impact of public attention on their field: a comparison', *Science, Technology & Human Values* Vol. 19 N. 1, pp. 23–46.

Rosenberg, B.H. (1993), 'North vs. South: Politics and the Biological Weapons Convention', *Politics and the Life Sciences* Vol. 12 N. 1, pp. 69–77.

Santos, R. V. and Coimbra, Jr C. (1996), 'Sangue, bioética e populações indígenas', *Parabólicas*, Brasil, Boletim do Instituto Sócio-Ambiental, N. 20, Ano 3, Julho.

Shapiro, D. (1994), 'A single national voice for bioethics: reflections from Europe', *Politics and the Life Sciences* Vol. 3, pp. 98–99.

Swinnerton-Dyer, P. (1995), 'The importance of academic freedom', *Nature* Vol. 373, pp. 186–188.

USA Congress, Office of Technology Assessment, (1989) *New Developments in Biotechnology: Patenting Life—Special Report* OTA-BA-370, Washington, DC: Government Printing Office.

USA Congress, Office of Technology Assessment (1991), *Biotechnology in a Global Economy* OTA-BA-494, Washington, D.C.:U.S. Government Printing Office.

Weinsteim, J.N.; Myers, T.G.; O'Connor, P.M.; Friend, S.H.; Fornace, Jr A.J.; Kohn, K.W.; Fojo, T.; Bates, S.E.; Rubinstein, L.V.; Anderson, N.L.; Buolamwini, J.K.; van Osdol, W.W.; Monks, A.P.; Scudiero, D. A.; Sausville, E.A.; Zaharevitz, D.W.; Bunow, B.;Viswanadhan, V.N.; Johnson, G.S.; Wittes, R.R.; Paull, K.D., (1997) 'An information-intensive approach to the molecular pharmacology of cancer', *Science*, Vol. 275, pp. 343–349.

16

ADOLESCENTS AND CARRIER TESTING

Attitudes and Ethical Presuppositions

Kris Dierickx

Centre for Biomedical Ethics and Law, K.U. Leuven, Belgium

INTRODUCTION

The insight in human heredity has been accelerated by the human genome project. By charting human genetic material, the project has helped to identify, localize and clarify the structure of an increasing number of genes which play a role in the origin and development of disease. This is leading to improved techniques for diagnosing disease, above all those in which genetic factors play a part. From the actual results one can predict that in the near future it will be possible to detect many of the most frequent genetic diseases early in pregnancy and at any moment thereafter. As more genes are identified, there is growing pressure to broaden existing testing opportunities and screening programmes. These programmes are increasing both the number of available tests and the volume of genetic information they generate. The emergence of the biotechnology industry increases the likelihood that these findings will be rapidly translated into widely available test kits and diagnostic products. The potential for generating all of this genetic information about individuals raises serious questions of informed consent, confidentiality and discrimination. Numerous factors effect the eventual use of genetic tests, not in the least the value of the test, the price, and the precision. Overshadowing these factors however, may be the attitude of the people towards genetic testing. How will they perceive such tests? How will they view and evaluate the implications?

Normative frameworks on genetic screening and testing in bioethics literature consider testing for recessive disorders at an appropriate moment as an important criterium (Andrews, 1994; Council of Europe, 1992; Danish Council, 1993; Nuffield Council 1993). Students are an important possible target group for information about genetic risks and tests. They are the adults of tomorrow and they are the potential users of genetic tests. This means that it is important to see what their knowledge, beliefs, opinions and attitudes concerning genetics and genetic tests are.

This article investigates the perception of adolescents toward carrier testing for cystic fibrosis. Cystic fibrosis is the most common life-shortening recessive disorder affecting

Genetic Information, edited by Thompson and Chadwick.
Kluwer Academic / Plenum Publishing, New York, 1999.

people of European descent. By the recessive nature of the disorder, it is meant that carrying the gene does not affect the carriers' health. Yet, if two carriers of the CF gene marry, their children will be at risk (1/4) of having cystic fibrosis. The genetic disorder expresses itself by chronic lung disease and pancreatic insufficiency. Most individuals born today with CF are expected to survive into their thirties. In Belgium about one individual in 22 carries the mutant gene, resulting in an incidence of 1 in 2000 births (Cuppens, 1997, p.720). The detection of the CF gene in 1989 raised the hope that heterozygote detection would lead to a definite answer about carrier status of any arbitrary individual. In the meantime more than 600 different CFTR mutations have been documented. This means that there are approximately half a million carriers in Belgium. Attitudes toward screening and testing depend on knowledge of the disorder, its burden (the type and severity of symptoms it causes), the availability and nature of treatment, ability of sufferers to participate in normal activities and enter the workforce, and longevity (Haan, 1993).

This article will focus on adolescents and genetic carrier tests, from two perspectives. First I will make an analysis of what the literature reports as being the advantages and disadvantages of carrier testing in this age group. What follows is the result of a study among Flemish senior high school students. The survey explores their knowledge about and attitudes towards genetic risks and testing. Finally, I will consider underlying presuppositions about health and genetic diseases and will reflect upon factors which can be applied in future information campaigns on genetics, and genetic testing in particular.

1. CARRIER TESTING OF ADOLESCENTS: PROS AND CONS

If one wants to introduce carrier testing or screening, the question is raised: To whom and when should it to be offered? There is potential for its application at different times in life, with concommitant advantages and disadvantages: before birth (embryonic testing) at birth (neonatal testing), in school (adolescent testing), at adult pre-pregnancy, premarital or family planning clinics (preconceptional testing), at pregnancy clinics (prenatal testing of the parents)(Dierickx, 1997a).

An overview of the pilot projects for CF shows a positive evaluation of adolescents in comparison with other groups who undergo genetic testing and screening (Brock, 1995). In secondary schools screening programs in the past had a high participation rate in Italy for thalassemia (Bianco, 1985) and in Canada for Tay-Sachs (Zeesman, 1984). In theory, carrier testing during school age offers a good opportunity of education and information about genetics and reproduction: it can be immediatly directed to the target group and be embedded in the education system, i.e. in courses of biology (Cobb, 1991). It also affords ease in sample collection in the school system. Another advantage of carrier testing on school age adolescents is that most of the subjects are not yet pregnant. In theory, this provides identified carriers maximal freedom to exercise an informed choice when faced with reproductive options.

Carrier testing of adolescents is not without problems, however. Belgium has compulsory education until the age of 18. On the one hand this provides an opportunity to screen a large group of students in their later school years. On the other hand, the majority of this group are minors and need the consent of the parents. Hypothetically, this has implications on the final decision because student participation is predicated upon parental consent. It should be noted that on the basis of the juridico-medical practice, it might be possible for the students to give a special type of informed consent. In the context of organ transplantation the legislature in Belgium agreed with the prelevation of a regenarable organ of minors on the basis of the combined decision right (Nys, 1995). In the Norwegian law on assisted reproduction and genetics, it is stated that genetic testing of a child under the age of 16 may only be carried out with the consent of the parents or guardians (Bulletin of Medical Ethics, 1994). The existing practice of consent, obligatory education in Belgium, and the legislation in some other countries appears to create the opportunity for the carrier testing of students between 16 and 18 after informed consent of the adolescent and their parents. Beside the problem of minors and informed consent, some state that knowledge about one's own carrier status is not immediately relevant. There can be a long period between the moment of the carrier test and the moment where the results of it become actually important (mating, reproduction). Some forget the results in the meantime[1]. Subsequently, a second, perhaps more sensitive test, may be necessary. Further, carrier testing can lead to stigmatization: it is possible that students identified as carriers, could be perceived as 'abnormal' or 'sick' by their peer group. Knowledge of carrier status before a partner is found may also induce assortive mating among other carriers because of a common stigma (Te Meerman, 1991). Assortive mating will most likely occur because those identified as non carriers would avoid mating with people identified as carriers. Besides this vulnerability to social stigmatization, there is also the risk of self stigmatization and the feelings of fear and unrest that may persist even years after the carrier test. The actual imperfect sensitivity of the carrier test and a false positive or false negative test result, a last disadvantage, may produce either an unwarranted sense of concern or a false sense of security.

2. INITIAL KNOWLEDGE AND ATTITUDES TOWARD CARRIER TESTING IMMEDIATELY AFTER BEING INFORMED

In Flanders a study was executed by the Centre of Human Genetics (K.U.Leuven) in cooperation with the Medical School Health Service. In the first part, initial general knowledge about CF was measured. Then, all the 162 students were provided with an informative text. After reading the text they had to answer questions about CF and wrong answers were always corrected. Only after this information process were the attitudes towards carrier testing for CF assessed. In the second part of the study, about six months later, they had to answer the same questionnaire, but without the informative text about CF2.

Before any information was given, the general knowledge about CF of Belgian senior high school students was rather poor. Less than one third (29%) of the initial sample has ever heard of it. This is considerably less than in British studies involving high school students, where between 59% and 84% reported they had heard of the disease (Williamson, 1989; Cobb, 1991; Watson, 1991). About one quarter (26%) of the Flemish students mentioned that CF was characterized by respiratory problems. The genetic cause was mentioned by only 5%. It is obvious from these data that education about the nature and course of CF will be required before young people can make more informed decisions about CF carrier testing.

After reading an informative text the understanding of the recessive nature and of carriership was rather good. Wrong answers were always corrected to ensure that the students were correctly informed. Since the perception of health risks can play an important role in decisions about health behavior (such as a carrier test for cystic fibrosis) it is interesting to see what the adolescents' perception of the risk of being a carrier of a recessive gene of CF is. In the study it was clear that they tended to underestimate their own risk on carriership: almost 50% of the students underestimated their own risk in comparison with other persons of the same age (Welkenhuysen, 1996c).

Looking at the attitudes towards carrier testing for CF, we can see that the proportion of subjects who do not want to know their carrier status at the time of the study (47%) is larger than the proportion of those who do (37%). But most of them (82%) are positive towards knowledge about their own carrier status 'in the future', or more precisely, before making reproductive decisions (59%).

What are the students' perceptions of the benefits and barriers of such a carrier test? More than a quarter of the initial sample stated that knowing their own carrier status at the time of the study had benefits. However, this proportion decreases significantly to less than 5% when considering carrier testing in the future. In contrast, more than half the initial sample (54%) perceived some benefit for a carrier test at the time of the study. The most important benefits included 'having certainty'(21%), and 'knowing the risks for future children' (17%). With regard to having a test in the future, 82% mentioned some benefit. Two benefits stood out in importance: (1) informed reproductive decisions (51%) and (2) preventing the birth of a child with CF (15%).

Concerning the barriers to knowledge about their own CF carrier status, a minority of 17% of the initial sample perceived no barriers at all at the time of the study. This proportion increased significantly to about one fourth (26%) for having such a test in the future. Complementary to these proportions, three quarters (75%) and more than half (58%) of the initial sample mentioned some barrier at the time of the study and in the future, respectively. The expectation of being worried about increased risk of having a CF child (or about transmitting the carrier status to future children) (32%) and the negative impact on self-image (12%) are most mentioned. Concerning barriers to testing in the future, the expectation of worry remained important (22%), together with increased anxiety about or during a pregnancy (19%). The negative impact of a positive test result on self-image remained a salient barrier for a small proportion of the initial sample (6%).

3. KNOWLEDGE AND ATTITUDES SIX MONTHS AFTER BEING INFORMED

The general knowledge about cystic fibrosis six months after reading the informative text improved. The proportion of the group which was mentioning at least one CF feature (56%) was higher than 6 months before (27%). However, the results from the questions concerning the (genetic) cause of CF were not much better than they had been 6 months before (9% v. 5%). Also the frequency of 'I do not know' and missing answers on other questions on CF was still high. The results of the questions concerning asymptomatic carriership were significantly worse 6 months after reading the informative text about CF than immediately after receiving the information. But, similar to the first part of the study, the total score for understanding was significantly higher for the subjects who were acquainted with a CF patient.

The answers concerning the attitudes towards CF carrier testing gave largely the same pattern of results as 6 months prior: the proportion of subjects who do not want to know their own carrier status at the time of the study (39%) was larger than the proportion that do (31%), while the proportion not wanting to know it in the future was fairly small (11%). The attitudes towards knowing one's CF carrier status at the time of the study was significantly more negative than the attitude towards knowing it in the future (55%).

When considering future carrier status information, the follow-up group answered largely in the same way. The perceived benefits and barriers of knowing their own status were also very similar, except for the following: six months after being informed the proportion of students who perceive no benefits at all in knowing their own carrier status at the time of the study was much smaller, while the proportion mentioning `having certainty, being informed' was about twice as large as immediately after reading the informative text.

4. PRESUPPOSITIONS AND RECOMMENDATIONS

1. Empirical studies show people have little or no foreknowledge about recessive affections, genetic tests, or screenings. A study among Belgian adults reveals that more than half of the participants (58%) had heard of cystic fibrosis, but only 38% were able to give a typical feature of CF (Decryenaere, 1991). About one in five (21%) were aware of the genetic cause of this condition. As previously mentioned, these figures are much lower for adolescents. Both the uninformed opinions of adolescents and the test results after reading an informative text illustrate the distinction between a genetic and a contagious disease is not always understood. Public education or the information of target groups will have to take into account the limited basic knowledge of the population in general and of students in particular. When providing information about genetics and genetic risks, one will also have to give special attention to differentiating between genetically determined conditions and contagious diseases.
2. The perception of (genetic) health risks—principally concerning the extent of one's own risk and the ability to control the given health problem—may have a mediating influence on preventive behavior towards such (genetic) risks. As for the risk perception of Flemish adolescents bearing a handicapped child or a child with a hereditary disease in the future, we see that about two out of three adolescents seldom or never think of this risk. Further, half of these assess this risk as very low or rather low. An information campaign concerning genetic risk must take this into account. When providing information, it seems best not to start from the questions or problems of young couples or people having a strong desire for children. This approach has little correspondence to the world context of most high school students. It seems better to give this information in the light of problems and questions with which young people in their own age group can be identified; e.g. questions about the illness of a sister, a brother, a friend... (Welkenhuysen, 1996a).
   Beyond this, less than a quarter of the students believes the birth of a child with a genetic disease or handicap can be prevented. The perception that genetic diseases are uncontrollable suggests that compliance to preventative behavior may prove difficult in this context. Thus it seems important to adapt the introduction and dissemmination of information about possibilities to detect certain abnormalities or risks for abnormality by prenatal diagnosis delicately. In this way the anxieties associated with genetic diseases can be lessened.
3. In this context it is important to be aware that adolescents are generally poorly informed about new diagnostic and preventive possibilities. Nevertheless, adolescents are not adverse to preventive measures. A majority wishes to get information about an eventual increased genetic risk before the pregnancy and almost

half of the group during the pregnancy. These figures are much lower than the percentage of Flemish adults that have a positive attitude towards prenatal diagnosis. The most important reason given by those who advocate preventive measures among students is that it helps to be emotionally prepared for a birth of a child with a genetic disease and it helps them to make informed decisions. In this context it is important to notice that the most mentioned argument against acquiring information during pregnancy is the opposition to abortion. Flemish adolescents are rather reserved on the subject of abortion. They are more reserved than Flemish adults. Connections with socio-demographical variables and features such as going to church, the desire to have children, and values, could not be found among adolescents. Students want to be free to decide what they want to do with information they receive; they don't want to be obliged by society to 'prevent' the birth of an affected child. An information campaign about genetics, genetic risks and the new genetic technology will have to take into account these variables as well.

4. Several studies have illustrated how difficult it is to write an informative and clear text. There are significant differences between the knowledge of students regarding asymptomatic carrier status in students who do not know CF patients and those who have personal contact with a CF patient, e.g. students who know a patient in their environment have an average score that is higher than the score of those who do not know any CF-patient. The fact that personal contact with a patient continues to effect the knowledge six months after being informed, combined with the significantly lower score of the group in the second phase, seems to show that reading an informative text and answering questions in the first phase has, in the long run, only a limited effect on the knowledge of adolescents. More is needed to provide a longer lasting influence which goes beyond informing the students via passive knowledge.
5. It is important to note the growing indecision concerning the intention to have a CF carrier test in the future, if the informative leaflet is no longer available. The informative text may have functioned as a cue to action and broken down some barriers. This suggests that the participation rate for carrier screening for cystic fibrosis would be much higher if the test were systematically offered when (for any reason) visiting a physician or a health care provider (e.g. a medical control in the school or the workplace, family planning clinics, etc.). The increased indecision indicates that making the choice to participate in genetic testing (e.g. a carrier test for cystic fibrosis) is a complex process. This process may engender (among others things) a fear of self-stigmatization and social discrimination. Together with the doubts and the lack of knowledge concerning human genetics in general, and cystic fibrosis in particular, this observation is an additional argument for caution exercised with sensitivity and prudence when designing and implementing a carrier screening program in secondary schools.

CONCLUSION

Medical genetics is in full expansion and makes fast progress: the expectations are high. For certain conditions there are already very reliable tests available and others are in preparation. It is expected that within a short time, important groups of the population will be eligible for a genetic test (Dierickx, 1997b). A worldwide study among geneticists shows that these specialists have various opinions about the most appropriate time for carrier screening and testing to be carried out. A comparison of the advantages and disadvantages as stated in the literature and the results in the Flemish study among adolescents indicate that some of the advantages are perhaps not so clear. The attitude of the Belgian adoslescents towards genetic carrier testing at the time of the study is not so positive: providing them with written information has, in the long run, a rather limited effect. The question can be posed if *offering carrier testing* and screening at school age is the best time for offering a test. It appears that—in theory—testing adults at reproductive age before pregnancy has more advantages: they are no minors; they are not (yet) pregnant; the interval between the time of the test and the moment of the pregnancy is smaller; all alternatives are still open: partner choice, adoption, preimplantation diagnosis, artificial reproduction with donor. On the other hand, there is no doubt that *offering information* about genetics, genetic riks, and genetic (carrier) testing and screening is an integral part of contemporary education. Young people are receptive to this kind of information. It is the duty of our society to acquaint the adults of tomorrow with the crucial evolutions within genetics and genetic testing.

NOTES

1. This is not a mere hypothetic risk (Zeesman, 1984): after 8 years, 10% if the screened adolescents had forgotten the test results.
2. For more details on the methodology and the following results cf. Decruyenaere, 1995; Welkenhuysen, 1996a; Welkenhuysen, 1996b.

REFERENCES

Andrews, L.B., Fullarton, J.E., Holtzman, N.A. & Motulsky, A.G. (ed.) (1994), *Assessing genetic risks. Implications for health and social policy*, National Academy Press, Washington D.C.

Bianco, I., Graziani, B. & Lerone, M. (1985), 'Prevention of thalassaemia major in Latium Italy', *Lancet*, Vol. ii, pp. 888–889.

Brock, D.J.H. (1995), 'Heterozygote screening for cystic fibrosis', *European journal of human genetics* Vol. 3, pp. 2–13.

Cobb, E., Holloway, S. Elton, R. & Raeburn, J.A. (1991), 'What do young people think about screening for cystic fibrosis?', *Journal of medical genetics*, Vol. 28, pp. 322–324.

Council of Europe (1992), *Recommendation on genetic testing and screening for health care purposes*, No. R (92) 3, Strasbourg.

Cuppens, H. (1997), *Mucoviscidose* [Cystic Fibrosis], in XXX, *Preventieve gezondheidszorg* [Preventive health care], Kluwer, Diegem, pp. 720–725.

Danish Council of Ethics (1993), *Ethics and mapping of the human genome. Genetic screening*, Kopenhagen.

Decruyenaere, M., Evers-Kiebooms, G. Denayer, L. & Van den Berghe, H. (1991), 'Cystic fibrosis: community knowledge and attitudes towards carrier screening and prenatal diagnosis', *Clinical genetics*, Vol. 41 pp. 189–196.

Decruyenare, M., Evers-Kiebooms, G., Welkenhuysen, M. Bande-Knops J. et al. (1995), 'Adolescents' opinions about genetic risk information, prenatal diagnosis, and pregnancy termination', *Journal of medical genetics*, Vol. 32, pp. 799–804.

Dierickx, K. (1997a), 'Ethische reflecties omtrent het tijdstip voor een screening op gendragerschap. Het voorbeeld van mucoviscidose' [Ethical reflections on the time of screening for gene carriership. The example of cystic fibrosis], *Tijdschrift voor geneeskunde* [Journal for medicine], Vol. 53, pp. 1493–1500.

Dierickx, K. (1997b), 'Preventie en predictieve genetische gezondheidszorg' [Prevention and predictive healthcare], in XXX, *Preventie en gezondheidspromotie* [Prevention and health promotion], Kluwer, Diegem, pp. 64–73.

Haan, E.A. (1993), 'Screening for carriers of genetic disease: points to consider', *The medical journal of Australia*, Vol. 158, pp. 419–421.

Nuffield Council on Bioethics (1993), *Genetic screening. Ethical issues*, London.

Nys, H. (1995), *La médecine et le droit* [Medicine and Law], Kluwer, Diegem.

Te Meerman, G.J. & Dankert-Roelse, J.E. (1991), 'Pros and cons of neonatal screening for cystic fibrosis', L.-C. Tsui (ed.), *The identification of the CF gene*, New York, pp. 83–95.

(1994), 'Norwegian Law on assisted reproduction and genetics', *Bulletin of medical ethics*, Vol. 99, pp. 8–11.

Watson, E.K., Williamson, R. & Chapple, J. (1991), 'Attitudes to carrier screening for cystic fibrosis: a survey of health care professionals, relatives of sufferers and other members of the public', *British journal of general practice*, Vol. 41, pp. 237–240.

Welkenhuysen, M., Evers-Kiebooms, G., Decruyenaere, M., Van den Berghe, H., et al. (1996a), 'Adolescents' attitudes towards carrier testing for cystic fibrosis and its relative stability over time', *European Journal of human genetics*, Vol. 4, pp. 52–62.

Welkenhuysen, M., Decruyenaere, M., Evers-Kiebooms, G., Van den Berghe, H. (1996b), *Jongeren en erfelijkheid. Hun beeldvorming over erfelijke ziekten, erfelijke risico's en genetische tests* [Adolescents and genetics. Their image about genetic diseases, genetic riks and genetic tests], Garant, Leuven/Apeldoorn.

Welkenhuysen, M., Evers-Kiebooms, G., Decruyenaere, M. & Van den Berghe, H. (1996c), 'Unrealistic optimism and genetic risk', *Psychology and health*, Vol. 11, pp. 479–492.

Wertz, D.C. & Fletcher, J.C. (1989), 'An international survey of attitudes of medical geneticists toward mass screening and access to results', *Public Health Review*, Vol. 104, pp. 35–44.

Williamson, R., Allison, M.E.D., Bentley; T.J. et al. (1989), 'Community attitudes to cystic fibrosis carrier testing in England: a pilot study', *Prenatal diagnostics*, Vol. 9, pp. 727–734.

Zeesman, S., Clow, C., Cartier, L. & Scriver, C. (1984), 'A private view of heterozygosity: eight-year-follow-up on carriers of Tay-Sachs gene detected by high school screening in Montreal', *American journal of medical genetics*, Vol. 18, pp. 769–778.

17

DOWN'S SYNDROME SCREENING

How Do They Know?

Ros Bramwell[1] and Susan Wade[2]

[1]Department of Midwifery Studies
University of Central Lancashire, Preston
[2]Barnsley District Hospital NHS Trust
Gawber Rd., Barnsley

1. INTRODUCTION

Maternal serum screening for Down's syndrome is an optional blood test available to pregnant women in the United Kingdom. It is often assumed by health professionals that they are the main providers of information upon which women make an *'informed decision'* regarding screening.

Considerable research has identified that the information women have about screening is insufficient to make an informed choice. An area neglected by research is the source of the information acquired: *How do they know?*

2. OBJECTIVE

To establish how, where and from whom women receive information concerning Down's syndrome screening.

3. METHOD

3.1. Setting

The study was conducted in a district general hospital in a provincial town with a relatively stable population of approximately 225,000. Historically, the district had a strong traditional emphasis on heavy industry. Currently, of the 22 electoral wards in the district, 17 have an unemployment rate higher than the national average. Lost mining jobs

Genetic Information, edited by Thompson and Chadwick.
Kluwer Academic / Plenum Publishing, New York, 1999.

(over 15,000 between 1984–1994) continue to influence the high unemployment rate. The average weekly wage is £14 below the Council of Europe decency threshold (which is one definition of the poverty line).

3.2. Design

Anonymous questionnaires completed by pregnant women.

3.3. Participants

208 women at 14–20 weeks of a singleton pregnancy from the hospital/community antenatal clinic consultation list. The sample was stratified by age, with at least 40 women in each category (under 20 years, 20–25 years, 30–35 years, 36 years and over). The response was 100% in all age groups. One woman was drawn from the selection because she expressed a communication / language difficulty. All mothers with a previous Down's syndrome pregnancy were identified from the GP's referral letter and excluded.

3.4. Procedure

The first 40 women in each age strata on the antenatal clinic list meeting the selection criteria were recruited to the study. Recruitment covered a two month period. Following the clinic consultation with the midwife and medical staff, women were invited to participate in the study. An explanation was given, and verbal consent obtained. The researcher was available to assist and reassure any participants who had problems reading or completing the form. A clearly marked box was available in the clinic for women to return their completed questionnaire.

4. RESULTS

Participants were aged 15 to 44, with a mean age of 28 (SD=7). As can be seen from table 1, 60% of pregnancies were planned, and approximately half the mothers were experiencing their first pregnancy. 77.8% of mothers were married / cohabiting (table 2). Table 3 shows the employment status of respondents and their partners. Maternal and paternal

Table 1. Parity, gravidity, and whether current pregnancy was planned

	Number	%
Number of pregnancies		
First pregnancy	90	43.5
Second pregnancy	53	25.6
Third or more pregnancy	64	30.9
Number of babies		
First baby	102	50.2
Second baby	58	28.6
Third or more baby	43	21.2
Was the pregnancy planned?		
Yes	125	60.4
No	82	39.6

Table 2. Marital status

Marital status	Number	%
Single	36	18.7
Living together	58	30.1
Married	92	47.7
Divorced	3	1.6
Separated	1	0.5
Other	3	7.6
Total	193	100.0

job titles were classified according to the Office of Population Census and Surveys Standard Occupational Classification (OPCS, 1991), and these are shown in table 4. Both of these distributions are a good reflection of the social and economic mix of the region, and it will be appreciated that this study has sampled a wide range of pregnant women.

Responses shown in table 5 demonstrate that 74.4% of women accepted screening, which is higher than the national uptake of 60% quoted by Wald et al. (1996). Of these, 86% had knowledge of their test results (table 5).

4.1. How Do They Know?

99 (48.5%) of respondents replied that they had first heard about this test before pregnancy. A high proportion of women reported that someone had spoken to them (80.7%) and that they had received written information (76.8%) about this screening test. As can be seen from table 6, it was from a variety of formal and informal sources that the respondents had first heard about the test and received written information. They had also spoken to a variety of health professionals, family and friends to gain further information (table 7), and indicated that they would use a variety of sources to gain further information (table 8). Health professionals were more important actual and potential sources of information for these women than family, friends and media, and it is interesting to note that by far the most important source of information is the midwife. This appears to be understood by health professionals themselves: Sadler (1997) found that the community midwives were regarded by the majority of health professionals being primarily responsible for counselling antenatal patients about serum screening for Down's syndrome, whilst obstetricians were the group least regarded as primarily responsible.

4.2 Comparison of 'Test-takers' and 'Non Test-takers'

Respondents were classified as 'test-takers' if they replied that they had already had a blood test taken for Down's syndrome, or intended to do so. Those who had not had such

Table 3. Maternal and paternal employment status

	Number	%
Maternal employment status		
Employed	108	66.3
Unemployed	55	33.7
Paternal employment status		
Employed	154	81.1
Unemployed	36	18.9

Table 4. Maternal and paternal socio-economic status by occupation

		Number	%
Maternal			
I	Professional occupations	2	1.2
II	Managerial and technical occupations	30	18.2
III NM	Skilled occupations non-manual	34	20.6
III M	Skilled occupations manual	28	17.0
IV	Partly skilled occupations	33	20.0
V	Unskilled occupations	38	23.0
Total		165	100.0
Paternal			
I	Professional occupations	8	4.8
II	Managerial and technical occupations	19	11.4
III NM	Skilled occupations non-manual	22	13.2
III M	Skilled occupations manual	72	43.1
IV	Partly skilled occupations manual	40	24.0
V	Unskilled occupations	6	3.6
Total		167	100.0

a test, and did not intend to, were classified as 'non test-takers'. The demographic profile of these respondents is compared in table 9. Non test-takers were very significantly more likely to be single (including divorced or separated) than living with or married to a partner (chi-sq (1) = 10.7, p<.01). This finding was consistent with the impression gained by the midwives working locally that male partners were often keener on taking up the opportunity for this test than the mothers, suggesting that some interesting dynamics may underlie decision-making by parents on genetic screening, and thus merits future research attention. It will be noted that there was a (non-significant) trend towards rejection of the test being more likely in women experiencing a first pregnancy or baby. This may be due to a greater concern amongst mothers with existing childcare responsibilities about the implications of having a baby with special needs.

As can be seen from table 10, there was a (non-significant) trend towards those who had or would choose to take the test being more likely to report that they felt they had received enough information about the test, and that they had talked to someone about the

Table 5. Uptake of Down's syndrome screening test

	Number	%
"Will you have the blood test for Down's syndrome screening?"		
Yes	145	74.4
No	36	18.5
Don't know	14	7.2
"Have you had the blood test for Down's syndrome screening?"		
Yes	114	57.0
No	77	38.5
Don't know	9	4.5
Knowledge of test result (in those who had had the test)		
Had result	80	86.0
Don't know/awaiting result	13	14.0

Table 6. Sources of information:
Where women first heard about the test and sources of written information

	First heard from		Where women received written information from	
	Number	%	Number	%
Hospital midwife	83	40.1	30	14.6
Community midwife	N/A	N/A	105	51.2
General practitioner	41	19.8	34	16.6
Friend	39	18.8	1	0.5
Hospital	24	11.6	8	3.9
Family	24	11.6	3	1.15
Health visitor	N/A	N/A	3	1.5
Magazine	20	9.7	9	4.4
Television	7	3.4	1	0.5
School	3	1.4	1	0.5
Other source	2	1.0	2	1.0
Total	243		197	

test. Each woman was also asked to give 'true or false' answers to some simple statements testing 'knowledge' about the screening test for Down's syndrome, and these provide interesting comparisons of test-takers and non test-takers. Quite a high proportion of both groups confused this screening test on blood samples with amniocentesis, and this clearly has implications for the granting of informed consent for this procedure. Such confusion was significantly (chi-sq $(1) = 4.3$, $p<.05$) more likely in those who stated they had not and would not undertake screening for Down's syndrome. It is evident that this must include some, at least, of the 69.4% who felt that they had received enough information about the test. One can only speculate as to whether this confusion had a causal role in the rejection of the test (due to the risk involved in amniocentesis being attributed to the blood test), or whether this inaccuracy of response and rejection of the test both reflect an attitude in the expectant mother that information on whether the baby has Down's syndrome before the birth is irrelevant and unwelcome. However, Green et al. (1993) found that the majority of women *do* want to know if there is something wrong with their baby.

Table 7. Who women talked to to gain information

Who talked to respondent	Number	%
Community midwife	126	60.6
Hospital midwife	40	19.2
General practitioner	14	6.7
Family	9	4.3
Hospital doctor	8	3.8
Friend	6	2.9
Other source	6	2.9
Hospital consultant	2	1.0
Nurse	1	0.5
School	1	0.5
Health visitor	1	0.5
Total	208	

Table 8. Where respondents said that they would ask or go if they wanted more information about the test

	Yes		No/maybe	
	Number	%	Number	%
Hospital midwife	182	71.9	32	28.1
Community midwife	151	91.5	14	8.5
General practitioner	64	82.1	14	17.9
Hospital consultant	51	58.0	37	42.0
Hospital doctor	41	50.6	40	49.4
Friend with a baby	14	32.6	29	67.4
Health visitor	11	27.5	29	72.5
Family	7	19.4	29	80.6
Family planning clinic	7	19.4	29	80.6
Friend	3	8.3	33	91.7
Library	2	5.6	34	94.4
Other	1	33.3	2	66.7
School	0	0.0	33	100.0

Clearly, the presentation of information on this test to women presents a difficult challenge to the health professional. Smith et al. (1994) found that 43% of midwives and 14% of obstetricians were able to answer less than half of a multiple choice questionnaire on prenatal screening correctly—although it should be emphasised that the questions in that study were considerably more complex than those asked of the women in this study. Khalid, Price and Barrow (1994) found that 40% of midwives in their study did not feel confident counselling women about screening.

Table 9. Demographic profile of test takers with non test takers

	Test takers		Non test takers	
	Number	%	Number	%
Number of babies				
First baby	71	46.7	28	59.6
Second or more baby	81	53.3	19	40.4
Number of pregnancies				
First pregnancy	62	40.5	27	54.0
Second or more pregnancy	91	59.5	23	46.0
Was the pregnancy planned				
Planned	94	61.4	29	58.0
Unplanned	59	38.6	21	42.0
Marital status				
Single	21	14.7	15	32.6
Living together	46	32.2	11	23.9
Married	70	49.0	19	41.3
Divorced	2	1.4	1	2.2
Separated	1	0.7	0	0.0
Other	3	2.1	0	0.0
Total	143	75.7	46	24.3
Maternal employment status				
Employed	78	65.0	27	69.2
Unemployed	42	35.0	12	30.8
Total	120	75.5	39	24.5

Table 10. Test-taking, information, and knowledge

	Test takers		Non test takers	
	Number	%	Number	%
First heard about the test				
Before pregnancy	75	49.3	21	43.8
After pregnancy	77	50.7	27	56.3
	152	76.0	48	24.0
Enough information about the test				
Yes	128	84.8	34	69.4
No	23	15.2	15	30.6
	151	75.5	49	24.5
Talked to someone about the test				
Yes	129	84.3	34	68.0
No	24	15.7	16	32.0
	153	75.4	50	24.6
Received written information about the test				
Yes	117	76.5	38	76.0
No	36	23.5	12	24.0
	153	75.4	50	24.6
Knowledge question 1: "Down's syndrome screening is a blood test"				
True	149	100.0	42	93.3
False	0	0.0	3	6.7
	149	76.8	45	23.2
Knowledge question 2: "The blood test is sometimes called an amniocentesis"				
True	32	23.5	17	45.9
False	102	75.0	18	48.6
Don't know	2	1.5	2	5.4
	136	78.6	37	21.4
Knowledge question 3: "The blood test result will say 'yes or no' to Down's syndrome"				
True	35	24.0	9	22.5
False	111	76.0	31	77.5
	146	78.5	40	21.5

5. DISCUSSION

Data on sources and accuracy of information accessed by women is crucial to planning health information for informed consent on uptake of testing.

This study showed that midwives and GPs are the main source of information for women, they are also the women's choice if additional information is required. This choice, however, does not necessarily ensure accurate knowledge. Evidence repeatedly suggests that women remain confused and unclear about the purpose of screening programmes and therefore the decisions that are being asked of them. Health professionals also recognise this, for instance, Thornton et al. (1995) found women's understanding of the test was seen as a problem by 81% of obstetricians and 50% of midwives.

Communicating knowledge is not the same as providing information. Health professionals must give women what they want and need to make an informed choice. To be effective, communicating information must be a two way process. Health professionals must listen and respond to the individual (see e.g.RCOG, 1993), but they may feel under pressure to rely on using a 'check list' of information to be given by a common perception that only in this way can they satisfy their professional accountability for informed choice. A more two way process may also require more time than that which is available in the booking-in interview.

Sampling the experience and knowledge of pregnant women themselves has provided useful insight into the processes of information acquisition and use, and suggests that future research could usefully investigate the agreement between health professionals' and pregnant women's perceptions of what information has been transmitted and received in the counselling process for prenatal screening.

ACKNOWLEDGMENT

Data preparation for this paper was undertaken by Russell Gurbutt

REFERENCES

Green, J., Stratham, H. & Snowden, C. (1993), *Pregnancy a testing time. Report of the Cambridge prenatal study.* Cambridge Centre for Family Research, University of Cambridge.

Khalid, L., Price, S. & Barrow, M. (1994), 'The attitudes of midwives to maternal serum screening for Down's syndrome' *Public Health,* Vol. 108, pp.131–136.

Office of Population Census and Surveys (1991) *Standard Occupational Classification,* Vol. 3. HMSO: London.

Royal College of Obstetricians and Gynaecologists (1993), *Report of the Audit Committee's working group on Communication Standards Obstetrics.*

Sadler, M. (1997), 'Serum screening for Down's syndrome: how much do health professionals know?' *British Journal of Obstetrics and Gynaecology,* Vol. 104, pp.176–179.

Smith, D.K., Slack, J., Shaw, R.W. & Marteau, T.M. (1994), 'Lack of knowledge in health professionals: a barrier to providing information to patients?' *Quality in Health Care,* Vol. 3, pp.75–78.

Van den Akker, O. (1993), 'Prophylactic benefits of antenatal screening: Helpful or harmful?' *British Journal of Midwifery,* Vol. 5, pp.220–223.

Wald, N.J., Hutly, W., Wald, K. & Kennard, A. (1996), Down's syndrome screening in UK. *Lancet*, Vol. 347, pp.330.

18

PUBLIC PERSPECTIVES OF THE NEW GENETICS

The Citizens' Jury Experiment

Rachel Iredale and Marcus Longley

Welsh Institute for Health and Social Care
University of Glamorgan, Pontypridd, CF37 1 DL

1. INTRODUCTION

This paper describes the process of the Citizens' Jury on Genetic Testing for Common Disorders which was held in Cardiff in November 1997. The Jury's principal recommendations are included and this experimental method of public consultation examined.

In health care systems that are publicly funded there is increasing pressure to be open about, and involve the public in its decision making processes. However, decision making requires access to information. "The informed citizen can ask a variety of questions of government and of public, voluntary and private providers of goods and services if, and only if, she or he has a sufficient base on which to build and phrase questions in such a way as to get data that will be useful in making a specific decision" (Neuberger, 1994:27). If a concerted effort is to be made by policy makers to encourage public involvement, it is important to ascertain what the public actually think about healthcare issues, why they think the way they do, and what forces have shaped their beliefs.

Recently numerous methods of overcoming what is frequently referred to as the 'democratic deficit', have evolved in order to open up possibilities for public participation in policy making and to encourage active citizenship. Methods for involving the general public include focus groups, deliberative opinion polls, consensus conferences, and most recently in the UK, Citizens' Juries, all of which rest on critiques of contemporary democracy (Fishkin, 1991, Joss and Durant, 1994, Kerr, Cunningham-Burley and Amos, 1996, Morgan, 1997). The importance of public involvement in decision making, the lack of qualitative research on public opinion, and the necessity to redress the democratic deficit have been themes which have informed the research on public participation at the Welsh Institute for Health and Social Care (WIHSC).

Genetic Information, edited by Thompson and Chadwick.
Kluwer Academic / Plenum Publishing, New York, 1999.

2. HISTORY OF CITIZENS' JURIES

A Citizens' Jury is based on the legal model for trials and is typically held for a period of four to five days in which a group of jurors question expert witnesses, deliberate on a question or series of questions and present their recommendations. Citizens' Juries involve using the lay populace in their capacity as citizens as opposed to users of services, consumers, or members of specific interest groups. They are based on the premise that ordinary people given enough time, support, resources and the opportunity are eminently capable of arriving at decisions about complex policy matters. According to the Institute for Public Policy Research (IPPR) (1996:1) "jurors are not merely a resource to be mined by researchers, nor actors in a public relations exercise. They are citizens engaged in a serious civic task who become lay experts as well as confident and competent decision makers".

The concept of Citizens' Juries originated almost 30 years ago in Germany and the US. The German model was developed by Professor Peter Dienel of the Research Institute for Citizen Participation and Planning Procedures at the University of Wuppertal in 1969 (Dienel, 1978). The American model was developed two years later by the millionaire Ned Crosby who established the Jefferson Centre for New Democratic Processes in Minneapolis (Crosby, 1996). Citizens' Juries were first piloted in the UK in 1996 by the IPPR, a left-of-centre think-tank, and the market research organisation, Opinion Leader Research (OLR) and drew on both German and American experiences (Leneghan, 1997).

2.1. FORMAT OF CITIZENS' JURIES

Citizens' Juries may take a number of different forms, but according to the IPPR they have in common some of the following characteristics:

- Time—Several days to consider the question
- Information—As much evidence as possible in the time available
- Scrutiny—Opportunity to cross examine witnesses and call for more evidence
- Deliberation—Opportunity for discussion amongst themselves and with witnesses
- Independence—The Jury is independent of the organising body
- Authority—Findings carry a weight of authority derived from the independence of the Jury and the integrity of the process

A Citizens' Jury differs from an ordinary legal trial in that much more interaction amongst jurors, and particularly between jurors and witnesses, takes place. Jurors engage in group work and discussions and have considerable opportunity to cross examine witnesses themselves after they have presented their evidence. Over a period of four days 10–15 witnesses may be called. Fifteen minutes is usually allowed for each witness session followed by forty five minutes or so for questions. The Jury deliberates over the evidence together and in small groups before reaching any decisions. With the help of a moderator a number of recommendations are agreed.

Not all issues are appropriate for a Citizens' Jury to address. The issues chosen must be of sufficient importance to justify the time and the costs involved, and also must potentially affect every citizen to some extent. According to Hogwood and Gunn (1984) an issue is likely to reach the political agenda only if one or more of the following circumstances apply: (a) the issue has reached crisis proportions and can no longer be ignored; (b) it has an emotive aspect or a human interest angle; (c) it seems likely to have a wide impact; (d) it might raise questions about power and legitimacy in society and, (e) the issue is fashionable in some way. Genetic testing clearly meets all of these criteria.

3. WIHSC RESEARCH

WIHSC is exploring public perspectives on genetics and health care and there have been two stages to this research: (1) focus groups with informed members of the public which took place from November 1996 to June 1997 and which led to (2) the Citizens' Jury on Genetic Testing for Common Disorders in November 1997. This was the first Citizens' Jury in Wales and the first in the UK on the often controversial topic of genetic testing. This research is confirming that the structure of lay thought, opinion and perception about science and technology in general, and the new genetics in particular, is complex and sophisticated.

The precise question for the Citizens' Jury on Genetic Testing was derived from a series of seven focus groups organised by WIHSC and held between November 1996 and June 1997. These groups involved a total of about 70 lay people from different parts of Wales, and each spent about two hours in semi-structured discussion of the broad topic of the new genetics and its impact on healthcare. At the end of each focus group, all of the participants were asked which issues they now felt were sufficiently important—or sufficiently complex—to be considered in greater depth by a Citizens' Jury. The question for the Citizens' Jury—together with the various sessions in the programme—was initially derived from these views. A Steering Committee was brought together to oversee the process and to offer guidance on issues such as publicity, selection of expert witnesses, recruitment of Jurors, contracting for responses to the Jury's report and generally maintaining the integrity of the entire exercise. Using the data from the focus groups, and after consultation with Steering Committee members and others, the following question was chosen for the Jury:

> What conditions should be fulfilled before genetic testing for people susceptible to common diseases becomes available on the NHS?

3.1. Recruitment

A variety of recruitment strategies for Citizens' Juries have been used in the UK already. Some Juries have used random samples drawn from the electoral register; others have been recruited by independent market research organisations by standing on street corners. WIHSC invited bids to undertake the recruitment of the Jury from a number of different market research organisations and university departments in the UK in June 1997, without specifying any particular method. The Jury was recruited by an independent market research company in Cardiff-Beaufort Research Limited—who used a multi-stage method to ensure that between 12 and 16 people would agree to participate in the Jury. The method used by Beaufort Research had never been used to recruit any other Jury in the UK before, but it is described briefly below.

In September 1997, a total of nineteen primary sampling locations (unitary authorities) were selected across Wales. At each sampling point a trained interviewer knocked on doors and asked each resident aged 18 to 84 if they would be interested in attending a discussion on health issues in November 1997. No mention was made of genetics or Citizens' Juries at this stage. The interest of potential Jurors was established by leaving a short self-

Table 1. Socio-demographic characteristics of jurors

North of Wales	4	Men	8
South of Wales	11	Women	7
18–24	3	Single	2
25–44	4	Married	9
45–64	6	Divorced/separated	2
65–84	2	Widowed	2
Working full-time	5	Welsh fluent	3
Working part-time	5	Welsh not fluent	1
Not working	5	Non Welsh speaker	11
Degree	2	AB social grade	1
Other qualification	7	C1	5
Secondary level	5	C2	5
Full time education	1	DE	4

completion questionnaire which respondents had to return to Beaufort Research. Each interviewer was set a quota of ten persons to find who were interested in attending a group discussion at which further details of the research would be explained. Certain categories of respondent were specifically excluded, e.g. employees of the University of Glamorgan. A total of 195 questionnaires were left, of which 131 (67%) were returned completed.

The next stage of the process involved inviting all those who were still interested to attend one of the two group meetings—in either the North or the South of Wales—at which the topic and the process were explained fully. At the end of these group meetings every attendee was asked to indicate whether they were still interested in participating. The eagerness of those present to be chosen as Jurors was encouraging. From a final pool of twenty four who were able to make a commitment to attend for the full four days of the Jury, fifteen people were eventually selected broadly in line with the demographic variables specified at the outset.

3.2. Jury Profile

Fifteen Jurors were selected by the recruiting agency to be as representative as possible of the Welsh population in terms of a number of key demographic variables, including age, sex, social class, education, employment status, marital status, and ability to speak Welsh.

3.3. Moderation

The moderator was an independent consultant contracted by WIHSC on the suggestion of the Steering Committee specifically for this Citizens' Jury and he had no prior specialist interest in genetics. The moderator's role was to bring together a group of people from diverse backgrounds, build their confidence, manage proceedings and facilitate both large and small group discussions in such a way that everyone felt that they had a say throughout the entire process. Essentially his job was to ensure that the Jury addressed the question properly and performed their task of making recommendations effectively.

3.4. Format

There was an introductory afternoon for Jurors on 1 November 1997, and the Jury proper took place on 8 November, and from 10–12 November 1997. Therefore, the Jury met for a period of four days during which Jurors were presented with the question, had a chance to examine evidence, interrogate expert witnesses, debate the issues and formulate a series of recommendations which could be presented to decision makers.

Evidence was received from fifteen expert witness representing a wide range of opinion on genetic testing, including perspectives from clinical genetics, sociology, general practice, psychiatry, nursing, NHS management, the private sector, the public policy maker, and the patient. After each presentation there was an opportunity to question the witnesses on points of clarification or opinion. Every afternoon was devoted to group work where the issues raised during the course of the day were debated. The final day of the Citizens' Jury was devoted to making recommendations on future policies for genetic testing for common disorders.

It was agreed that there are two types of genetic testing for common disorders: (a) Presymptomatic testing: applicable to a mendelian high risk subset, e.g. BRCA1/2 (high risk sub-group); (b) Susceptibility testing: applicable to disorders with a series of interacting genetic and environmental factors (multi-factorial). The Jury was concerned with both of these types of genetic testing, but more attention was paid to the second type as the issues raised go beyond current NHS practice, and no policies have yet been formulated in this area.

With the help of the Moderator the Jury worked through the following stages to enable them to address the question: (a) understanding genetic testing for common disorders, (b) outlining concerns and hopes for susceptibility testing, (c) specifying conditions (rules) that should be fulfilled and (d) making recommendations for policy makers.

3.5. Recommendations of the Citizens' Jury on Genetic Testing

Given that the Jury only met for a period of four days, it was unrealistic to expect Jurors themselves to write the report containing their principal recommendations. As the Jury progressed, Jurors were asked to put together all the evidence they had heard and come up with a series of recommendations for future policies for genetic testing for common disorders. All of these recommendations were discussed and agreed before the Jury disbanded. Staff at WIHSC then drafted the report on behalf of the Jurors, which was circulated to them, and their comments incorporated as necessary.

The Jury supported current NHS practice on testing patients for *single gene* disorders, and endorsed the continued provision and development of such services. In relation to genetic testing for susceptibility to *common* disorders, the recommendations are summarised in Table 2.

The Jury presented their recommendations to the government's Human Genetics Advisory Commission (HGAC) on 15 December 1997. After the meeting with the HGAC the Jury's recommendations were circulated to a number of key decision making bodies in the UK, including Health Authorities, Trusts, Community Health Councils, Royal Colleges and patient organisations. These bodies are being asked to respond formally to the Jury's recommendations. All of these responses will be put together into a report and published in the summer of 1998.

3.6. Evaluation

Given that Citizens' Juries are a relatively new method of consulting with the public in the UK, they are still being evaluated and compared with other more traditional methods. This Citizens' Jury was independently evaluated by a small team of researchers from

Table 2. Summary of recommendations of the citizens' jury on genetic testing

General Principles
That planning for the future of genetic testing should be much higher on the agendas of policy makers and professionals at all levels within the NHS and government
That there should be no general population screening for genetic susceptibility to common diseases: genetic testing should only be offered to families known to be at high risk
That the NHS should take the leading role in the provision of genetic testing services
That the timing and pace of new developments should be controlled, following expert advice
That ethnic, cultural, and religious considerations should be respected at all stages in the planning and provision of services
That the interests of future generations should always be considered when drawing up policies in this area
Ensuring Equity of Access to Genetic Services
That everyone with a family history of genetic disease should have equitable access to high quality genetic testing services at the primary care level
That every patient should have equitable access to specialised genetic services, through a well-understood system of referral from the GP
Achieving the Right Balance of Funding
That money be 'earmarked' to ensure adequate funding for genetic testing services for families at high risk, both at the primary care and specialist levels
That the 'earmarked' funding be distributed according to need, and that this distribution be regularly reviewed
That research into the genetic causes of common disorders should continue
Improving Genetics in Primary Healthcare
That a family history should be taken from every new patient registering with a primary healthcare team in order to identify those at high risk of a genetic condition
That good communication between service providers is necessary, perhaps by sharing information in the form of 'best practice' examples
That every primary healthcare team should have equal access to information and advice from specialist genetic services
That one member of the primary healthcare team should have lead responsibility for genetics
Ensuring Adequate Counselling For All
That a genetic testing service should always include both pre- and post-test counselling
That genetic counselling should always be initiated by the primary healthcare team before referral to a specialist genetic centre
That all members of the primary health care team involved in genetic testing receive training in non-directive counselling
Maintaining the Regulatory Framework
That no regulation in addition to the recent *Code of Practice and Guidance on Human Genetic Testing Services Supplied Direct to the Public* is required *at present*
That developments in genetic testing should be kept under close scrutiny by the Advisory Committee on Genetic Testing, and that the government should not hesitate to introduce legislation immediately if it is required
That consideration be given to the establishment of an independent 'Ombudsman' for genetic testing to investigate individual complaints and to advise on acceptable levels of pricing and profits
That individuals should remain free to use private genetic testing services offered within current regulation
Expanding Public and Professional Knowledge
That the general public and politicians need to have a better understanding of the implications of genetic testing
That the media should be provided with balanced and objective information about genetic advances
That genetics should have a more prominent part in the National Curriculum, and also in community education
That adequate resources be allocated to the appropriate training and up-dating of all NHS staff who will be involved in genetic testing in the future
That basic training in genetics for NHS staff should meet nationally agreed standards. This could perhaps be achieved through a 'National Curriculum' on genetic testing for healthcare professionals at the pre-qualification stage
Keeping the Public Involved
That it is vital to stimulate an informed public debate on the healthcare implications of the new genetics in order to maintain public confidence, and therefore that the Human Genetics Advisory Commission pay particular regard to their third term of reference.

the Universities of the West of England and Birmingham, in conjunction with researchers from another department within the University of Glamorgan.

WIHSC will also be evaluating the Citizens' Jury process. One purpose of this Citizens' Jury was to inject into the policy-making process a coherent summary of informed public opinion on the specific question posed. Success in this instance was therefore measured by the extent to which the jurors were able to address the issues involved, and then frame relevant recommendations which would command the attention of those responsible for policy making in this area. A second purpose was to contribute to the Institute's broader research programme on public attitudes to the new genetics in healthcare. It is hoped that a fuller analysis of the Citizens' Jury will provide a unique insight into the perspectives of lay people, and the values and priorities which they bring to bear on these issues.

4. CRITICISMS OF CITIZENS' JURIES

Arguments in favour of the jury method in general include that it is a safeguard of liberty, it can be an essential check on injustice and that it is the best means for establishing the truth. Arguments in favour of Citizens' Juries reflect these points, and also emphasise improving reflexivity, fostering notions of active citizenship and providing the public perspective on topical issues. Critics of Citizens' Juries tend to generally emphasise the following: (1) expense, (2) lack of representativeness and (3) that Jurors may lack the ability to understand all the evidence.

The question for the Jury must be one of sufficient importance to justify the costs involved and the significant amount of time input from Jurors. Citizens' Juries typically cost between £16–20,000 and concern is regularly expressed about value for money for an exercise that may have little or no impact on decision making. It has been suggested that the money might be better spent for example, by employing a community worker, providing education or directly improving services. However, if a question of significant magnitude is chosen, where the consequences of particular decisions are great, it can be argued that commitment of such sums actually represent a good investment. The full costs of the Jury have not yet been assessed, but will be calculated once the process is complete. Much of the total cost went directly to the Jurors, who received an allowance of £300 for taking part, and whose travel and other expenses were also met. Other costs included the use of the venue and the salaries of WIHSC staff involved. Additional costs were incurred because of the facilities provided for observers, such as closed circuit television.

A Citizens' Jury, given that it involves 12–16 people, can never be truly 'representative' of the population as a whole, if understood in conventional quantitative terms. For a Citizens' Jury, as with their analogue in the legal process, it can be argued that strict representativeness is less important than the fact that Jurors have no personal or selfish interest in the subject and have an opportunity to deliberate in a considered and unfettered manner. It is important that the Jurors are recruited in a way that is rigorous and independent of any other vested interests involved.

Another common criticism against Citizens' Juries is that Jurors may lack the ability to understand all the evidence presented to them by the witnesses. This may partly be a consequence of the fact that many policy makers (and even scientists) are not themselves entirely convinced of the capacity of the public to understand such a complex subject as genetics, and then to make sensible decisions on it. According to Turney (1996) the lament about the public's inadequate understanding of science is often heard, and some scientific experts argue that the general public are ill-equipped to understand the complex issues that surround biomedicine. Other experts have cautioned explicitly against the inclusion of public opinion in the decision making:

> It can be argued that "public opinion" is a shallow concept and that in relation to "esoteric" subjects like genetic engineering where the public is generally not knowledgeable, opinions are formed at the time the questions are asked—converting them almost into "non-opinions", in that they had not been thought about, or held deeply (Lemkow, 1993:10).

The focus group research undertaken by WIHSC demonstrated that genetic testing is an issue that people can grasp reasonably well in a short period of time. Allied with the fact that the Jury question was derived from the focus groups gave confidence from the outset that Jurors would be eminently capable of understanding many of the esoteric issues usually perceived to be the province of experts. Of the 75 or so observers who viewed the Citizens' Jury over the four days, 50 filled out a short questionnaire asking them for their opinions on the proceedings. One question was whether they thought that the topic of genetics was too difficult to be understood by a lay audience. 88% of observers claimed that it was not too difficult, 8% were undecided and only 4% of observers thought the topic was too difficult for the Jury.

5. CONCLUSIONS

This Citizens' Jury had two main purposes. The first was to inject into the policy-making process a coherent summary of informed public opinion on the specific question posed. Success in this instance was therefore measured by the extent to which the jurors were able to address the issues involved, and then frame relevant recommendations which commanded the attention of those responsible for policy making in this area. The second purpose was to contribute to the Institute's broader research programme on public attitudes to the new genetics in healthcare. Although the Jury is still being evaluated, a number of plausible hypotheses can be constructed at this stage:

The first is that ordinary members of the public are both willing and capable of getting involved in potentially complex debates such as those surrounding the new genetics. Their capacity to offer their opinion about future healthcare policies was not contingent upon their level of scientific or technical expertise, but rather upon the role which they perceived they were playing—responsible citizens acting for the good of the community—and the contribution which they felt, in the words of one Juror, they as 'common sense people asking common sense questions' were willing to make to the policy process. As Williams and Calnan (1996:1616) argue, "at best, one can only hope to be an expert in a very small number of areas of contemporary social life—an expert in one area may be a lay person in another..." Yet we can all be expert citizens, given the opportunity to exercise our civic duties.

The second might be that this Citizens' Jury has demonstrated how the key stakeholders are effectively prepared to accept the legitimacy of the public voice, let the general public join the dialogue and are amenable to working *together* to broaden the debate on the new genetics and to influence public policy. Not only did clinicians and scientists willingly share their particular expertise on genetics and related issues with Jurors, but there was also a large number of interested parties observing the proceedings who expressed an interest in responding to their recommendations and feeding the results into the policy process. Although the Jury proved to be educational, informative and empowering for all those involved, this attempt to engender new democratic processes will need to be sustained.

Finally, the emergence of a critical perspective by the lay populace will pose a significant challenge to the advisory bodies on human genetics that have been established by government in this country. The creation of such bodies is a useful first step—a belated recognition that consideration at least should be given to the regulatory framework in which human genetics operates. But however well they function, many still seem to operate within the very system of government which is failing to address public concerns over issues such as environmental pollution, BSE and the many food hygiene scares, and even the very real possibility of human cloning. Such bodies clearly have an important role to play, but the onus now rests with them and other policy makers to embrace the Citizens' Jury concept—and other forms of public consultation—to ensure that ordinary members of the general public can come in and sit down at the policy making table, and even be permitted to choose the issues for debate. The challenge for the future is to take forward innovative public consultation exercises like the Citizens' Jury process, develop models of best practice and apply them to other controversial areas of science and technology in order to ensure public confidence and trust in the ability of the policy process to address such issues effectively.

REFERENCES

Crosby, N. (1996) *How Should We Live Together? The Philosophical Underpinnings of the Citizens' Jury® Process*. Unpublished Paper: Centre for New Democratic Processes.

Dienel, P. (1978) *Plannungszelle*, Opladen: West Deutsche Verlag.

Fishkin, J. S. (1991) *Democracy and Deliberation: New Directions for Democratic Form*, New Haven: Yale University Press.

Hogwood, B.W. and Gunn, L.A. (1984) *Policy Analysis for the Real World*, Oxford: Oxford University Press.

Institute for Public Policy Research (1996) *Citizens' Juries: Towards Best Practice*, London: IPPR.

Joss, S. and Durant, J. (1994) *Consensus Conferences A Review of the Danish, Dutch and UK Approaches to this Special Form of Technology Assessment, and an Assessment of the Options for a Proposed Swiss Consensus Conference*, London: The Science Museum.

Kerr, A., Cunningham-Burley, S. and Amos, A. (1996) 'The New Genetics and Health: Exploring Lay Perceptions'. A paper presented at the British Sociological Association Medical Sociology Conference, University of Edinburgh, 20–22 September 1996.

Lemkow, L. (1992) *Public Attitudes to Genetic Engineering: Some European Perspectives*, Dublin: European Foundation for the Improvement of Living and Working Conditions.

Lenaghan, J. (1997) 'Citizens' Juries: Towards Best Practice', *British Journal of Health Care Management*, Vol.3, No.1, pp 20–22.

Morgan, D.L. (1997) *Focus Groups as Qualitative Research*, 2nd Edn. Qualitative Research Methods Series Vol. 16. California: Sage.

Neuberger, J. (1994) 'What Sort of Information Should Be Available to the Public in an Open Society', in Marinker, M. (ed.) *Controversies in Health Care Policies*, London: BMJ Publishing Group.

Turney, J. (1996) 'Public Understanding of Science', *Lancet*, Vol.347, 20 April, pp 1087–1090.

Williams, S.J. and Calnan, M. (1996) 'The 'Limits' of Medicalization?: Modern Medicine and the Lay Populace in 'Late' Modernity, *Social Science and Medicine*, Vol.42, No.12, pp 1609–1620.

GENETICS AND JOURNALISM

A View from the United States

Joan Stephenson

Medical News and Perspectives
Journal of the American Medical Association
515 N. State St., Chicago, Illinois 60610

1. INTRODUCTION

Charles Petit, a science writer for *US News & World Report*, recently lobbed a mock news story at his colleagues to illustrate journalistic clichés perpetrated by science writers. Although Petit's spoof was intended to lampoon a formulaic approach to science reporting, he makes some telling points about how journalists serve up science news to the public:

> Science writers are nearing a breakthrough, perhaps a major breakthrough, in their age-old quest to unlock the secrets, even the ultimate secrets, of cliché-free prose, researchers reported yesterday. Using cutting-edge, state-of-the-art, high-tech and other [hyphen]-laden methodologies, the science journalists sifted obscure clues to reach their tentative conclusions. 'This is statistically significant,' one senior researcher said. 'It is an important step forward,' said another. 'This is science in action,' they agreed.
>
> The research was reported in *Science Magazine*, a prestigious journal, and in *Nature* too, a leading British journal. Other researchers welcomed the report, but were cautious. They called for more research.
>
> Science writers covered all the (usual) bases, quoting John Pike of the Federation of American Scientists, climatologist Stephen Schneider of Stanford University, bioethicist Arthur Caplan, live astronomer Steve Maran, dead astronomer Carl Sagan, outspoken physicist Robert Park, and neo-Luddite anti-technology gadfly Jeremy Rifkin. Stephen Jay Gould would have added class, but was unavailable for comment.
>
> Clichés are a window into the past, even if they are redshifted like the whistle on a passing train that changes pitch when it goes by, an analogy that itself is a window into the past. They

> offer a glimpse of the future, too. They add to growing evidence of the cataclysm that may have killed the dinosaurs. Debate is sure to continue.
>
> And while the latest results do not offer a cure, they point the way to better understanding of the underlying basic cellular causes to the ancient affliction.
>
> 'We may never know all the answers but this is an important piece of the puzzle,' said everybody (Petit, 1997).

Petit fingers several common shortcomings found all too often in science reporting today, including an overly-positive tone, engaging in pack journalism by giving a disproportionate amount of attention to a few journals and a relatively small pool of experts, and trying to sell the relevance of a science story by hinting that clinical applications are likely to follow.

In any case, whether the prose is clichéd or not, the stories that science journalists choose to tell and how they frame these stories shape public awareness and understanding of genetics.

2. WHAT DOES THE PUBLIC KNOW ABOUT SCIENCE?

Those who write for the lay public always must keep the audience and their level of understanding in mind. After a while, it is tempting to assume that readers, listeners, and viewers at least know a few basics.

That, as studies show, is a risky assumption.

Jon D. Miller, of the International Center for the Advancement of Scientific Literacy at the Chicago Academy of Sciences, has tracked the public's awareness and understanding of science since 1979 for the National Science Foundation, in the United States. In a 1996 report, he and coauthor Linda Pifer noted that:

- About 40% of the Americans surveyed indicated a high overall level of interest in scientific discoveries and the use of technologies.
- Nearly 70% of Americans said they had a high level of interest in medical discoveries.
- One in 9 Americans thinks he or she is very well informed about science and technology. People who have undergone more years of formal education and more math and science courses are significantly more likely to believe they are better informed than others (Miller and Pifer, 1996).

These and findings from other polls suggest a high level of awareness among these respondents. But how sophisticated is their understanding of the science?

The 1996 NSF study indicates that Americans' interest in science exceeds their grasp. There is a disheartening gap between our interest and our understanding of basic concepts. Miller and Pifer found:

- No more than 1 in 10 Americans can define a molecule or give a minimal explanation of what bacteria are.
- Just 1 in 5 can give a minimally acceptable definition of DNA—for example, that it has something to do with inherited traits, or is the blueprint for our genes. In other words, only 20% have a level of understanding of DNA that is sophisticated enough to read and understand a news story on a new genetic finding.

Americans also do not have a very good idea how science works, that is, an understanding of the scientific method. Miller and his colleagues found that only 2% understand science as the development and testing of theory. Moreover, only 23% of Americans have a good enough understanding of the nature of scientific inquiry, such as the rationale of using control groups, to make informed judgments about the scientific basis of results reported in the media.

But perhaps ignorance is bliss. Miller and Pifer found that Americans with little understanding of how science works seem to have an unduly optimistic view that science can cure every disease and solve any experimental problem. In general, more than 70% of Americans believe the benefits of scientific research outweigh any current or potential drawbacks, although there is less consensus over certain technologies, including genetic engineering. Those who are college graduates are much more likely to have a positive view of the gene jockeys' efforts.

So where do science journalists fit into this picture?

Miller and his colleagues have found that the vast majority of Americans get most of their information about health and medicine from the print and broadcast media. The Internet is growing as a source of medical information—and misinformation. The majority of Americans already use a computer at home, work, or both, and many are leaping onto the Web. The next generation of so-called Web TVs, which will simplify Web access for people who are computerphobic, is thought to be only a few years away and may vastly increase access to both accurate, balanced information and misleading nonsense about genetics and health issues.

3. SHAPING THE NEWS: HOW SCIENCE AND MEDICAL JOURNALISTS SELECT STORIES AND SOURCES

How do science and medical journalists select what stories to cover and shape it into news? Often, they turn to familiar sources: the professional journals.

As sociologist Peter Conrad of Brandeis University has pointed out, science and medical journalists tend to focus on a few journals, and these are overwhelmingly represented in the news: *Science*, *Nature*, *The New England Journal of Medicine* (*NEJM*) , the *Journal of the American Medical Association* (*JAMA*), *Proceedings of the National Academy of Sciences*, *Nature Genetics*, and *The Lancet* (Conrad, forthcoming). For example, in a 1995 study of the press coverage of the link between alcohol and breast cancer over a 13-year period, 88% of the news stories came from studies reported in *JAMA* or *NEJM* (Houn *et al*, 1995).

There are a couple of reasons for this pack journalism. First, science reporters believe these journals, which are desirable places to be published on the part of researchers and have reputation for rigorous peer review, usually publish the most important research. A less benign reason is competition. News is not merely what is important or novel—it is also what one's competitors are writing about.

An increasingly slick effort by journals to package the news by sending out engaging self-promotional news digests with lay-language summaries of upcoming stories also encourages running with the pack.

According to National Public Radio science reporter Richard Harris, *Nature* started the trend in the mid-to-late 1980s, as a way of getting more play in the US press (Harris, 1997). Not to be left out, *Science* started its own promoting of upcoming articles, and others also jumped on the bandwagon. *JAMA*, for example, sends out a press package each week, including video news releases for television reporters, highlighting stories that are often given plenty of ink and air time.

On the plus side, such advance promotion gives reporters an opportunity to prepare such stories with greater care. But it is also not surprising that many science writers rely on the journals' own publicity machines for stories because deadline pressures make a predictable source of story ideas very attractive.

Science reporters also shape news coverage of genetics by who they choose to interview and quote in their coverage of a news story. People who are well-known in a given area often are quoted because their opinion appears to add a voice of authority—a self-perpetuating process. Frequently quoted in the press, they become even better known and more sought-after by reporters.

In other cases, experts who make time for reporters, who explain things clearly without jargon or endless qualifiers, who are reliable providers of colorful quotes or sound bites, get more than their Warhol unit, or 15 minutes of fame.

4. NEGATIVE STUDY BIAS: WHAT JOURNALISTS IGNORE

Another way in which the press shapes public perception of genetics is their bias against reporting negative studies. Peter Conrad calls this the "disconfirmation dilemma" (Conrad, forthcoming). Articles reporting significant new genetic findings typically get major play in newspapers and magazines, but if later research does not replicate the findings or calls into question the validity of the first study, these follow-up stories are buried in the paper, if they are written about at all.

Conrad, who has been tracking genetics coverage in 5 major US papers and 3 major newsmagazines followed the reporting of two such stories, a 1987 study about the Old Order Amish and a gene for manic depression and a 1990 study about a link between alcoholism and the dopamine D2 receptor. Both stories received major attention in the print media. But while the *New York Times* reported later studies discounting the findings of the manic depression gene story, the other papers and the newsmagazines either did not report the new evidence at all or only did so years later as part of another article.

By neglecting to write about these negative studies, the press unwittingly contributes to the persistence in the public mind of ideas that are no longer valid. One example of this, Conrad notes, is the widespread reporting in the late 1960s of an apparent link between men with an extra Y chromosome and criminal behavior. Although this supposed link was shown NOT to be valid in the 1970s, people still refer to XYY as "the criminal gene."

Another problem arises when science writers oversimplify in their effort to translate complex material. For example, the influence of genes is usually complex, but science writers sometimes present new findings—"*the* breast cancer gene," "*the* gay gene," "*the* obesity gene" as if a single gene were directly responsible for the trait in question.

What is often lost is the notion that the appearance of a particular disease or behavioral trait or condition usually depends on the contribution of many genes and environmental factors. This kind of oversimplification is most common in headlines—which, incidentally, are generally written by editors, not the science writer. But it also shows up in the stories themselves.

Critics such as Dorothy Nelkin and M. Susan Lindee have charged that this kind of reporting can promote a deterministic view of genetics—that all-powerful genes are the fundamental cause of the traits with which they are associated (Nelkin and Lindee, 1995).

It also may imply that the mere identification of a gene associated with a condition explains something meaningful about the disease process.

5. PUTTING NEWS STORIES IN A SOCIAL AND ETHICAL CONTEXT

Finally, a complaint some people have made about media coverage of genetics stories is that science writers may fail to put the findings in a social and ethical context. This is an area where experts who are familiar with the issues can help, by educating science reporters and the public they reach. Bioethicists might consider taking a more activist role in educating journalists by identifying their areas of expertise to the public information officers at their institutions and agreeing to talk with the press about bioethical angles of genetics stories. Many reporters, for example, send queries to a service called Profnet,[1] which works to put them in touch with experts who can help them interpret news stories dealing with scientific discoveries and their implications for society.

And if the press comes calling for an interview or help with a story, one more factor comes into play to determine which expert sources are likely to be quoted: deadlines. Sources who return calls of reporters on tight deadlines and communicate information in a clear, jargon-free manner often are the ones who have a powerful opportunity to improve the people's understanding of issues that will affect their lives and their well-being.

NOTES

1. Profnet. 100 North Country Road, Suite C, Setauket, NY 11733–1353. USA.1-800-PROFNET (in the US) or 01-516-941-3736 (from outside the US). E-mail address: profnet@profnet.com. The Internet URL is http://www.profnet.com/whatisprofnet.html.

REFERENCES

Conrad, P. (forthcoming publication), 'Media Images, Genetics and Culture: Potential Impacts on Reporting Scientific Findings in Bioethics', in Hoffmaster, B. and Wylie, A. (eds.), *Bioethics in Context.*

Harris, R. (1997), 'Journal journalism,' *Current Biology*, Vol. 7, R458.

Houn, F., Bober, M., Heuerta, E., Hursting, S., Lemon, S., and Weed, D. (1995), 'The Association between Alcohol and Breast Cancer: Popular Press Coverage of Research,' *American Journal of Public Health*, Vol. 85, pp. 1082–1086.

Miller, J. and Pifer, L. (1996), 'Science and Technology: Public Attitudes and Understanding,' *Science and Engineering Indicators*—1996, National Science Board, U.S. Government Printing Office, Washington, DC, chapter 7.

Nelkin, D. and Lindee, M. S., (1995), *The DNA Mystique: The Gene as a Cultural Icon.* W. H. Freeman: New York.

Petit, C., (1997), 'Secrets of Clichés Uncovered,' National Association of Science Writers mailing list, nasw-talk, http://nasw.org/lists/nasw-talk/hyper/ old/1997/1509.html

20

GENETIC INFORMATION AND "GENETIC IDENTITY"

Richard E. Ashcroft*

University of Bristol, Centre for Ethics in Medicine
73 St Michael's Hill, Bristol BS2 8BH

1. THREE MAIN USES OF GENETIC INFORMATION

What is genetic information for? One way in which genetic information is used is to identify individuals in the sense of picking them out reliably and uniquely. This is the way information is used in so-called genetic finger-printing and in tests of paternity. Another way genetic information is used is in diagnosis. Suppose I have a certain pattern of symptoms which may indicate one disease, but which may indicate something else. Suppose further that the prognoses of the diseases are different, or that the indicated treatments for the two diseases differ, so that it matters that the diagnosis be correct. It may be known that one of the diseases has a genetic basis, while the other does not (or has a different one). Then one could use genetic information in a diagnostic test to rule out one or both diseases. A third use for genetic information is in predictive screening. Does a patient have a gene implicated in a particular disease, such that they are at increased risk of developing that disease (or are certain to do so)? Is a patient a carrier for a gene which means that any children he or she may have with another carrier for the gene are at risk of a certain genetic disease?

When and why is genetic information sought? Genetic information may be sought for identification purposes, as in genetic fingerprinting. In such cases it is taken that each individual's genome is unique to them (with the possible exception of identical twins, triplets and so on). So if human tissues are found at a crime scene, the genetic material they contain can be used to identify who was present at the site if a matching sample can be obtained from an individual or from a database. Genetic information functions in this case as a sort of natural sign. The genetic material points to the identity of a person.[1]

* 0117 928 9843 (phone), 0117 927 9814 (fax)

Genetic Information, edited by Thompson and Chadwick.
Kluwer Academic / Plenum Publishing, New York, 1999.

The second case is the use of genetic information in diagnostic testing. In general that information is only gathered when a person, having fallen ill, defines themselves as a patient and requests diagnosis and treatment. The doctor seeks for diagnostic signs, among which genetic information may be sought as a marker, in order to make a causal judgement. Without other diagnostic signs, the genetic information on its own is unlikely to be reliable.[2] In this situation the usual moral and emotional hazards of diagnosis arise—just as they do when a patient passes from unspecified sickness to being a "cancer patient". Ethically, the specifically genetic nature of the cause may not be relevant, except insofar as the patient theorises their condition historically.[3]

My topic in this paper is the third of the uses of genetic information: its use in screening for conditions in which some identifiable genetic factor plays a causal role. When is information of this kind gathered? There are three main occasions. The first is when a couple are planning to have a baby, and suspect that they may be carrying genetic factors which put the putative child at risk of some disability or disease; this may be because they are themselves sufferers, or because there is a "history" in one or both of their families, or because they already have a child with the condition in question. The second is when the mother is already pregnant, and she (or the father or both) wish, following medical advice, to have the foetus tested for some of the more common or more serious genetic disorders. The third is after the child has been born (at some point in childhood or later), where parents or child want to know whether the child is at risk of expressing a genetic disorder.

2. A DECISION-ANALYTIC MODEL OF GENETIC SCREENING

Consider genetic testing *in utero* (Campbell, 1992). Once the mother becomes aware she is pregnant, typically a range and sequence of tests are made available to her. Some are invasive (maternal blood serum tests, chorionic villus sampling, amniocentesis), and some are not (ultrasound scanning), and there are varying degrees of risk attached to each as well as varying degrees of expense. Assume for simplicity that expense constrains the range of tests available, but that the tests available are free to the mother and not rationed. So the main determinants affecting the choice the mother makes in respect of any given test are the risk attached to the test (e.g., the probability of miscarriage), the desire for the information, and the utility of that information. The latter two determinants are linked but separate, since a mother may not desire information because she will not act on it (if she is against all abortions in principle, for instance), or she may desire it even though she will not act on it (for the sake of preparedness or peace of mind).

According to a simple theory of rationality, this can be described using a decision tree. Consider amniocentesis, and (in this section only) assume for simplicity that we are only interested in Down's syndrome. With a certain probability p, the test will cause miscarriage; so with probability 1-p the foetus will be unharmed and information obtained. Using epidemiological data we can determine the probability this information will indicate that the child has Down's; call this probability q. So with probability 1-q the information will tell us that the child does not have Down's.

Here the mother is faced with two decisions: to test or not to test, and to abort or not to abort, conditional on a positive test for Down's. What should she do? This depends on the utilities she assigns (or we ascribe to her in the light of her behaviour) to the different outcomes. It may be that this is her (actual or perceived) last chance to conceive—perhaps this pregnancy arises out of IVF treatment. So, she may judge that the risk of miscarriage caused by amniocentesis is not worth taking. The negative utility of a miscarriage is in far excess of the utility of any information she may get if she does not suffer a miscarriage. Or it may be that she prefers to be childless rather than have a child with Down's (for whatever reason), so much so that the risk of the test is worth taking.[4]

3. QUALITY-OF-LIFE MEASURES AND SCREENING CHOICES

In this next section and thereafter, consider genetic testing for a wide range of conditions, including Down's, Cystic Fibrosis, Spina Bifida, Huntington's Chorea, Phenylketonuria, and other conditions or traits which may prove to have a genetic basis. The decision procedure discussed will be structurally identical. All of the utilities that get assigned to the different outcomes in this decision procedure in any given case and whatever the condition tested for reflect judgements about "quality of life". This is made complicated by the fact that two (or more) different quality-of-life judgements are made. In the short run time frame, the judgement is effectively about the mother's short run quality of life—as influenced by the utility she will derive from the different information she may receive. In the long run time frame, the judgement may also involve judgements about the mother's quality of life (for instance possible regret and remorse after a miscarriage as a life-defining factor, or the difficulties attendant on raising a sick or disabled child). But more important (for most people) is the judgement about the quality of life of the child as predicted by the possible outcomes.

Disentangling these time frames and the self- and other-regarding reasons is complicated—perhaps impossible—and in practice it may be thought of as no one's business but the parents', except inasmuch as we have an interest in helping them understand how to make a decision in a manner that helps them get into reflective equilibrium regarding their choice. But, there is an interesting question about this model of choice which does have a public character. Can these utilities be objective? At first blush, it seems impossible. But why not set the utilities by using objective quality-of-life data? The rationale for this is clear enough—some conditions we regard as being tolerable, some as intolerable. We can rank these. Many conditions fall into a grey area in which there is disagreement. But the point is not to set binding conditions, but to give a normal standard which parents can choose to depart from, but which helps in setting their "initial values".

There are different ways of approaching this task of setting standard weights to the outcomes of the test. The simplest is to determine weights which we take to measure bulk quality of life. Ranking possible lives in this way is a very crude approach: it takes no account of length of life, or variations in quality over time. So an alternative is to use an approach like the QALY method, which assigns the measure 1 to one year of standard healthy life, and then assigns fractional values to years of ill health or disability (and perhaps values greater than one to years of supranormal functionality?). One can make this quite sophisticated by building in discounting measures to reflect the preferences we may have for optimum health in different phases of life. As well as modelling preferences for health at different phases of life, the QALY method can reflect different "quality curves" that may be associated with different conditions, to reflect the way a condition induces different levels of impairment, loss of function or pain and suffering over time.

The sophistication of this way of deriving weights for different years of life under different conditions is not in question. But is it any more sophisticated in its basic assumptions about quality of life than the coarse method? Not really: we evaluate each individual year using the coarse method (rather than a total life), albeit the comparative method adds finesse. In addition, the apparent standardisation dissolves when we use the most sophisticated versions of the QALY, because we have to incorporate time preferences (which are very personal to each individual), and because what counts as contributing to quality or to relative imperfection goes unanalysed. It simply results from a preference ranking, either individual or social. Most seriously, however, the outcome of all this adjustment, comparison, ranking and preference evaluation is still a brute number assigned to the different outcomes in the outcome space of the decision tree. And what the mother is faced with is a choice of actions regarding tests and regarding abortions (or post-natal therapies).

Aside from the risks and harms introduced by the testing itself, the choice bears on the question what quality of life will my child have? Or, what quality of life can I choose for a child of mine to have? If I have an abortion, could I have another, less "damaged" child? Is no child better than a child severely disabled or ill? Can I inflict this suffering on a child of mine? All of these questions have slightly different moral undertones, and much could be said in analysis of them. But the assumption I want to analyse in this paper is that which this species of consequentialism makes in linking genetic information to quality of life. Being in possession of a certain gene seems, on this account, to cause a certain quality of life (or quality of life profile). This would be a very surprising thing for a gene to cause, since consensus is far from being established about genes causing particular behaviours; and quality of life is, on any account, less "physical" than behaviour.

This account seems in practice to involve a kind of genetic determinism (which can be called practical genetic determinism, in contrast with theoretical genetic determinism with which biology has to do). In the rest of this paper I want to explore the nature of this determinism, to show that it can be blocked as the basis for general policy-making about genetic screening. As I have shown, this model fits quite naturally with a simple kind of consequentialist moral reasoning in decision-making. My argument will not question consequentialism as such, although I think it shows that the scope of the set of "consequences" needs rethinking. The approach I contrast with this type of genetic determinism involves a different style of moral reasoning, which may be of more use in thinking about the choices involved in genetic screening.

4. GENETIC IDENTITY

What seems to be at stake in the decision-analytic model is a theory of how what the future person will be like is based on some defining property of their physical identity. Built into this account of what the future individual will be like is an evaluation of the value of that life to that person. This prediction of the future individual's subjective quality of life is based on two sources of what we could term "evaluative data": objective quality of life data, and judgements about "what makes life worth living"—or not worth living—founded on the predicting individuals' experiences and attitudes to reports of others' experiences. To unpick the genetic determinism I identified above, we should begin by unpicking this theory of future quality of life. While it is possible to simply dispute the QALY theory and its analogues, or to propose opposite evaluations of experiences, this strategy does not offer resolution of the debate, nor does it respect the problems the QALY theory and personal reflection on values are honest attempts to address.

We have seen that genetic information can be used both as an identifier of individuals and as a marker for some medical conditions. From here it is a natural philosophical move to think about genetic information in connection with personal identity. A body of literature exists which takes exactly this tack, relating the possibilities for genetic therapies to metaphysical theories of personal identity (Persson, 1995 and 1997, Elliot, 1997). Another body of literature exists which invokes such theories of personal identity in discussions of refinements of the decision-analytic model I described above, through an approach to weighting future QALYs to weighting attachments and preferences for future states of persons (or future persons *tout court*) (Lockwood, 1994). But it seems to me that these discussions reflect on the wrong sense of "identity". We can talk about identity of persons in the sense of what makes two tokens of a species identical (so, for persons, we look for conditions of mental connectedness or continuity). Yet in ordinary speech, my identity is a loose synonym of my personality, that is what makes me, me. And in discussions of quality of life in the future, for future persons, or for future states of existing persons, it is this sense of identity which counts. To distinguish the theory of "what makes me, me" from theories of metaphysical identity, the philosophers Lycan and Boer developed a theory of "knowing who someone is", which as a special case includes "knowing who I am" (Boër and Lycan, 1986, Morton, 1990). This theory proceeds roughly as follows; for short, call such a theory a KW-theory (knowing-who-theory).

A KW-theory tries to specify salient properties of an individual which allows someone who knows these properties to know who that individual is, in the sense that we ask "Who is the Prime Minister now? -Tony Blair. -Who? -Oh, the one who is always smiling and talking about hard choices—Oh now I know who you mean!" When we consider our own lives, we sometimes are able to identify characteristics and activities that we do and capacities we have, such that we can give a list which gives us a reliable sense of knowing who we are, what is important about us. In some cases, we are able to point to one or more say—without that (or those), I would not be me, or, without that or those my life would no longer be worth living. This is an important step in many bioethics arguments concerning euthanasia, and for a powerful discussion of how powerful this sense of identity and its destruction can be one should consult William May's account of the Dax Cowart case (May, 1991, ch.1). The importance of this here is that arguments with steps like this bear crucially on future quality of life judgements in the decision-analytic model, and on arguments about choosing to be genetically screened when one is aware of a possibility that I may be at risk for Huntington's chorea, for instance.

So suppose that one can make a list of properties, which we can call a P-list. We may differ about what sort of things should go on a list, or whether it could ever be completed, but for a KW-theory all we need is reliability, not certainty. We can use a P-list to explain features of a person's behaviour. And we can use a P-list to make predictions. But most interestingly we can use it counterfactually: there may be, as noted, P-properties which if they were deleted from the P-list, we would have to conclude that the identity of the person was now different. So there *is* a connection with the metaphysical theory of identity, but it is a weak one. If I say, if I could not Φ, then I would not be me any more that might mean, I-before and I-after are really different people (as May argues in Dax's case). Or it might mean that I am no longer attached to my existence and find my life not worth living (as Dax said of himself). Being different people clearly involves physical or psychological differences, which may in turn involve metaphysical difference or non-identity, but this metaphysical non-identity is not the surface topic of discourse. The metaphysics goes on at the wrong level of description and analysis.

When I can consider my P-list, and consider the way deletions or additions to it might change what I can do or how I feel or my ability to enter into certain relationships with others, I can say various things. I might say, I cannot do some (given) activity I prize, and while I do regard my life as worthwhile for me, I would not wish this situation on another person. Or I might say, I am happy in my state or have the ability to become so, and

I would be happy for someone else with the same P-list to come into being. Note that the P-list need not be comprehensive; it need only pick out properties which are "important" in some personal sense. Also, it is not necessarily objective—your P-list for me (i.e., for knowing who I am) need not even overlap with mine, although it probably will.

The challenge of quality of life thinking in genetic screening is to come up with P-lists for individuals which do have some objectivity. Objectivity is important because in pre-natal diagnosis one is choosing for another, or perhaps choosing *which* other will come into being (possibly none).

5. REDUCTIONS OF P-LISTS

P-lists can be quite long and heterogeneous. Perhaps I can reduce my P-list to some other list of properties which explain the properties on the P-list? For instance, I am red-headed and irritable, so perhaps there is a common cause. Constructing reductions for a P-list will involve drawing up another list, call it a Q-list, the elements of which explain the elements of the P-list (Blackburn, 1984, pp. 182–190). There need be no overlap between the two lists. Q-lists typically contain physical properties with causal powers, and perhaps other sorts of disposition-like properties. And where the P-list is drawn up with a view to evaluative importance, the Q-list need not be. For instance, being red-haired is on my P-list. The gene-pair which causes red hair is on my Q-list, but it is not on my P-list, since I don't particularly care how it is that I come to have red hair, but being red-haired is significant to me. In philosophical language, some of the P-properties and Q-properties may be of the same kind, but where they are not, the work of reduction is either causal explanation, or an account of how the P-properties *supervene* on the Q-properties.

The activity of identifying relevant properties and then constructing reductions is scientific in the broadest sense of the term, and in particular we call the activity of making explanatory reductions "theory-construction". So, in discussion of the informal activity of person-definition and evaluation, we have two readily available theories, the genetic theory, and the virtue theory. In most cases these two theories appear to be consistent, but in some situations this consistency is not at all obvious, and so we cannot assume that an inter-theoretic reduction is possible.

6. POSITIVE AND NORMED REDUCTION THEORIES

Theories like these are theories of the evaluatively essential features of individuals. Call them *normed theories*. In contrast, consider *positive theories*. Such a theory could have a similar structure, that is, a P-list of positive identifying properties and a Q-list of explanatory properties, such that the property sets allow the discriminatory identification of individuals. For example, we have positive theories regarding the uniqueness of fingerprints, and of the individual's DNA code.

Normed theories seek to reduce subjectively identifying P-lists. The force of "subjective" here is that the P-list represents properties (or descriptions of properties or property-bundles) under an evaluation: P-lists identify properties which identify me and are also candidates from properties which make my life worth living or not worth living. From such a set we can pick out properties which allow me to say *my* life is not worth living (with or without) it; or that all things considered I prefer to live, but would not choose to live my life over again with these P; or that no one should be obliged to live such a life.

The use of Q-lists is to permit short form identifications (in positive theories) or to explain and predict corresponding P-lists. In particular, it allows recognition that some Ps which, taken apart, may allow us to think that one is undesirable and could usefully be dropped or forestalled, are, taken together, linked through some individual Q or conjunction thereof.

Are positive and normed theories different? Is a positive theory of personal identification possible? A trivial answer is: no, because any theory of identification will always involve some judgement of relevance. This is trivial because while vacuously true at the level of the P-list, it need not be so at the level of the Q-list. For example, a genetic theory has the advantage that whatever we may use as a list of identifying phenotypic properties, the genome itself is a unique chemical-alphabetic string. Relevance is irrelevant to that. It reappears in the practicalities of sampling actual genomes to compare them; but that sort of "cognitive" relevance is not what was intended. So, yes, there are positive theories. At the P-level, the trick is to take the genomic Q-list as the P-list itself. But in practice, are such theories independent of normed theorising? I want to argue that they are not.

7. DYNAMIC AND STATIC REDUCTION THEORIES

In normed theories, we can distinguish *dynamic* and *static* types. *Dynamic* normed theories allow that identity is processual and malleable, so that the contents of the Q- and P-lists can change, as can the relationship between them. The variations in the P-list are constrained by the Q-list and their changes; and we expect some sort of lawlike regularities linking these regularities. Most *moral* normed theories are dynamic in this way: I may be *lonely* now, but this can be changed by exercise of another property or disposition I have, *friendliness*. Some moral normed theories are not dynamic, or less so: my loneliness may be related to my actually being unfriendly, with these characteristics standing in a reinforcing relationship, and I may take it that this unfriendliness is actually a fixed trait. Only in fantasy are moral theories absolutely dynamic or absolutely static; absolutely static moral normed theories seem to rule out responsibility or self-possession, while absolutely dynamic moral normed theories involve various puzzles about the corrigibility of ordered desires and preferences (i.e., could I want whatever I want to want, and so on).

Most genetic normed theories appear not to be dynamic however. Specification of the genome appears to account for expression of the genes in the phenotype, up to environment. The Q-list or genome appears to *regulate* the P-list, to the extent that permanent choices or expectations must be framed—changes in diet to prevent the effects of PKU to be expressed, expectation of relatively early decline in Huntington's chorea, early death and limited mental development in Down's, and so on. These things either will happen, or have a significantly raised risk of happening (which psychologically may be the same thing). Note that "genetic" in this context may refer to the genome itself, but it may equally well refer to exposure to environmental risk while an embryo or in the early years of life—as, for instance, when a foetus is developing in the womb of a mother who smokes.

Dynamic genetic theories can be devised, however. Two are available in the medical or philosophical literature. The first we may call a *genetic therapy* theory, in which the Q-list is replaced, piecewise, by a modified genome Q*. This raises the interesting philosophical puzzle, which many writers have considered, of whether the resultant individual is the same person. In brief: if gene therapy is localised to some organ or part thereof, the problem is no more or less than the problem of whether I am the same person that I was seven or eight years ago when all my cells were different. However, if gene therapy takes place in early foetal development *in utero*, so that the genome in all cells is changed, it may be argued that the *individual* is changed—but, so most argue, the *person* is not, simply because the person does not exist yet. So this kind of dynamic genetic normed theory is at most a special case of the second kind, which we turn to.

The second kind of theory is a *possible people* theory (Hare, 1993). A dynamic genetic positive theory involves making normed selections over the set of possible individuals that could result from my gametes and those of my partner or donor. Of course, comparing such individuals is largely like comparing different, but equal, actual people. Some broad classes of individuals could be ruled out, perhaps, but not on the basis of choosing genomes *alone*. Selection would have to be done on the basis of the associated P-lists, and on the predictive inference from Q-list to P-list. And even this dynamic theory is dynamic only in the restricted sense of variation of the "initial conditions". All the rest is biochemistry and the environment.

Now this makes any genetic normed theory, even if dynamic, into a species of determinism. But it is not determinism of a very fearsome kind, as it stands, because the P-properties it seems to envisage are largely physical properties of individuals, while in moral normed theories we were interested in moral properties, about which the genetic normed theory apparently says nothing. But it conceals a determinism of a more worrying kind, as I will now discuss.

The real world version of this possible people dynamic genetic normed theorising is, in effect, *genetic screening*. Genetic screening is usually aimed at detecting particular genes (i.e., under most genetic positive and normed theories, particular members of the genomic Q-list), which are associated with particular favoured or disfavoured phenotypic traits (members of a P-list). The aim of a given screening test depends on the trait, and on the mechanism connecting it to some conjunction of members of the Q-list, in which conjunction some particular genetic string is thought to be critical. When people talk about genetic information, it is not simply the information that one has a certain string of genetic code (or that one is likely to pass this string of code on), but the information that a certain trait on a P-list is likely to be expressed.

This is where the linkage of genetic positive and normed theories becomes important. In medicine as in life, we are interested in normed theories for the most part. But we use screening information *as if* we refer only to the Q-list, that is, as if we were using only the minimal positive theory in which we take the P- and Q-lists to be the same. In this, we may only be considering the genetic information as a short-form identifier, as in fingerprinting. Note that many critics of what we could call "geneticism" take short form identification to be in itself problematic, because it allows monitoring of individuals through collation of genetic information in databanks. In this regard, we can see how the thought is plausible that genetic positive theories are, while positive, implicated in a social technology of monitoring and control.

8. GENETIC REDUCTION THEORIES AND BIOPOWER

This is a simple form of what Michel Foucault called biopower—the regulation and government of individuals through a twofold process of defining them through biological characteristics and then monitoring them through tracking these characteristics and controlling them (Foucault, 1980, ch.9). Foucault got this idea through consideration of the advent in the nineteenth century of public health initiatives (through which the state intervenes and creates society in defining and monitoring "normal" behaviours, not only "deviant" or "criminal" ones) and police detection techniques including fingerprinting (which arises out of anthropometry). In medicine, genetic screening could be used to contribute to a new public health politics simply by tracking genetic traits systematically. This is already done in a small, local way when a genetic disorder is identified in one individual and then tracked through a family. At present this is done only in a piecemeal way, because the choice to be screened rests with the patients, whose autonomy is deemed final. One can choose not to know, and confidentiality requirements require the information not to be passed on by the screener to others, even if they are "at risk". This situation is under pressure, however, from insurers, who may require such information in order to set premiums, and from some tendencies in public health, that regard communication of risks as more important than individual privacy.

The significance of genetic positive theories rests on two non-positive features: the use to which the information is put, and the theory of reduction invoked to connect phenomenology and genetic information. Is it possible to have an abstract theory of reduction of the P-list to genetic Q-list? Or, put the other way around, does the genome project make sense under the description that it is a theory to completely specify a generic genomic Q-list and derive a generic positive P-list? That is, is there a class of individual properties that we can derive from the generic Q-list which do not depend on normed selection? This is immensely unlikely.

Consider the reduction of chemistry to physics: in principle it can be done (and the principle is all that matters), but in practice we cannot do chemistry by doing physics—in part because we cannot solve the three body problem in classical physics, never mind in quantum physics, so we cannot get an exact theory of the helium atom, never mind the rest of the table. On the other hand, we can, by doing physics, get a reasonable guide to the sorts of chemical entity we might expect, and some their associated phenomenology. Consider also proposed physicalistic reductions of mental to physical phenomena. The reductions propose that we consider the class of mental phenomena and work out what the physical mechanisms must be, or at least, be like. In a "final theory", it may be that we could use the reduction predictively. Critics of such reductions may use the strategy of showing that no such predictive reduction could work, in order to discredit the project of explanatory reduction. But no one else is much interested in the idea, save perhaps artificial intelligence workers. Both these examples suggest that the idea of a pure genetic normed theory is at most a project, not a reality.

Parts of the theory do exist. Successful genetic reductions of phenotypic characteristics are fairly common. But if we consider the sorts of reduction that are proposed, we find different sorts: PKU disease, in which presence of a "wrong" gene means that a certain enzyme is not synthesised, so that a certain amino acid cannot be metabolised, so that it builds up, toxically, and so that another chemical is not made which is required in cortical development. Here we have a step by step mechanism, in which we can be confident of a successful reduction which is both explanatory and predictive. In some breast cancers, which appear to "run in families" a certain gene seems to be implicated in the process, which regulates (or fails to regulate) the "programmed cell death", the failure of which causes the cancer. But the mechanism of this regulation is not well understood, its onset is not immediate (onset of breast cancer can be very young, but it can also be very late in life), and not all women with the gene express the cancer, nor do all women without the gene fail to express it. The association between the cancer and the gene has some explanatory power in some cases, but the predictive power is—as yet—small. Finally, consider traits like homosexuality or aggression. Here some researchers believe that there must be a genetic component to these traits. But the identification of these traits is, to say the least, description-sensitive. The mechanisms that could lead from gene—or interaction of genes—to behavioural expression are very obscure, and likely to remain so. Again, consider reductions in the philosophy of mind. Here, we know what characterises mental phenomena (say, intentionality), and we know what physical system we want to use as a source of explanatory factors. Neither of these features are available for behavioural traits. Reduction in these cases, at the present time, is at best a promissory note, and most commonly an ideological programme (Schüklenk, Stein, Kerin and Byne, 1997).

One strategy for reductionists—which I would commend—would be to try to characterise the types of phenomena which are, in principle, explicable by genetic reduction. We could then decide whether some given trait fell into one of these types, and then the possibility of a genetic positive theory which talked about them would at least make sense. But imagine that homosexuality did fall into one of these types. Two things would then arise. First, we would be considering a topic in which the normed and positive theories had the same structure, and the only difference between them would be the implicit or explicit evaluation of the trait under the normed theory.

Second, because, as discussed earlier, genetic normed and positive theories are effectively static, we would be implying that the expression of the trait would be necessary (even if not immediate). The positive theory makes of the trait something objective (evaluation independent). Hence the subject with the gene can be regarded as Being Homosexual, even if no behaviour is ever expressed. In other words, through a genetic positive theory of behavioural and other traits, the subject is constituted: and the truth of the subject is their list of reducible traits. In discussions of disability, precisely this constitution of the subject is resisted. Am I a deaf person first, with other characteristics later? Or a person, who, among other characteristics they possess, is deaf? Politically and tactically, I may choose the former description. But that does not mean that I wish to be fixed under it. Disabilities can often also be understood as maladaptations of society and its physical arrangements to certain individuals (Rioux, 1997). So even apparently objective physical traits, which look like natural candidates for reduction in a positive theory, must also be understood as candidates for reduction in a normed theory.

But which normed theory? It may be that a positive theory is available that reduces homosexuality. This need not be a genetic theory, but even supposing there is one, this does not entail that the normed theory should be a genetic normed theory too. At present, there is a common belief that if a certain trait has a genetic reduction then it is essentially "genetic". So, in thinking about its importance to actual or future subjects, the appropriate stance is to adopt a genetic normed theory about it.

As I said before, genetic normed and positive theories tend to be static. The reason for this is that genetic theories inherit the model of "coding" instructions which was used to derive many of the properties of DNA in the late 1950s, at the same time as modern computing was getting off the ground (Rose, 1991). In effect, a genetic mechanism is regulated by something rather like an "instruction", albeit one implemented physically and through self-assembly rather than stepwise. The idea of an unambiguous programme which regulates the cells of the body and their development proved useful in research and in understanding the mechanisms at work in disorders like PKU (how much progress in genetic research has actually been provoked by research into disorders?). So if we have a genetic reduction available, we have a cultural pressure on us to think of it as the effective implementation of a programme, which cannot be stopped, short of reprogramming or system shutdown. More complex disorders arise out of the effects of different kinds of programme interacting, as in cancers. Persons with such disorders are subjects with bugs in their programmes. This is not a standpoint which would curry much favour with evolutionary biologists!

So, under a genetic normed theory, we are encouraged to see individuals as constituted by their programming. When we learn something about this programming, that is, when we discover something about our genetic information in the narrow sense, we make the reasonable inference that we have found out something immutable about our P-list under the normed theory. That is, that we have learned something about the constitution of our *selves*, that is, about the set of properties we evaluatively select in thinking about what makes me, me, and in determining what makes life worth living (or not) for me now, or for me "over again", or for individuals (actual or possible) like me.

Moreover, the static genetic normed theory, because of its "necessitarian" features does seem to carry over into decision-making about the life worth living in general. Most normed theories do not easily allow us to judge that if some P makes my life intolerable, then it would make yours intolerable too. It can do in some cases, perhaps, and this would be true in certain genetic disorders where the mechanism is well-understood, predictively reliable and fatal. But in most cases these conditions are not met. And the conclusion that we are obliged, morally, to find out about our basic properties (under the genetic theory) and to tell others (insurers, the state, lovers) does not follow.

9. CONCLUSION

So genetic normed theories have a moral property: fatalism. This is not meant loosely. Genetic normed theories imply that individuals have a genetic fate, over which they have little control. If I have such a "genetic" disorder, I can do little about it, save getting used to the idea of my essential imperfection and early decline. In itself this may be wise; but not on its own. We need, in any chronic illness or disability, to come to terms with it and remake our lives so that they become (or remain) worthwhile. This involves living, not as externally defined subjects, but as reflexively and morally dignified selves. Genetic positive theories are useful in gaining philosophical self-knowledge: but genetic normed theories are counsels of despair. In those cases, and only those cases, where a genetic reduction is possible, we may make use of the information—so long as it does not make us. In fine, contrary to usual reductionist expectation, we cannot reduce moral normed theories to genetic positive theories; we should seek to combine them.

In concluding, let us return to the quality of life judgement that can be made in prenatal screening (choosing for a potential (?) other) or in adult screening (choosing for oneself). In the adult case, the lesson is not to assume that in your genes lies your fate. Your future life has some features which are genetically given: but the quality of your life still remains under your control to the extent that you can choose your reduction theory, your Q-list, and choose your P-list, that is, which things you take to be important about yourself. Morally, if your context is given, how you act in that context is what matters. For the prenatal situation, we have seen how to block the collapse of options into necessities, by showing how the link between gene and the normative content of quality of life data can be cut. That link is contingent on the adoption of a static, normed, genetic reduction theory, which we have shown to be only one way of considering the moral and human dimensions of the choice.[5]

NOTES

1. It may be possible to interpret the genetic information as a proper name—a "rigid designator" in Kripke's terminology. Some of the ethical debate about the use of genetic information in criminal investigations (especially in the maintaining of databases) could perhaps be seen as exploring our anxieties about the revelation of our "true names", the secret name which identifies the truth about us, and the possession of which gives power over us. Whether or not my genome is a name in Kripkean sense, the idea that it tells some—most—of the important truths about me is what I shall dispute in this paper. (Kripke, 1984)
2. Genetic information has in this instance the character of a sign, with what Paul Grice called a "natural meaning" (as smoke is a natural sign of fire). Like any other kind of sign, its interpretation is difficult without some context—other signs are needed to give a context, as is an account of how the signification works. (Grice, 1957)
3. For instance, in adjusting their perception of their "normal" state from "being healthy" to "being sick", and in posing the question "why me?". The moral significance of genetic disease enters into this fashioning of a personal "sick role". As well as gathering information in the case where the genetic factor in a disease is known, such information may be sought in a case where the diagnosis of a disease is known, but it is not known whether a genetic factor is responsible in general. Information may be gathered for epidemiological research.
4. Here I am simplifying matters greatly by ignoring the possibility of errors in testing (false positives and negatives) and by ignoring the wealth of psychological studies showing that our understanding of risk and probability is very haphazard (Gärdenfors and Sahlin, 1988). I take it that we can build this understanding into the utility assignments. In other words, I propose that an apparent misunderstanding of probabilities expressed in a judgement we might make can be taken to reflect judgements about the relative utilities of the outcomes in the "outcome space" (instead of judgements about the probabilities in the "probability space"). The internal paradoxes of rationality cannot be removed in this way.
5. I am very grateful for constructive criticism from the following: Alastair Campbell, Serafina Cuomo, Jane Hutton, Adam Morton, and Paul McNeill. The usual disclaimer applies. Earlier versions of this paper were read to the Centre for Ethics in Medicine, and to the Department of Philosophy, both University of Bristol.

REFERENCES

Blackburn, S. (1984) *Spreading the Word: Groundings in the Philosophy of Language* Oxford University Press: Oxford

Boër, S.E. and Lycan, W.G. (1986) *Knowing Who* MIT Press: Cambridge, Massachusetts

Campbell, A.V. (1992), 'Ethical and legal issues' in Elwood, J.M., Little, J., Elwood, J.H. (eds.) *Epidemiology and Control of Neural Tube Defects* Oxford University Press: Oxford

Elliot, R. (1997) 'Genetic therapy, person-regarding reasons and the determination of identity' *Bioethics*, Vol. 11 No.2

Foucault, M. (1980) *Power/Knowledge* Harvester: Brighton

Gärdenfors, P. and Sahlin, N.-E. (eds.) (1988) *Decision, Probability and Utility: Selected Readings* Cambridge University Press: Cambridge

Grice, H.P. (1957) 'Meaning', *Philosophical Review*, Vol.lxvi No.3

Hare, R.M. (1993) 'Possible people' in his *Essays on Bioethics* Oxford University Press; Oxford

Kripke, S. (1984) *Naming and Necessity*, Blackwell: Oxford

Lockwood, M. (1994) 'Identity matters' in Fulford, K.W.M., Gillett, G., Soskice, J.M. (eds.) *Medicine and Moral Reasoning* Cambridge University Press; Cambridge

May, W.F. (1991) *The Patient's Ordeal* Indiana University Press: Bloomington, Indiana

Persson, I. (1995) 'Genetic therapy, identity and the person-regarding reasons', *Bioethics*, Vol. 9 No.1

Persson, I (1997) 'Genetic therapy, person-regarding reasons and the determination of identity—a reply to Robert Elliot', *Bioethics*, Vol. 11 no.2

Schüklenk, U., Stein, E., Kerin, J., Byne, W. (1997) 'The ethics of genetic research on sexual orientation', *Hastings Center Report*, Vol. 27 No.4

21

GENETIC INFORMATION AND KNOWING WHEN YOU WILL DIE

Margaret P. Battin

Department of Philosophy
University of Utah
Salt Lake City, Utah 84112

1. THE SIZE OF OUR LIVES

Most of us, whether we like it or not, already have some rough idea of the likely size and shape of our lives—how long we will live, in what state of health, and what we will eventually die of. We live in an age in which life expectancy patterns for populations and subgroups of populations are known and predictable. We know the frequencies of the principal causes of death and the ages at which they are most likely to prove fatal. We know accident rates, risk ratios, and predictive factors for high-risk behaviors. Most important, we know that the age and cause of death of our own parents is the best predictor of our own mortality. Of course, we cannot for the most part tell if we as individuals will actually contract a specific disease, or become the victim of an accident, or succumb to some other cause of death. Furthermore, we often deceive ourselves about the ways in which our own health maintenance habits (or lack thereof) influence our expectable lifespan. Nevertheless, in general, in an extremely rough, often inchoate, not fully recognized way, we have a sense of what to expect about our own deaths and the periods of morbidity that may precede them: what is likely to happen to us, more or less, and about when, at what age, and for what reasons it will occur. If our ancestors all lived into their 90's and died of "old age," that is, of conditions that occur primarily at very advanced ages, we have a pretty good chance of doing so too; if they died of heart attacks in their 50's or cancer in their 60's, our anxieties mount when we reach these ages.

Many factors—improved actuarial computation techniques, better recordkeeping of mortality and cause-of-death statistics, better inter- and intra-population comparison data, better prediction of the emergence of new pathogens like viruses and flu strains, and so on, are likely to contribute to a change in this picture. Most of them are likely to make the picture sharper and clearer. But the biggest factor in changing this picture will be the increased possibilities of genetic analysis, coupled with clinical and epidemiological data

Genetic Information, edited by Thompson and Chadwick.
Kluwer Academic / Plenum Publishing, New York, 1999.

that make possible both population-wide and individual prognostication. As it becomes increasingly informative to trace an individual's genetic legacy and thus to identify inherited disorders and diseases, physiological characteristics, and disease susceptibilities, it will become increasingly possible to make more and more accurate predictions about how long a person is likely to live, in what condition and with what degree of function, and when and how that individual is likely to die. Of course, as the possibility of prediction increases so will the possibility of treatment, but people will still eventually die—in more predictable, forseeable ways.

This is the matter I want to explore. While I think human awareness of eventual illness and death will involve gradual change, barely perceptible to most individuals, I also think we must recognize that this is a process of change already well underway. This change will constitute a substantial transition from the present, and a huge departure from the past. What lies at the center of this change, I want to show, is the increasingly informative possibility of genetic prognostication (something of which we are already partly aware) about the size—the length, health characteristics, and cause of demise—of an individual's life.

I'm not the only one who wants to explore this possibility; Hollywood does too. Just the issue I want to explore has been very alarmingly raised—though I think trivialized—in the science-fiction film *GATTACA,* released in 1997.[1] I take the box-office success of this film as a symptom that the general public may be sensitive to this issue too, at least at a superficial level, and I even imagine that this sensitivity may itself be a sign of the changes I want to describe.

2. DEATH IN THE FUTURE: A CONJECTURE

Let us, then, jump to the future—the fairly near future, a world we already see on the horizon, not so awfully far away—maybe 20 years, maybe 50, maybe 100 years away at most. It will be a world, we may foresee, in which the human genome has been fully mapped, comprehensive correlation with clinical and epidemiological data has been carried out, and in which genetic diagnosis and treatment have become routine, largely non-controversial parts of the medical armamentarium. It will be a world in which familial patterns of disease are readily identifiable, and outcomes of both untreated and treated disease are known. It will be a world in which patterns of susceptibility to workplace toxins, environmental influences, ambient parasites, bacteria, and viruses, and other external factors, wherever these are controlled or influenced genetically, are also understood. In this not-so-distant future world, for example, it is possible to predict which miners are most likely to get black lung disease, which taxi drivers are most vulnerable to air pollution in urban environments, which women will probably suffer toxemia in pregnancy, who will get the flu, and which smokers will die early deaths. When new strains of viruses or new mutuations in human genetic material appear, enhanced capabilities for detection and tracking will make it possible to understand these phenomena and their probable consequences for specific individuals quickly. Of course, all these things will involve an enormous amount of data collection and correlation with the findings of genetic research, but this will be the business of medicine, carried out with the aid of such specialized research fields as population genetics, epidemiology, and other areas of applied research. We might call it the complete "gene scan," routinely and easily made available for any individual human being.

Data too will be available for accidental risks, disaggregated finely over very small subgroups. While it will never be possible to anticipate which accidents will befall exactly what individuals, risk ratios for various groups, we may assume, will be known. We already know, for example, what subgroup is most likely to suffer fatal ski injuries (males ages 17–24), how often epileptics have fatal automobile accidents (very seldom, compared to adolescent males), how often condom breakage during sexual intercourse leads to the transmission of HIV. If genetic information becomes available about behavioral traits associated with, for instance, various forms of risk-taking behavior, including sports and adventure risks, sexual risks, risks in impulsive and aggressive behaviors, this picture may also grow sharper and clearer.

Thus the future picture I want to explore is one in which many factors, but especially developments in genetics, make it possible to prognosticate far more accurately about the cause and timing of an individual's death. What I want to ask is whether this will be a bad thing—yet another feature of some brave new world (like *GATTACA*) in which privacies are invaded and human meaning undermined—or whether it will be a good thing, still another product of advances in science which improve and enhance the human condition. I don't think this is an easy question to answer, and what I can say here will be but the barest sketch. But now let us return to the present, partway between the past and the future we are here to explore.

3. THE COMPLEXITIES OF GENETIC PROGNOSTICATION

Some genetic prognostication about the size of life, though comparatively crude, is already possible. The child diagnosed with Tay-Sachs disease, for example, can be predicted to live only a couple of years—to age three or four, at the most. At the opposite extreme, the person whose parents died of Alzheimer's disease with onset in the 80's can with some degree of likelihood expect a similar demise at about a similar age, at least if his or her genetic makeup includes the relevant gene or genes from one—but only one—of these parents; if he or she has the gene or genes from both parents the likelihood is of onset and death at much earlier ages. Here, the likelihood which is genetically predictable includes both contracting Alzheimer's at that late age and not contracting any other fatal disease at an earlier age: if the parents did not, the offspring is not likely to do so either. People testing positive for Huntington's can, on the whole, expect the onset of symptoms in their mid-30's to mid-40's, with death about 10–15 years later. Women with mutations in the BRCA1 gene associated with breast cancer have a 50/50 chance of developing the disease by age 50, and an 85% chance by age 70, though some women never contract the disease at all. With familial polyposis, colon cancer is most commonly diagnosed in the 30's and 40's. Inherited cancers tend to be of earlier onset than those associated with sporadic mutations and those in which there is no known genetic abnormality at all.

With this handful of exceptions, genetic prognostication about the size of life remains at the moment enormously crude. Furthermore, these exceptions are in themselves still quite crude. Many genetic conditions are of differing penetrance, expressed in severe symptoms in one individual but mild or no symptoms in another. It is often difficult to distinguish between early-onset versions of a disorder and late-onset forms, and whether, indeed, they are different forms at all. The predictive value of genetic information often depends on information about other affected family members, information which is not always available. The activity of some genes may counteract that of others, as do oncogenes and tumor-suppression genes, and there may be other as-yet-undiscovered patterns. We have only a rudimentary understanding of "two-hit" conditions, in which a genetic disposition results in disease only in the presence of specific environmental insults. We do not fully understand the function of genes which, in heterozygotes, protect against some conditions, like malaria, but in homozygotes, cause others, like sickle cell anemia. Treatment may already substantially affect the outcomes of some genetic diseases, so that there are fewer cases available for observing its natural history. And, of course, our present capacities for genetic prognostication are limited by the fact that the human genome is not yet fully mapped: not only have we not yet identified the genes involved in all single-gene mutation disorders, but disorders involving several interacting genes, multifactorial disorders exacerbated by or associated with environmental factors, behavioral and personality traits, and many other factors have not yet been discovered.

But, let us assume, they will be. That is what a *complete* mapping of the human genome, coupled with comprehensive medical and epidemiologic study, would provide: this is the complete "gene scan." We cannot of course at the present moment know how far advances in genetic science, together with developments in medicine and epidemiology, will take us, or at what point we will know, so to speak, everything there is to know about the human genome and its implications for human health—or, indeed, whether we will ever reach that point. But we will certainly know a *great* deal more in the near future than we do now, and it is crucial to anticipate what these changes in scientific and clinical knowledge will contribute to changes in human experience.

4. HOW IT USED TO BE

At all periods of human history up until the middle of the previous century, the length and character of the end stages of an individual human life were far less predictable—indeed, virtually unpredictable—except perhaps for someone born in already compromised health and with obvious deficits: then, clearly, life would be short, though some individuals born apparently healthy would also live short lives while others much longer ones. In these earlier periods of human history, life expectancy fluctuated between about 20 and 40, depending primarily on the occurrence of famines and epidemics of infectious disease, particularly evident in high rates of infant mortality. These were interrelated: famine resulted in malnutrition and hence increased vulnerability of the individual to infectious and parasitic disease; epidemics of infectious and parasitic disease affected food supplies, including crops and animals; and epidemics affecting crops and animal food supplies, together with variations in the weather, increased the chance of famine for humans. The specific occupations or roles of individuals might enhance their risks: soldiers suffered an increased risk of death due to infected battle wounds (a bigger killer in most wars than outright trauma), and women in childbirth were far more vulnerable to generalized sepsis. To some degree, the aristocracy in nearly every society was protected from some of the risks the peasantry faced, at least by virtue of better nutrition, better sanitation, and freedom from the risks of physical labor, but the aristocracy were not entirely protected: they too succumbed to infectious and parasitic disease. In all these previous eras of human history, death could occur at any time, largely without much warning, and largely without the physician being able to do much to prevent it. While it might have been possible to predict life expectancies for population groups and subgroups in various circumstances (though no societies had adequate datakeeping methods for doing so), there was little one could predict about the fate of a given individual. If a person got an infection, he or she might or might not survive; if the plague was raging, some individuals but not others who were exposed would live through it. There was little or no way to predict which ones. Similarly, some soldiers but not others would die of seemingly equally infected battle wounds; some women but not others would become septic in childbirth. Beyond apprais-ing such matters as nutritional status, age, and the presence or absence of other concurrent diseases, it would be hard to predict of any individual with any potentially but not uni-formly fatal disease or condition just what the outcome would be, survival or death.

Some of this mortality, we may assume, was associated with heritable diseases. Much more of it, we may also assume, was associated with varying degrees of susceptibil-ity to endemic bacteria, toxins, viruses, and parasites, and these varying degrees of suscep-tibility, we may suspect, will also have been at least to some degree heritable patterns. But while many of these societies recognized notions of family traits—similarities of facial appearance and physical characteristics within a lineage—and also had extensive empiri-cal experience in the practice of animal husbandry, they had no well-developed notion of genetic science. Heritable conditions and diseases in humans could be recognized as "run-ning in the family" only after they became symptomatic; and the notion of a genetic pat-tern fixed at conception was completely unknown. Adopted children and people whose parents had been killed in accidents or war early in life had little way of knowing their own family health traits at all.

Even with a conception of family traits, people in societies prior to the development of genetic science would be prone to various sorts of error in predicting the likely course of their own lives. Each individual has two genetic parents, four genetic grandparents, eight genetic greatgrandparents, typically with different disease and mortality histories. Without the mechanisms of genetic diagnosis, that individual has little reliable way of tell-ing which "family traits," from which side of the family, he or she receives. Of course, there is often an enormous amount of speculation within a family about which ancestors an offspring "takes after" that might provide the basis for speculation about the size of that offspring's life, but this of course would trade on the unfounded assumption that a child with resemblances to a given parent in facial appearance or other externally observable characteristics will therefore inherit the same health traits as that ancestor, including as yet asymptomatic diseases and disease suceptibilities. This is a form of rudimentary genetic prognostication, but it is not very reliable at all.

But that picture is changing, for two principal reasons. First, beginning around the middle of the 19th century, with the development of the germ theory of disease and associ-ated improvements in public sanitation, antisepsis, immunization, and, in the 20th century, the development of antibiotics and other resources for controlling infectious and parasitic disease both in humans and in their food sources, the principal causes of death have changed. Now, in the developed world, most people die of degenerative diseases—heart disease, cancer, various types of organ failure, neurological diseases, etc., which are not so much the product of external causes like bacteria, viruses, and parasites, but of internal breakdown. To be sure, some of this breakdown is associated with or caused by external factors, as liver disease is with alcohol consumption, or lung disease with smoking and en-vironmental conditions. Some we're not sure about, for example, the possible associations of Alzheimer's Disease with a virus. But by and large, the contemporary causes of death are different in kind from those of our earlier world. To put it most simply, in earlier peri-ods of human history, most people were killed by germs and worms; now most people in the developed world fall apart from malfunctioning of their own systems.

In contrast to the infectious and parasitic diseases of which people typically died in all earlier ages of human history, the diseases of which we die now tend to have onsets at comparatively predictable ages and exhibit comparatively predictable courses. With germs and worms, you could never really predict when they would kill you, though certain situations—agricultural labor, military service, childbirth—provided much greater invitation. In contrast, the principal contemporary causes of death in the developed world are degenerative diseases—heart and circulatory disease, cancer, organ system disease, neurological diseases—are far more predictable: once a "terminal" illness is underway, we have some idea about the way it will go and about the duration of the downhill course. Furthermore, this is true for most contemporary dying in the developed world: heart disease and cancer, both clusters of degenerative conditions with comparatively predictable characteristic courses, account for about two-thirds of deaths. Only a minority of deaths in the modern world are completely unforseen and unpredictable.

That picture is also changing because of our greater alertness to the importance of genetic informaton. As Bob Cook-Deegan suggests (personal communication), we are probably the first generation to care about what happens to our parents not just from affection and filial loyalty but in part because we realize in an informed way that it is predictive of our own personal futures. The greater predictability of degenerative disease over infectious disease is further enhanced as genetic diagnosis makes it increasingly possible to detect heritable versions of these conditions and susceptibilities to disease. For the first time in human history, we now sometimes reliably know what to expect even in advance of the appearance of symptoms. Thus, not only do we die of more predictable diseases at more predictable ages than did human beings in the past, but genetic diagnosis also makes it possible in some cases to say which disease specific individuals are likely to die of, and thus, what the timing and conditions of their deaths are likely to be. This is not, of course, guaranteed; but it is *likely*. In short, we already know far more about the likely circumstances of the ends of our own lives than have human beings at any earlier age. This, I believe, has already subtly begun to change the way we see death.

5. THE COMPLEXITIES OF GENETIC PROGNOSTICATION, CONTINUED

But although we now know far more about the probabilities in the character and timing of our own deaths than have human beings in the past, we must not oversimplify this claim. There is an enormous amount we do not and cannot yet know.

Take, for instance, what we can now predict about the lifespan of someone who tests positive for the Huntington's gene mutation. In the first place, we must admit that this claim itself is not certain; there is a small chance—though less than 1%—of lab error. Second, we cannot say for certain in a more general metaphysical sense that this person will *ever* develop the symptoms of Huntington's disease; there is always the possibility that life will be interrupted in some other way, say by an unrelated fatal disease, accident, or other cause of death, even though this individual would have contracted HD if he or she had remained alive.

Even the predictions we do make are rough at best. Although we can predict, for the person testing positive for the genetic mutation associated with Huntington's disease, an onset of symptoms somewhere between the ages of 35–45, a subsequent life expectancy somewhere around 12–15 years, a course that declines into pronounced disability due to chronic progressive chorea and dementia, and death of secondary infection, aspiration, or heart failure, none of these predictions are exact. For one thing, they are all based on statistical distributions. For example, although in the bulk of cases, the onset of symptoms occurs between 35–45 (on the earlier side if the affected parent is male, on the later side if the affected parent is female), there is a bell-shaped tail extending away from the mode on either side, with a few cases far earlier—children as young as 2, and some later, at 50, 60 and even a few at 80. Life expectancy after onset is also a statistical claim: although average life expectancy is about 12–15 years, some people die of Huntington's (not other causes) within a few years of the first appearance of symptoms, and some live for quite a number of years, especially in late-onset families; some are relatively mildly affected and only become symptomatic at later ages, increasing the likelihood that other factors will be the cause of death. Typical life expectancy for HD children is shorter, about 8–12 years, but this too is a bell-shaped modal prediction with tails on both sides. The nature of the symptoms is also variable: although for most people with HD the chorea or tremor is the most pronounced physical symptom, in some individuals the disease involves progressive rigidity, not choreic movement. And death can occur in a variety of ways, not just those most frequent forms listed above.

To be sure, some of this variation itself may be predictable. For example, although some Huntington's families are comparatively late-onset families and others early-onset ones, and although some families exhibit lesser, others greater amounts of variation within the family, this variability is apparently associated with expanded repeat size of the mutated gene pattern on the chromosome. While variability remains wide, where there is a CAG repeat size of more than 60, for example, the disease tends to have become symptomatic before age 40, while individuals with lower numbers of repeats tend to have onsets at later ages.

In short, just having the Huntington's gene doesn't entail that a person will develop specific symptoms at a certain age and die a fixed number of years later. Nevertheless, HD is one of the best current examples of a heritable condition in which the identification of a gene mutation makes it possible to say—not precisely, but roughly, very roughly—how long a person will live and what he or she will die of—that is, what the approximate size of his or her life will be.

This picture is far from exact. While Huntington's is currently one of the most predictable known genetic diseases, the predictions now possible are so far from exact that most HD testing centers caution against using data to make specific predictions about the life expectancy of a given individual at all. Nevertheless, even if this prognostic picture is far from exact, it is far more specific than the kind of prognostic picture an individual could have made prior to 1983, the year the markers were identified and linkage data became available. Before 1983, the best guess you could make, if you had a parent with HD, was that you had a 50/50 chance of having HD yourself and a slightly greater chance of earlier onset if that parent was your father rather than your mother; but if you did not have an HD parent, you had virtually no chance of developing HD. That was *all* you could know. And this prognostic picture of just a little more than a decade ago was in turn already far more specific than what was available in the more remote past, before it was known that Huntington's involves an autosomal dominant gene. We can speculate about how an autosomal dominant disease like Huntington's might have appeared to a prescientific, pre-Mendelian tribe (though the disease was not described until 1872): a scourge that afflicts some of the children of a person afflicted by it—perhaps the manifestation of divine revenge for wrong behavior or a curse upon a particular lineage—but skips other children of the same parent. With just this information, the children of an affected parent might themselves wonder whether they would get it, but would have no explanation for why it might be themselves and not their siblings. To a group that knew nothing of genetics, autosomal recessive conditions would surely have appeared even more erratic, more mysterious in their manifestations, explanable only as a curse or scourge the gods could inflict on anyone at any time, but without much warning at all. Even now, people symptomatic for recessive conditions often protest, "but it never showed up in my family before."

Genetic prognostication, even in its inexact present, must take into account other factors as well: for example, the changing risks as avoidance of conception, selective abortion, and somatic and/or germline therapy are used. For example, another of the currently best predictive examples is Tay-Sachs disease, a heritable condition in which symptoms lead to death within a fairly narrow range of ages: a Tay-Sachs child typically dies by 3 or 4. However, due to widespread, well-accepted screening, Tay-Sachs is very rapidly disappearing as a disease among the population formerly at highest risk, Ashkenazi Jews, though the prevalence of the genetic mutation is not reduced by parental choices to abort or not to reproduce at all, since their affected children would in any case not themselves have reproduced. While it is not likely that this disease will disappear altogether from human experience, its prevalence is already dropping dramatically, and with it, one of the conditions in which the predictability of the size of life was highest.

Improvements in treatment will influence predictability as well. For example, variability of outcome appears to be increasing among patients with cystic fibrosis. Not long ago virtually all CF victims died by the time they were teenagers; now, many live longer lifespans. To some degree, this may reflect greater prevalence of mutations associatied with milder forms of the disease, but it certainly also is a function of substantial gains in effective treatment. Similarly, untreated hemochromatosis characteristically produces organ deterioration beginning in the 30's and 40's and death in the early 50's in males (though iron overload is not typically a problem in females until after menopause), but with early, presymptomatic treatment, life expectancy rises to normal. Predictability may seem to be reduced by this greater spread of possible outcomes for diseases like cystic fibrosis and hemochromatosis, though what we are actually seeing is in part the spread in outcome among untreated, routinely treated, and expertly treated disease: cases in each group would be increasingly predictable.

In the long run, advances in genetic diagnosis, together with advances in treatment of genetic diseases and disorders, may work to make the size of life more nearly uniform for all individuals. Heritable conditions involving very young onset age and uniform lethality, like Tay-Sachs, anencephaly, and some of the trisomies will, I think, be virtually eliminated as diseases altogether, as screening programs and risk analysis dissuade parents from conceiving infants who will with certainty be affected in these ways. Improved therapy will delay the onset, increase the survival rate, and in other ways increase life expectancy for those with later-onset or later-mortality conditions. Of course, we cannot know whether advances in genetic diagnosis and therapy will also work to increase the life expectancy of the oldest old, so that human lifespans will still vary substantially in size, or whether the length of life—increased for those who would have died earlier of heritable conditions—will approach and eventually equal those who now die at the oldest ages. We also cannot know how our healthcare institutions will change, and what their degree of commitment will be to treatment of these conditions. But whether or not individuals' lifespans continue to grow more nearly equal (as they already have in the industrialized nations), the predictability of them, whether short or long, is certain to increase.

6. THE FUTURE: MORE PREDICTABLE DEATH

Will it be a good thing to inhabit a future world in which, routinely, we know our complete genetic makeup and hence—as is my specific conjecture here—have a far more concrete idea of the likely size and shape of our lives? We will never know in advance a specific date of death, of course, but we may well know when it is likely to occur within a far narrower range. Will this be a harm, either to specific individuals, to their family members and social groupings, or to society as a whole? And will it be a loss both of human innocence and of something central in human experience, something at the root of deepest human meaning? Or will the future mean relief from uncertainty, a waning of background anxiety over death, or the end of chronic, though suppressed, terror over impending doom? If we are now partway between a past in which human beings had little way of guessing when they would die and a future in which human beings will routinely almost all know more or less what to expect, is this progress in human experience, or a loss?

A science-fiction social-issues film like *GATTACA* tries to answer just this question, but does it in a perfectly predictable way. Like many other such films, including for example *Clockwork Orange* and *Soylent Green, GATTACA* works on a tacit "slippery-slope" model, taking a possible future technological change (whether sophisticated operant conditioning or genetic diagnosis and prenatal genetic engineering) and positing extreme negative social consequences that would ensue—usually, supression of individual liberty, the development of intrusive social constraints, and tyranny by the technological elite. Such films' plots work by focusing on one or two innocents who resist such a society, and in the process, typically, fall in love. What is glossed over in such films are the questions about why such a society might develop, and what other consequences might be forseen for the particular technological changes imagined. Hollywood needs challenge, threat, crisis, and the small forces of good against a huge, entrenched evil to make a powerful plot; but philosophers may want to look at the issues in a more open and informed way.

Would the world we can forsee be a better one, or not? This is a complex question, with a number of interrelated issues to be explored.

Some initial objections might be entered to assembling information about a person's overall genetic makeup or to divulging it to people in a routine way. For example, it might be argued that divulging such information to people would be undesirable because it would alter their risk-taking and health-maintenance behavior: believing they will live to 90, they will not protect themselves in their 20's. But such behavior would involve a logical error: that someone expects to die of Alzheimer's disease in her 90's does not mean that she cannot die of a skydiving accident in her 20's. On the other hand, at least some health-maintenance behavior would, presumably, be altered favorably to try to arrest the course of a known though not yet symptomatic disease: people testing positive for hemochromatosis, for example, would refrain from eating iron-rich foods.

It might also be argued that routine assembling and divulging of such information would undercut people's insurability. Certainly, information about genetic risk can currently constitute a real liability if it is disclosed to a health insurer. But in a world in which assessment of individuals' entire genetic makeup is routine—perhaps routinely disclosed, for example, to schools, employers, the military, the justice system—insurance as we now know it cannot exist, since both the prospective insureds and the insurers would have the same information, and cream-skimming—selective coverage of people with better health risks, but denial of coverage to those with worse risks—would be impossible. People with long expected lifespans and low expected morbidity would not need much health insurance, except coverage for accidents, though insurers would be willing to sell it to them; people with short expected lifespans and/or long periods of expected morbidity would seek health insurance but find it impossible to purchase at any reasonable fee. Thus, the entire system of health insurance would be undercut. Presumably, health insurance would be replaced by some form of universal-access program of care, and life insurance, trading as it does on risk, would cease to exist at all, to be replaced by individual savings or perhaps welfare programs for surviving dependents of those with shorter lives.

Greater predictability of medical and behavioral characteristics of individuals will raise many other questions as well. Workplace exclusion policies are already heavily debated, both publicly and in the courts: should employers be able to exclude from jobs workers who exhibit genetically demonstrable higher susceptibility to diseases associated with various workplace toxins and environmental factors? May mining companies refuse to hire miners at greater risk for black lung disease, or taxi licensure boards refuse medallions to drivers more vulnerable to air pollution? May the school system or the military or the criminal justice system track individuals on the basis of genetically-based behavioral predictions? Or suppose musical talent turns out to be genetically based: minimal endowment in this area may provide a reason not to save a child a place at the conservatory, but would it be sufficient reason to exclude him or her from the high school band? And there is *GATTACA's* practical question: may people with greater risk of heart conditions be excluded from eligibility for a space-flight program?

These questions and objections aside, the real issue concerns what sorts of changes might occur in the human experiential condition. How will we feel about knowing so nearly the likely character of our final illnesses and the time of our dying? Of course, not all deaths are due to heritable diseases, disease susceptibilities, or causes influenced by inherited or sporadic genetic mutations: some people will still be hit by lightning or be run over by trucks. But except for unpredicted trauma, including some but not all homicide and suicide (some of which may be associated with genetically based behavioral traits, like aggressiveness), we may conjecture that genetic factors control or at least influence most deaths. This includes both deaths from infectious and parasitic disease and also deaths from degenerative disease, but since degenerative disease is itself more predictable in time of onset and in its terminal course once begun, predictability is far greater here. Thus, newly available genetic information, together with advances in background epidemiology and other data-correlating sciences, will give rise to a dramatically new prognosticative picture. We cannot be sure of the details; but we can be sure that huge changes are coming, and indeed are already underway.

How will we feel about the far greater predictability this information will bring? Predicting how people will respond and behave in such a world is, of course, sheer conjecture; but it is not entirely uninformed. Here are some things to consider about how people now react to genetic diagnoses that may suggest how people will respond to new capacities for prognostication in the future. For one thing, as one genetic counselor currently puts it, people measure their plight in terms of everybody else.[2] If a person believes (albeit falsely) that most people live to the average life expectancy, now nearly 73, but discovers that because of a genetic disorder he is likely to die at 30, he experiences this as deprivation compared to everybody else—he "loses" more than forty years. Furthermore, it seems to him that it is only he who has discovered that his life will be short; other people do not have such specific knowledge about when or of what they are likely to die. He is doomed, it seems, and they are not. He also loses the capacity they have to believe they will live long lives, even if it is not likely to turn out to be true.

How people react to the news about a future disability or death also depends on how everybody else reacts, or how that a person believes others will react: if they expect others to react to the news about genetic disease with horror, disgust, rejection, accusations of taint, and so on, that person's own reaction is likely to be that much worse. Furthermore, many people (and their doctors) do not process risk information very well: what does 1:200 mean? 1:5000? 1:2?, if you still cannot tell what this means for sure about your own specific case? Some people avoid genetic testing because, they fear, this would take away hope (they sometimes forget that the test result could be negative); others, even when they have a positive test result, continue to hope—not that they will not develop the disease, but that they will have a very-late-onset case, or a mild case, or that a cure is developed. It may also be that people seem primarily concerned with potential symptoms, disability, and the experiences of other affected family members when they are confronted with the possibility of genetic disorders, but these may also mask concerns about illness as the process which leads to death. We see all these responses to what is regarded as "premature" death; we also see them even for many individuals who live well past the average life expectancy, since even the latter deaths seem unpredictable and often, thus, somehow unfair when they do occur.

7. AN IMPOSSIBLE FUTURE: KNOWING THE DATE OF ONE'S DEATH

Check one box:

☐ I'd like to know the exact date, time, and cause of my death now.
☐ I would not like to know the date, time, or cause of my death in advance, at least not now.

Barring astonishing developments in clairvoyance, human awareness of death—regardless of the sophistication of genetic and other prognostication—will never be so good as to be able to predict the precise date, time, and cause of an individual's death. But what people say in response to the forced-choice question above provides some indication of how they might respond to substantially increased possibilities of genetic prognostication about the likely size of one's life. Here's what some people say:

1. Yes, then I could take care of my business and plan for the end.
2. No, it would ruin my life; I'd be worrying about it all the time.
3. I'd only like to know if the date isn't within 5 years.
4. I wouldn't like to know now; but I would like to know if, in general, everybody else knows the dates of their own deaths.
5. I'd rather know about my own parents' deaths than my own.
6. I don't want to know, and I want to convince myself that it doesn't matter.
7. I'd like to know only if my death would be sudden, so that I can say goodbye to people first. [Note that accidental deaths not associated with a genetically based behavioral trait are the one sort that would not be predictable on the basis of genetic information.]
8. I wouldn't want to *know*, but I would want to have the predictive capacity to change my circumstances.
9. Yes, I'd want to know, just because it's *my* death.
10. I wouldn't like to know, because I'd give it away by acting differently even if I didn't actually tell anyone, and they'd treat me differently as a result.
11. I don't care so much about when I die, but now, and about what this would do to my relatives.
12. I don't care about the date of my *death*. What I care about is the date of my incapacitation—would I be in a coma for 20 years before death?

The question to which these answers are responses, "would you want to know the exact date, time, and cause of your death now?" is a crystal-ball question, an impossible question to which we will never be able to provide real answers. But the responses people give to it allow us to discern what we can expect would be two quite different patterns of response, to be explored below, to the prospect of knowing much more nearly the *likely* time-range and cause of one's death: not just 'yes' answers and 'no' ones, but two basically different types of answers: instrumental and noninstrumental forms. This difference between instrumental and noninstrumental responses to the prospect of knowing the crystal-ball date and cause of one's death allows us to begin to answer the question of how our world will change—and whether these changes will be good or bad—as increasing capacities for genetic prognostication actually do make it far more possible than at any period in human history to know the approximate date and cause of one's own death—not just shortly beforehand, but for most of one's life, even, perhaps, from childhood on.

8. THE GENETIC FUTURE: KNOWING THE SIZE OF OUR LIVES

Knowing this size of our lives, at least approximately so, will, we can predict, elicit both instrumental and noninstrumental responses. But they are very different.

8.1. Instrumental Responses

Instrumental responses to the prospect of knowing something point to the uses that can be made of that information. If the information concerns the approximate timing and cause of one's death, known far in advance, we can imagine at least three areas in which this information might be used:

8.1.1. Prevention and Treatment Options. In some cases (for example, hemochromatosis), presymptomatic knowledge of a genetic disease permits the choice of prevention and treatment options—here, the prevention of iron accumulation, or treatment by flushing out of iron accumulation—that can radically alter one's prospects for life. Of course, one does not have to know the approximate timing of one's death in order to select prevention or treatment options, but one does have to know one's disease status, and with this, in the genetic future we are exploring, comes at least some knowledge of the approximate timing of one's death.

8.1.2. Planning. Pragmatic planning of many everyday matters may hinge on knowing the approximate timing and cause of one's death: for example, whether to purchase life insurance, if risk-rated insurance remains available at all (important if you have dependents and expect to die young; a waste of money if you expect to die in old age); whether to take out disability insurance or long-term care insurance (important in conditions like Alzheimer's and Huntington's); whether to take out a 30-year mortgage, whether to sign up for a long tour of duty in a country with an inadequate health-care system, whether to explore with one's physician the possibility of eventual assistance in suicide, and so on. Of course, it may be hard to predict whether knowing in advance will enhance or undermine an individual's psychological capacity for planning, and whether in particular those with predicted shorter lives are more likely to live in unplanned, "fast-lane" ways just because they expect to die young. It may also be hard to predict whether individuals will respond in rational or irrational ways to advance notice about their own deaths. But some individuals will see it, as an economist might put it, as an opportunity for allocating resources prudentially over one's expected lifespan, and hence maximizing benefit without risking loss.

8.1.3. Communicating. Prognostic knowledge of the likely cause and timing of death may facilitate communication among family members, both in order to begin saying goodbye and to prepare family members for the kinds of terminal illnesses to be expected. These patterns already occur among, for example, people at risk for Huntington's; their experience (though not necessarily the fear of such a difficult end) might be far more widely replicated among individuals generally. Communicating about these matters might of course generate its own moral dilemmas: if you fall in love, are you obligated to tell the person you love that you expect a lifespan of, say, only five or ten years? Should you undertake parenthood if you do not expect to live another twenty, until a child reaches adulthood? Should you enter a partnership or make a contract you do not expect to be able to maintain over time? Of course, these are dilemmas already faced by people with some genetically-based conditions; in the future, they would be faced by nearly everybody as they neared the end of their projected lifespans, on a far broader scale.

8.1.4. Life Choices. Basic life choices also might ride on knowing the approximate size of one's life. These life choices can involve such basic matters as career choices, mate selection, emotional investment in long-term projects and goals, and so on. For example, the smart kid who expects a comparatively short life might choose to be a mathematician, on the grounds that most advances in mathematics are made by people at very young ages—the late teens, early 20's; the same smart kid who forsees a very long life might instead choose a field (perhaps philosophy), where maturity of thought counts for a great deal. Similarly, knowing the approximate size of one's life might invite one to think more carefully about selecting a mate with the same sorts of prospects, about reproductive choices and whether and when to have children, and so on. And one will, presumably, select life-commitments and life-projects in accord with some more realistic notion of whether one will be able to see them through, and not be interrupted by an "untimely" death. These basic life choices are perhaps best conceptualized as a more profound sort of planning and communicating, but a sort that is only possible now in the most rudimentary, unreliable sort of way.

8.2. Noninstrumental Responses

Noninstrumental responses to knowing information about the likely end of one's life are far harder to understand. These are situations in which knowing information is significant, though not because it is to be put to any practical use. Where the information is information about one's own genetic makeup and so information about the likely cause and timing of one's death—this may contribute, in some intangible way, to one's sense of self, of who one is, what sort of person one is, what one's life is like. Self-knowledge is often painful, and this form of self-knowledge may be so as well. But it may also contribute to what we think of as the quintessential way of being human in the world. It is far harder to talk about the value of being able to comprehend and reflect about one's own circumstances as a human being, and there is even disagreement about whether it is a value or a disvalue (though the maxim "better Socrates dissatisfied than a pig satisfied" still rings in our ears), but it is the really important issue here. Advances in genetic science will presumably also make it possible for us to perform the same sort of prognostication about the length of life for individual animals and plants of various species, but animals and plants cannot comprehend this information and do not, we assume, have a sense of self, something usually held to be possible only for humans. For humans, the possibilities of genetic prognostication, I think, will bring something new to the sense of self an individual has, something we now have only in the most inchoate ways: the sense of myself as a person with a certain size of life, whether short or long.

How will knowing oneself as a person with a probably short lifespan, or a probably long lifespan, change us? Will knowing this from childhood on be more or less like knowing—something which will also become possible—whether one will be tall or short, something which, for the most part, an individual eventually must just accept? After all, the capacity to know in advance is likely to precede the capacity to do anything about changing it. Do we adapt better to the circumstances of our lives if we know what they are going to be?

The barest evidence from studies of responses of people taking diagnostic tests for Huntington's disease invites us to conjecture that knowing one's genetic makeup and hence knowing in advance the probable timing and cause of one's death will lead to a decrease in anxiety and stress. Of subjects measured before and after undergoing diagnostic analysis, those who received good news (negative for the Huntington's mutation) and those who received bad news (positive for the mutation) both experienced decreases in anxiety and stress; subjects whose results were inconclusive remained at higher anxiety and stress levels (Wiggins, 1992). There are reasons not to draw too heavily on this study: for one thing, it was flawed by the design of the control group, which included some subjects who were not tested at all, and, for another, it examined people who got news comparatively suddenly, after a test, about a matter they had been in the dark about for their whole lives. And in any case, it is a big leap from this one small study of one disease condition to a generalization about what it would be like for human beings to be routinely aware of their own genetic makeup and of the approximate size of their lives. Yet it is a tempting leap anyway: we may speculate, at least, that human beings would experience less anxiety about death, less open-ended, free-floating *Angst*. At the same time we can predict that that knowledge will be painful for some, namely those expecting short lives. And anxiety may increase for all individuals as the age-range in which they can expect the onset of the disease or condition that is likely to kill them approaches. And, of course, there can be no guarantees; the prognostications that genetic science increasingly makes possible do not ensure that some accident, violence, or unrelated disease will not interrupt life earlier, or that there will not be mistakes in prognostication. But it may well undercut our tendency to avoid issues of death altogether, to refuse to prepare for it in any but the most superficial ways, and to treat death as an existential unknown, when in fact it is becoming something that can be increasingly foreseen.

It is impossible to deny that we are already involved in a process of profound change, from the human past—say, the 16th century—in which a person could have very little realistic idea of when or how he or she would die, through the present, in which more of us have some rudimentary, inchoate idea of our own fates, into a future in which it will be possible to prognosticate increasingly accurately about the cause and timing of each individual's death. Of course, we cannot now say with certainty how accurate or complete our capacities for genetically-based prognostication will eventually be. But we can already begin to ask the ethically important questions: will this be a bad thing—yet another feature of same brave new world, the *GATTACA* dystopia, in which privacies are invaded and human meaning undermined, or will it be a good thing, yet another product of advances in science which improve and enhance the human condition? I myself see it as closer to the latter, but then I am someone who would choose to know now, if I could, the date and cause of my own death. Yet many—maybe most—readers of this paper will have answered the test question above the other way; this suggests they will see the answer to the issue here the other way around too. But whether one would want to know or not, whether one sees this change as good or bad, we cannot deny that it is a change that is coming in the human future, and that, compared to the human past, we are already partway along in a huge change, but one for which it is difficult for us to see that it is already well underway.

NOTES

1. It may interest some readers to know that, except for direct references to the film, this paper was written before GATTACA appeared.
2. I thank Bonnie Baty for this point.
3. For comments on earlier versions of this paper, I'd like to thank several of my temporary colleagues in March 1996 at the Green Center for the Study of Science and Society, University of Texas at Dallas, especially Robert Cook-Deegan and Gail Geller, as well as students and participants in a seminar at the Green Center; I'd also like to thank a number of my colleagues at the University of Utah, including Bonnie Baty, Jeff Botkin, David Green, Leslie Francis, Jan Van Riper, Chris Grammes, Tom Stillinger, and Lenny Moss.

REFERENCES

McKusick VA, Francomano CA, Antonarakis SE. Mendelian Inheritance in Man. (Online, http://www3.ncbi.nlm.nih.gov/, continuously updated). Catalogs of Autosomal Dominant, Autosomal Recessive, and X-linked Phenotypes. Print edition: Baltimore: Johns Hopkins University Press.

Wiggins S, Whyte P, Huggins M, et al. (1992), 'The psychological consequences of predictive testing for Huntington's disease', *The New England Journal of Medicine*, vol. 327, pp. 1401–1405.

22

INFLUENCES OF GENETIC TESTING ON A PERSON'S FREEDOM

Elisabeth Hildt

Center for Ethics in the Sciences and Humanities
University of Tübingen
Keplerstr. 17, D-72074 Tübingen

Autonomy and freedom of action are important characteristics of personhood which are of relevance in almost all areas of human existence. Also in the context of medical ethics the concepts of autonomy, freedom of action, and freedom of choice, which are all tightly linked to one another, play an important role. The concept of freedom can be interpreted in a wide sense so that it encompasses freedom of the will and freedom of choice, but also, more generally speaking, rationality, self-consciousness and intentionality. Also restricted room for manoeuvring, i.e. a limited capacity to make decisions and plans materialise, can be considered a reduction in a person's freedom.

On the one hand, an important aim of medicine is to maintain a patient's freedom and room for manoeuvring at the mental as well as the physical level as long as possible and for as far as possible. On the other hand, there is a considerable danger that, in the context of new developments in biomedical research, situations will arise whose consequences run counter to these aims. Especially in human genetics it seems that developments are emerging in which in addition to the diagnostic and therapeutic goals there is also a strong influence on the freedom of action of the persons concerned. In addition, there might be far-reaching affects on the way people see themselves.

In recent years genetic testing has to an increasing extent been performed in order to reveal whether a person carries certain genes which—by themselves or in combination with other factors—are responsible for certain disorders. Genetic knowledge may be of great benefit to the person who has undergone genetic testing. It may enable a person to make important decisions concerning his or her future life course, such as whether or not to have a child, whether or not to choose a certain career, or whether or not to engage in certain activities. By allowing a person to exercise self-determination on the basis of self-knowledge, genetic testing may clearly increase a person's autonomy and freedom.

In principle, the knowledge that can be derived from genetic testing is very similar to predictive knowledge that arises in other medical contexts. There are two important characteristics of predictive genetic testing, however:

Genetic Information, edited by Thompson and Chadwick.
Kluwer Academic / Plenum Publishing, New York, 1999.

I. Genetic knowlege is of great relevance to reproductive choice. In particular, the interrelation between genetic knowledge and the possibility of terminating a pregnancy for genetic reasons leads to complex social and ethical repercussions.
II. Very often, totally asymptomatic persons have to cope with the knowledge that there is a certain probability that they might develop a specified disorder sooner or later. In contrast, in other medical contexts predictive knowledge is much more tightly linked to symptoms, so that most often persons who already suffer from symptoms are confronted with a prognosis (Chadwick, 1997).

Most often, genetic testing is considered for one of the following reasons:

a. A person may wish to know details of his or her genetic make-up in order to be able to adjust his or her life-plans according to this knowledge. Persons who are at risk of being a carrier of the genetic abnormality leading to Huntington's disease might be an example of such a person.
b. People may wish testing in order to take preventive measures based on the test result, if necessary. For example, a woman who knows that she is a carrier of a breast cancer gene (BRCA1 gene) may choose to undergo regular medical examinations for early detection of breast cancer or she might wish to undergo a prophylactic operation.
c. In the context of family planning a couple may want to know whether they are carriers of a certain genetic disorder in order to know details about the risk that their offspring might be affected.

In addition to these possibilities of genetic testing, the following methods exist:

d. prenatal genetic testing with whose help a pregnant woman or a couple wants to know whether the foetus is a carrier of a genetic abnormality that may lead to such a severe disorder that it might justify abortion. Genetic testing may also encourage a couple to have a baby, however. In addition, genetic knowlede may also provide the chance to start medical treatment very early.

It seems obvious that those who undergo genetic testing expect that genetic knowledge allows for free and efficient planning of their future life course. Genetic testing clearly increases a person's freedom and autonomy in manifold respects. Only very seldom, however, is there an adequate therapy available which could effectively treat the disorder detected by predictive genetic testing. Due to this lack of adequate therapy, negative implications for a person's freedom may also arise from genetic testing, such as affects on an individual's sense of self and on his or her life-hopes; loss of alternatives for action; influences due to social pressures; loss of independence, etc. These aspects may considerably decrease the benefit a person derives from genetic testing. In what follows I want to discuss some of these aspects in greater detail.

1. KNOWLEDGE AND RATIONALITY

Genetic testing serves to resolve the uncertainty over whether a person is or is not a carrier of a certain genetic abnormality. In principle, after genetic testing a person can rationally start planning the future. This is especially the case with a diagnosis which shows that the person is not a carrier of the gene in question. After a test result like this, people may feel that they have got their future and freedom back again.

In the case of a test result revealing a certain genetic abnormality, the consequences for a person's freedom seem to be highly ambiguous, however (Quaid, 1994). On the one hand, a person clearly gains certain advantages from this knowledge with regard to life planning and family planning. He or she can adjust plans to the purported health expectations. On the other hand, with genetic knowledge a lot of uncertainties arise: it cannot be predicted when the up-to-now asymptomatic person will start suffering from symptoms, how severe these symptoms will be, what the course of the disease in this individual case will look like exactly, etc. The mere knowledge of genetic status often does not lead to any concrete certainty about the future course of life. Difficulties arise in cases in which, by means of a genetic diagnosis, nothing but an elevated risk or a predisposition to a multifactorial condition has been stated. Thus, difficulties in adequately assessing and interpreting the results may lead to enormous drawbacks.

According to the preference or desire satisfaction theory it is good for persons to aim at a maximum satisfaction of desires and preferences. In order to satisfy one's desires and preferences effectively, it is necessary for an autonomous person to know the facts that are relevant for deciding the issue in question (Husted, 1997). From this point of view it is clear that—at least for members of genetic risk groups—autonomous and rational decision-making with regard to one's future life course requires genetic knowledge.

As Diana Fritz Cates points out, genetic knowledge can even be considered to be an important presupposition for the exercise of virtue, for it tells us "who we are [...], what we have control over and what we must suffer as beyond our control" (Cates, 1994, p.60). According to this position, persons who do not want to know details about their genetic make-up deliberately obscure things about themselves that affect their exercise of virtue. By this, they become less capable of shaping their moral agency.

Certainly, based on the test result a person may be able to adjust life-plans and life-expectations, to take some health care measures, to prepare the family-members for the future, or to make some financial preparations. Aspects like these may clearly be very important for the overall organisation of life. However, genetic testing may also have more indirect implications on a person's life-style and the way he or she sees him- or herself. To what extent does this kind of genetic self-knowledge influence a person's freedom at a deeper, more personal level?

The way a young woman at risk of Huntington's disease (HD) argued for her decision not to use predictive testing has been quoted in an article written by Kimberly Quaid (Quaid, 1994, p.10):

> I decided that since I was in my midtwenties there was too long a gap between knowing and onset. I was not certain I could lead a full or productive life with the knowledge that I would develop HD. [...] There is no treatment or cure for HD. What good would it do me to know now? There was nothing I could do to change the inevitable one way or the other. Would I really modify my behaviour or lead my life any differently? A yes answer to that question would surely nullify the meaning of my present life.

This example shows that it is not only the amount of information available that matters in this context. Instead, genetic knowledge and decision-making have to be considered within a complex system in which a person's individual history, values, life-style and sense of integrity also play an important role.

Genetic information may provide the chance for living a more informed life—this chance should not be neglected. There is a considerable danger, however, that a person will adjust his or her whole life to this kind of information and to the purported future threat that goes along with it. In fixing this seemingly inevitable future personal fate not only may the value of the present reality be lost, but also projects done in the past, which before testing the person has considered as being worth aiming at, may present themselves in another context after the test.

In theoretical discussions on the importance of genetic knowledge it is often asumed that on the basis of genetic information autonomous persons will automatically be able to rationally choose what is best for them. Genetic knowledge is considered to be empowering. However, it is often forgotten that genetic information may also lead to severe psychological effects which clearly negatively influence a person's ability to engage in free and rational decision-making. With regard to genetic testing, information clearly is not the only aspect that matters.

I would like to draw your attention to David Hume who held that it is not only reason but also a person's sentiments which play an important role in pursuing one's goals. While the sentiments, affections, or desires set an agent's goals, with the help of adequate information reason pursues these goals as efficiently as possible (Lindley, 1986). According to this point of view, a person who chooses not to undergo genetic testing can be considered to exert self-government on the basis of his or her desire not to know. A choice like this is not derived from false beliefs: it is not an irrational one. Also, a decision that is made without knowing genetic information need not be irrational since decision-making done in this way can be considered the rational pursuit of the choice to live without genetic knowledge. The fact that someone lives according to a life-plan which is not based on the concept of genetic self-knowledge does not necessarily imply that it is irrational to live according to this life-plan.

2. LIFE-HOPES

Life-hopes are an individual's attitude to his or her future. As Ted Honderich (Honderich, 1993, p.82) puts it: "A hope is a desire for something, involving an approving valuation of it, bound up with feeling, and such that it is not certain that the thing will come about." Thus, in life-hopes it is not knowledge and rationality but a person's desires and personal feelings that play the most important role.

Genetic testing may negatively influence a person's life-hopes. Being aware of a kind of genetic determination which may consist of being struck by a disease such as Huntington's disease in some years time may destroy a person's life-plans and life-hopes to a significant degree. General resignation, depression, or other negative mental consequences have to be expected.

A person's life-hopes and his or her plans for action are closely linked and highly interdependent, not least since normally it is to a considerable extent by way of a person's own actions that his or her life-hopes will materialise. Somebody who is in uncertainty concerning his or her genetic status may be able to imagine two possible future lives: one with the disease and one without it. Despite clearly existing doubts and fears the person may have hope and may successfully engage in actions relevant for fulfilling his or her life-plans and life-hopes.

Certainly, in the case of a low-risk test result, the future seems to be totally open again. However, persons who came to know that they are carrying a genetic abnormality that leads to a severe disorder have been reported as feeling intensely that they no longer have a future (Quaid, 1994). Although these persons—at least as long as they are asymptomatic—are in principle free to do what they want to do, a diagnosis like this may lead to a subjective, self-imposed reduction of the individual's room for manoeuvring. A reduction like this may be mainly due to the modified self-perception a person has after genetic testing and to the comparatively low chances he or she expects to have of fulfilling projects and commitments (Brock, 1994).

A person's life-hopes intensely affect the plans and projects in which he or she engages. Decisions with regard to important plans and projects clearly depend on the extent to which the person expects he or she will be able to fulfill them. Due to the test results, a person may jump to the conclusion that it is useless to engage in longer-lasting projects and to plan for the future. For example, a student may choose to drop out of university. In view of the manifold uncertainties linked to genetic knowledge there is a considerable danger that genetic testing leads to enormous self-imposed restrictions in a person's freedom. Especially in those cases in which the course the disorder takes differs significantly from the expected course, life may prove to have been significantly restricted for no real reason. This may be the case when the symptoms appear much later than expected, or when they are less severe.

3. ALTERNATIVES FOR ACTION

In genetic testing, respect for the individual's autonomy plays an important role. In principle, persons can freely decide whether or not to undergo genetic testing and how to make use of the test results. In most Western countries there is a strong consensus that genetic counselling should be non-directive, i.e. the counsellor should not influence the counselled person in the process of decision-making.

Going along with this respect for autonomy there is a growing tendency to consider people responsible for their individual health. Especially in the context of genetic testing the situation may arise that, once a person has come to know details about his or her genetic make up, he or she will be considered responsible for the choices made. The argument runs as follows: If there is a possibility to avoid future burdens and future suffering and if despite genetic knowledge a person did choose to do nothing to avoid it, he or she clearly has to bear the consequences (ten Have, 1997).

There is a considerable danger that a person who has undergone genetic testing might be forced into doing actions which only serve to avoid the blame that he or she has acted irresponsibly. For a person who strives to avoid the blame of irresponsibility the range of alternatives for action may narrow down considerably. The person may be more or less forced to choose one option. In a situation like this a person cannot be considered to be fully autonomous any longer. As Joseph Raz puts it (Raz, 1986, p.378): "Autonomy requires that many morally acceptable options be available to a person" and "A choice between good and evil is not good enough" (Raz, 1986, p.379). For example, situations like this may arise with the disclosure of genetic information to family-members or to the person one loves and wants to marry. Disclosure might be felt to be a duty, especially with regard to those relatives who plan to have children. A person who carries a gene responsible for a certain genetic disorder may not have a free choice any more between two good options: telling or not telling the family-members about his or her genetic status, but a choice between a good option, namely informing at least all those family members who are involved in reproductive decision-making, and a bad option, namely going on without telling anybody about genetic testing. According to this point of view, in a situation like this a person is not given an adequate range of options, therefore the choice made cannot be considered a free one.

With regard to a person's whole life, there is a substantial danger that what results after genetic testing is a life not freely chosen any more, but a life in which freedom and self-determination have been given up to a considerable degree in order to fulfill the high burden of responsibility in light of genetic information.

4. INDEPENDENCE

Closely linked to the issue of alternatives for action is another issue which also greatly compromises an individual's freedom. This relates to the fact that the choice to be made is a choice more or less directly forced upon the individual by social influences and pressures.

This aspect is further underlined by Gary Watson's reflections on free agency (Watson, 1982). According to Watson's point of view in an unfree action the agent is led to action not by his valuational system which chooses (among alternative states of affairs) on the basis of his value judgements, but by motivations such as appetites, passionate desires or acculturated desires. For Gary Watson acculturated desires are desires which do not (primarily) derive from the agent's own judgements, but from social influences and acculturation. In my opinion, this category of acculturated desires is very helpful for the discussion of problematic issues arising in the context of genetic testing: after genetic testing a person may be motivated to act in ways which do not rely on his or her own system of values but which are somehow socially imposed upon the individual. According to this point of view one could say that the person is moved to action by acculturated desires. What results is an unfree action.

One may even question whether it would be preferable for rational individuals to have certain options, on the condition that these options can only be obtained by a person at the price of a widespread distribution through society (Dworkin, 1988). In my opinion, the field of genetic testing, especially prenatal genetic testing, can be considered an example of this issue which shows that it is especially the widespread use of this technology that may lead to significant negative implications (that do not arise if one considers an isolated case). The more prenatal genetic diagnosis is used, the more people may feel that there is an efficient possibility available to avoid certain genetic disorders. For members of genetic risk groups who want to have a child of their own, a growing pressure may exist to have prenatal genetic diagnosis and—if considered necessary—to undergo termination of pregnancy for genetic reasons.

I want to stress that up to now mainly indirect social influences on a person who has undergone genetic testing have been mentioned. The influences are indirect in so far as the person might expect social sanctions to occur and by force of this anticipation of social sanctions may act in a certain way. These aspects are only indirectly connected with direct pressures on the individual that might arise after the results of genetic testing have become known to other persons or institutions.

Society may be very much interested in certain results obtained by genetic testing for financial reasons. If a test result should ever become known to other persons or institutions, there might be manifold social consequences such as prejudice, stigmatisations and social isolation, not least of all because insurance companies, educational establishments and potential future employers are very much interested in this kind of information (Nelkin and Tancredi, 1994). For example, a person may not get a job he or she is qualified for, or may have difficulties with health insurances. An individual person's freedom is highly compromised by the ever-present risk that other persons or institutions might get to know details concerning his or her genetic make-up. It seems clear that with regard to freedom what is of particular importance is whether the person who has undergone genetic testing is able to control who will know the results. It is very difficult to achieve such a strict control, however, especially because there is a clear conflict between individual and public interests here.

In conclusion: If one considers those aspects related to freedom of action mentioned above, there is clearly no unequivocal answer to the question of whether or not genetic testing increases the freedom of the persons involved. On the one hand, genetic testing clearly increases a person's freedom by allowing decision-making on the basis of genetic knowledge. On the other hand, this increase in freedom is highly compromised by indirect influences arising within the context of genetic testing, not least since an individual's freedom not only depends on the number of options theoretically available, but also on the real possibilities of choosing them and the social consequences that go along with these options.

ACKNOWLEDGMENT

I would like to thank Eve-Marie Engels and Marcus Düwell for helpful discussions.

REFERENCES

Brock, D.W. (1994), 'The Human Genome Project and Identity', in Weir, R.F., Lawrence, S.C. and Fales, E. (eds), *Genes and Human Self-knowledge*, Iowa University Press: Iowa City, pp. 18-33.

Cates, D.F. (1994), 'Commentary on Genetic Knowledge and Self-understanding', in Weir, R.F., Lawrence, S.C. and Fales, E. (eds), *Genes and Human Self-knowledge*, Iowa University Press: Iowa City, pp. 59-62.

Chadwick, R. (1997), 'The Philosophy of the Right to Know and the Right Not to Know', in Chadwick, R., Levitt, M. and Shickle, D. (eds), *The Right to Know and the Right Not to Know*, Ashgate: Aldershot, pp. 13-22.

Dworkin, G. (1988), *The Theory and Practice of Autonomy*, Cambridge University Press: Cambridge.

Honderich, T. (1993), *How Free Are You?*, Oxford University Press: Oxford.

Husted, J. (1997), 'Autonomy and a Right Not to Know', in Chadwick, R., Levitt, M. and Shickle, D. (eds), *The Right to Know and the Right Not to Know*, Ashgate: Aldershot, pp. 55-68.

Lindley, R. (1986), *Autonomy*, Macmillan: Hampshire.

Nelkin, D. and Tancredi, L. (1994), *The Social Power of Biological Information*, Chicago.

Quaid, K. (1994), 'A Few Words from a 'Wise' Woman', in Weir, R.F., Lawrence, S.C. and Fales, E. (eds), *Genes and Human Self-knowledge*, University of Iowa Press: Iowa City, pp. 3-17.

Raz, J. (1986), *The Morality of Freedom*, Oxford University Press: Oxford.

ten Have, H. (1997), 'Genetic Information and Human Existence,' in Chadwick, R., Levitt, M. and Shickle, D. (eds), *The Right to Know and the Right Not to Know*, Ashgate: Aldershot, pp. 87-95.

Watson, G. (1982), 'Free Agency', in idem. (ed.), *Free Will*, New York, pp. 96-110.

23

GENETIC KNOWLEDGE

The Contribution of Sociologies

Simon Dyson

De Montfort University
Leicester, United Kingdom

1. INTRODUCTION

This chapter will consider the issues of public awareness and knowledge of genetics with specific reference to the haemoglobin disorders sickle cell anaemia and beta-thalassaemia. The paper aims to highlight problems in the whole notion of public awareness of genetics by examining the insights which may be derived from sociologies. In doing so it will draw critically upon the author's own work in assessing community and professional awareness of these conditions and their carrier states.

A recurrent feature of the processes of raising public awareness with regard to disorders such as sickle cell anaemia and thalassaemia is the use of multiple choice questionnaires. These have been used for assessing knowledge of genetics in general (Decruyenaere et al, 1992); for use in assessing knowledge of haemoglobin disorders in nurses (France-Dawson, 1990; Lorenzi, 1993) and midwives (Dyson et al, 1996) for estimating awareness of those affected by their haemoglobinopathy (Jennings, 1990; Midence et al, 1994); and for describing knowledge in screened populations (Rowley et al, 1979; Dyson, 1997) or opportunistic community samples (Dyson et al, 1993a; Dyson et al, 1993b).

Irrespective of how far these projects do or do not explicitly subscribe to empiricist epistemological assumptions, their limitations, especially their role in the social construction of 'lay ignorance' (Kerr et al, 1997) need to be acknowledged. In this respect, sociologies have a contribution to make in pointing out problems with untheorized notions of community genetic information. This chapter reviews the potential insights afforded by three different epistemological traditions, namely subjectivism; feminist, disability rights and anti-racist strands of critical realism; and postmodernism. These are each of course heterogeneous traditions. Moreover, many sociological researchers try to work at the points of tension between the positions.

Genetic Information, edited by Thompson and Chadwick.
Kluwer Academic / Plenum Publishing, New York, 1999.

2. SUBJECTIVISM

To the extent to which researchers draw upon subjectivist insights, they are concerned with social meanings. In particular they are concerned with the social construction of (genetic) knowledge. This means not only that genetic knowledge is apprehended contextually depending upon its social relevance (Kerr et al, 1997), but also that there will be different demarcations of what is socially meaningful about a particular topic such as genetics.

Genetic knowledge assessment imposes a single dimension onto the social world, without any alternative, in that multiple choice formats do not allow for any alternative property to exhibit itself, and constitute what has been called a problem of imposition (Cicourel, 1964). Furthermore, meanings assigned to language, objects and events may vary between respondents, as well as between respondents and researchers. This correspondence problem (Cicourel, 1964) can be related to research into public awareness of genetics. Firstly, does the researcher use the same criteria as respondents in demarcating what is meaningful about, for example, the haemoglobin disorders? Secondly, is the language used by the researcher understood in the same way by respondents and vice-versa? Thirdly, in interpreting the same social environment do the cultural values of respondents and researchers correspond? The problem of imposition and the three correspondence problems each deserve some consideration in turn.

2.1. Problems of Imposition

The notion of a problem of imposition refers to the fact that researchers, by imposing a particular set of questions, say about haemoglobin disorders, onto the social world, actively construct the possibility of haemoglobinopathy ignorance and simultaneously exclude a range of possible questions. Genetic counselling and knowledge of the risk ratios presupposes a particular use of criteria to identify what is important about genetics. These criteria are based on notions of rational action by social actors who can use knowledge to orientate themselves in an increasingly calculable world. But the notion of probability may be used differently by other cosmologies. Different questions are pertinent to the realms of science and magic (Evans-Pritchard, 1950). In genetics what may be important to the respondents is not theoretical probability but empirical probability, about which genetic science can tell us nothing. In other words, we can know that two carriers have a 1 in 4 chance for each pregnancy of having an affected child, but not why one couple have three affected children in a row, and another couple have several unaffected children. This question is in the realm variously conceptualized as religion, magic or fate, and represents a whole arena of potential interest to the respondent ignored by the questions imposed.

2.2. Problems of Correspondence

The second set of questions concerns the correspondence of language, both between researcher and respondent and between different respondents. The most critical possibility would be in asking people ‘what is sickle cell?’ or ‘what is beta-thalassaemia’. If they said they did not know, this might be because the condition, or at least its symptoms, was only known to them by a lay term or folk concept. This is theoretically possible, but subjectivists do not seem concerned to establish the extent of any intersubjective disagreement (Hindess, 1973).

Genetic conditions mitigate against the development of folk constructs (Rapp, 1988), though these do exist where the prevalence of sickle cell is high (Ebomoyi, 1988). In Britain the more plausible scenario is one where either a child with sickle cell remains undiagnosed, in which case their mortality is likely to be higher (Vichinsky et al, 1988; Powars, 1989), and is unlikely to be given any label, even a folk construct. Or, through contact with the health services a diagnostic label of sickle cell is given, though this may be belated and/or rejected and/or not communicated to the child (Hill, 1994). In developing countries it seems likely that a beta-thalassaemia death would be indistinguishable from other deaths, unless a diagnosis had been made. The author's own research in Britain has utilised several community workers representing Gujarati, Punjabi, Pakistani, Bangladeshi, Chinese, African and Caribbean communities. Neither training days nor debriefing interviews established such lay constructs for the haemoglobin disorders.

More specific breakdowns of intersubjectivity, such as the elderly Gujarati respondent who used the word "thalassaemia major" to evoke the name of the former British Prime Minister in the mistaken but apparent belief that he was being assessed for nationality as part of an immigration check, move us to a third area of concern, namely the different cultural interpretations of the same situation. Statistical risk presumes an abstract mathematical world that may not be shared by those counselled (Rapp, 1988). Multiple choice format schedules may not be understood by all respondents and by the researcher in the same manner. Certainly, in one community survey conducted by the author (Dyson et al, 1993b) a Bengali interviewer apparently interviewed a group of respondents together which at least suggests a conception of knowledge as residing in individuals and to be elicited on a one-to-one basis to be ethnocentric.

Furthermore, the interview situation may have a number of different meanings to respondents, including the intrusion of officialdom; fear of stigma; opportunity to complain about other illnesses as well as an opportunity to gain information (Dyson, 1995). Not all of these constructions threaten an assessment of how much people know about the haemoglobinopathy in question. Moreover, recent work on the co-production of accounts of health (Radley and Billig, 1996) mitigate against accepting the principle of error from an alleged 'true rate' of knowledge.

2.3. Limits to Subjectivism

If the data generated by multiple choice formats are so subject to caveats, and the interview situation and cultural meanings of genetic information are so variable at so many different levels, then is there any purpose in persisting with attempts to assess community awareness of the haemoglobin disorders, when no "real" level of knowledge can be unproblematically stated? Indeed, such surveys of knowledge might be conceived as part of the process of imposing a professional discourse onto communities. But uncovering the ethnographic meanings of the haemoglobinopathies to communities risks revealing lay concepts and strategies to the health professional with no guarantee that information will not be used against lay communities, the more effectively to co-opt them to professional discourses.

It seems that whilst giving information about the haemoglobinopathies is value-laden in that it imposes new invidious reproductive choices onto communities (Rothman, 1994), a general level of community awareness at least creates the necessary, if not the sufficient, conditions for a moral debate. That debate is certainly foreclosed if people, mostly women from minority ethnic groups, are presented with new, devastating, information as isolated individuals at an ante-natal clinic. It is this concern which particularly motivates some realist critics of the new genetics to which we now turn.

3. CRITICAL REALISM

Critical realist authors share a concern that 'knowledge' may reflect oppressive social relations rather than ahistorical scientific truths and that a transformation of knowledge may depend upon a transformation of such social relations. Here we consider authors who are concerned particularly with relations of gender, ethnicity or disability.

3.1. Gender

For feminist critics, writing from the basis that social relations are generated by both capitalism and patriarchy, the new emphasis on genetic information and ante-natal diagnosis is part of both the medicalization of social life in general (Zola, 1977; Crawford, 1980) and the extension of control of male-dominated medicine over women and their bodies in particular (Oakley, 1980; 1984).

Access to chorionic villus sampling is said to sometimes depend upon the woman agreeing to a termination if the foetus is affected (Stacey, 1996). Questions have been raised as to whether or not women have been consulted about their wishes with regard to pre-natal diagnostic tests (Farrant, 1985). It has been suggested that patriarchal concern to protect paternity is at the heart of prenatal diagnosis and that genetic counselling discourses subtly pressure women to the conclusion that termination was "what they had to do" or that "they had no choice" (Rothman, 1993; 1994). This is because genetic counselling does not occur in a vacuum, but in a gender-ordered society where a male partner may threaten to leave if the child born has a disability. The real 'choice' for a woman is not whether to have a child with a genetic condition or not, but whether to be a lone parent of such a child in a gender discriminatory, child-unfriendly, ablist society.

The ideology of prenatal diagnosis and counselling reaches its height in claims that negative test results reassure (this does not mean the child will not have a condition other than a haemoglobinopathy which is socially stigmatized) or that tests for those who under no circumstances would consider an abortion will help a mother prepare for the birth of an affected child (Rothman, 1994). However, it may be argued that whether or not advance notice of a child with a major haemoglobinopathy to come is beneficial may be a matter of empirical investigation. The absence of routine neonatal screening renders an undiagnosed child/infant with sickle cell vulnerable (Vichinsky et al, 1988). In Britain, whether preparing a case for housing transfer; for investigating procedures of statementing for special educational needs; for making occupational decisions; for financial planning; for making self-help contacts and reading relevant educational materials: all these preparations may or may not help some, none or all women.

3.2. 'Race' and Ethnicity

An analysis based on assumptions of an underlying structure of racism would emphasise the influence of racist immigration laws on the genetic variation in a community. For example, do laws restricting rights of British Moslems to marriage partners from the Indian Sub-Continent effectively narrow the gene pool of potential partners in communities practising consanguinity? An anti-racist analysis would also address the interplay of racism and class in education, housing and employment in determining material life chances and thereby the material basis to the situational relevancy of genetic information.

Moreover, it would be unwise not to consider that the ascription of genetic risk to racialized groups may engender stigma in a way in which such information appears not to in the white population (Watson et al, 1992; Burn, 1993) particularly as one such historical example apparently exists (Stamatoyannopoulos, 1974). Although in the case of beta-thalassaemia and a Pakistani Moslem community in Britain stigma is not socially constructed, this is because of very specific cultural factors, namely that the pattern of consanguinity in marriage means that to ascribe stigma would be to stigmatize one's own family (Darr, 1990). Indeed, where genetic information disturbs the one valued social role open to black women living in poverty, namely raising children, genetic information relating to sickle cell may be rejected or blurred (Hill, 1994). A political economy analysis also reveals implications for racial discrimination. Firstly, the manufacture of pharmaceutical drugs, such as desferrioxamine for beta-thalassaemia major, in the countries of the North raises a specific instance of the general exploitation of financially poorer countries by multinational drugs companies (Melrose, 1982). Secondly, haemoglobinopathies are carried by nearly 5% of the population globally (WHO, 1994), but since most of these are in developing countries venture capital is more likely to support research into the genetic basis of heart disease and cancers. Thirdly, the South Asian diaspora to five continents may mean there are extended families who have independently in several countries discovered their familial genetic status by the unexpected birth of a child with a haemoglobin disorder.

Anti-racist approaches would also be critical of explanations which cited lack of cultural understandings for failure to assimilate or act upon genetic information. Such authors have criticized the 'ethnic' culturalist approach in medicine and health education for failing to acknowledge racism; for portraying ethnic groups as homogenous and for construing minority ethnic cultures as deviant, deficient and resistant to change (Pearson, 1986 & 1989; Ahmad, 1994 & 1996). Moreover, material and structural factors may apply differentially between different ethnic groups; be experienced in differing ways and to differing degrees by the mediation of culture; and culture may itself be forged in the historical and social context of material factors (Smaje, 1996). Such is the case of low income families caring for a child with sickle cell in the US, where "...the strong cultural norm of motherhood among low income black women mitigates against easy acceptance of reproductive implications of having sickle cell trait." (Hill, 1994, p. 68).

3.3. Disability

Disability rights authors might be said to work with notions of underlying structures which generate the everyday discriminations against peoples with disabilities. These discriminations operate in the very construction of what is to constitute not only a social handicap and functional disability, but in the case of genetics, what is to constitute an 'impairment.' (Shakespeare, 1995).

Thus research into genetic information is conceived of as itself socially situated. People with genetic conditions are not usually consulted about research questions, methods of data collection or strategies of data analysis. For some, the context of an ablist society requires certain challenges to these traditional and oppressive relations of research (Oliver, 1992). Official UK research on disability reveals an ablist bias through questions such as "Does your disability affect your work in any way at present?" Alternative formulations of the question could locate the problems in society not the individual, for example

"Do you have problems at work because of the physical environment or the attitudes of others?" (Oliver, 1992).

By extension a question to ask a young woman with sickle cell is not "Does your sickle cell affect your pregnancy?" but "Does the (racist/sexist/ablist) structure of health service delivery affect your pregnancy?" For the course of pregnancy often runs well for women with sickle cell disorders. Moreover, their infants, even those small-for-gestational-age, appear healthy. This is attributable to nothing more than good obstetric care where health care workers are knowledgable about sickle cell disorders. Moreover, advice on pregnancy avoidance, even for women with the SS genotype, is not justified by the scientific evidence (Smith et al, 1996).

The implication appears to be that poor obstetric care and/or comparisons with general population birth outcomes (rather than the already compromised birth outcomes of black peoples oppressed by poverty and racism) could themselves create any apparent disadvantage of black mothers with sickle cell disorders. If community genetics awareness surveys were used to dissuade mothers with sickle cell anaemia from having children this would then amount to a cultural assault of three dimensions.

Firstly, black women have historically reported disproportionate pressure to undergo abortions, use dangerous contraceptives and to curtail their fertility (Davis, 1982; Phoenix, 1990). Secondly, it deflects attention from the knowledge and attitudes of health care providers and the extent to which they are prepared (in both senses of the term) to provide culturally competent care for sickle cell anaemia. And thirdly, there is the enormous symbolic value of children, including children born with sickle cell anaemia to women of African-American descent in the US (Hill, 1994).

4. POSTMODERNISM

Researchers drawing upon postmodernist concerns problematize the individual human subject which is seen as an effect of power. Moreover, the emphasis on the creation of knowledge through language (rather than by access to reality) leads to the analysis of scientific discourses and the constitution of scientific technologies and communities. This de-centring of the human subject and the social construction of knowledge come together in concerns about the body and the creation of self-identity.

4.1. Foucauldian Analyses

Historically specific power relations create the clinical subject (Foucault, 1973); the individual psyche (Foucault, 1971); the family (Donzelot, 1979) and the social realm of public health itself (Armstrong, 1993). The notion is of a series of disciplinary powers extending their "gaze" from the individual clinical subject to the social spaces between subjects. The latter is achieved by the increased use of epidemiological and social survey methods (Armstrong, 1983).

For this line of argument there is a real sense in which respondents to community genetics surveys are both subjects and resisters of surveillance. The in-depth interviews of sufferers which subjectivists might insist constituted giving the less powerful a voice would be construed in a post-structuralist sense as an even more thorough surveillance in which subjects effectively survey themselves. The move to the social also moves the focus away from those who are ill to those who are potentially so. Populations are "at risk", and the emphasis is on health education in schools (Department of Health, 1993), outreach into the community (WHO, 1988), the more extensive communication of female genetic counsellors (Zare et al, 1984), and approaches which seek to incorporate the extended kinship network into genetic counselling (Punales-Morejon and Penchaszadeh, 1992).

Thus, genetic public health reaches still further to control the tentative bodies of the foetus in the womb, and hypothetical bodies potentially, but not yet actually, in existence. In short, genetic public health could be viewed as a process culminating in the surveillance of hypothetical beings through prenatal diagnosis and genetic counselling. The global Human Genome project may stand accused of violating individual 'rights', but Foucauldian analyses would conceive these 'rights' as themselves created in the first instance by the power of genetic surveillance. The possibility for the invoking of individual rights as part of resistance to such surveillance is in the process of being countered by the discourse of the new genetics, in which individual autonomy is set against the responsibility to others (Wood-Harper and Harris, 1996) such as members of the family who may have a right to know their risk and/or the rights of others not to have preventable health economics costs incurred (see Clarke et al, 1992) as a result of the individual wish not to know. Disciplinary power, having first created a consciousness of the individual and the possibility of knowing individuals, now appears to downgrade those creations to mere expressions of historical lineage or global families.

4.2. Discourse and Scientific Communities

The analysis of scientific discourses (Gilbert and Mulkay, 1984; Bijker et al 1989) focuses on the *social processes* by which, for our purposes, such technologies as biotechnology, genetic modification, and information about carrier status become defined as 'morally neutral' (British Medical Association, 1992). Genetic scientists use permeable discursive boundaries to simultaneously claim expertise and to minimise their social liabilities (Kerr et al, 1997).The notion of a 'symmetrical account' (Pinch and Bijker, 1989) which uses the same order of concepts to explain the relative 'success' and 'failure' of the technologies could also be applied to the new genetics. This would focus on the different interpretations available to genetic scientists so as to demonstrate that scientific evidence is shaped by social interests and not by nature alone (Gilbert and Mulkay, 1984). The account could proceed to examine the processes of 'closure', that is how competing accounts are defined out whether by rhetoric or by redefinition of the problem. And finally, the notion of different types of development of technologies based on 'technological frames' (Bijker, 1989) could be applied to developments in genetics. For example, how might genetic information be socially constructed to enroll interest groups to a particular technology? One such case might be neonatal screening where only the advent of penicillin prophylaxis provided "a therapeutic rationale for screening newborns for sickle cell disease" (Andrews et al, 1994, p 42) leading to widespread screening but not necessarily widespread provision of the treatment which had provided the post-hoc justification for the screening.

4.3. Body and Society

According to other lines of postmodernist thinking we are now living in a society in which the major political and personal problems of the age are expressed via the conduit of the body (Turner, 1996). Certainly it is possible to discern in the new genetics processes of normalization and rationalization of the human body by new medical technologies (1992).

Other authors are concerned with how social bodies become *civilized*. The pacification of ever larger areas of territory that resulted in nation-states carried with it a civilising process of the manners of interpersonal relations (Elias, 1978[1939] & 1982[1939]). Etiquette in sexual relations already extends to a declaration of genetic status in obtaining the blessing of the Greek Orthodox church for the marriage of couples in Cyprus (Angastiniotis et al, 1986) and the blessing of the community matchmaker in Orthodox Jewish communities (Merz, 1987).

The rise in importance of interpersonal sexual intimacy in the creation of self-identity has been noted (Giddens, 1992) and there is evidence that genetic information is used creatively in constructing identity by avoiding matches unwanted for non-genetic reasons (Ghanei et al, 1997). Since the notion of the body as constituting a form of physical capital usually refers specifically to body surfaces (Bourdieu, 1984; Schilling, 1993) the wider role of unseen carrier states in the genesis of social identities remains uncertain. However, since we are a death-denying society in which our own (impossible) survival can only be affirmed in the (premature) death of others and our own avoidance of premature death from particular causes (Bauman, 1992), genetic information would seem to have the potential to extend the scope of the concept of survival as part of the social construction of our identities.

5. CONCLUSION

This chapter has begun with an acknowledgement of the limitations of multiple choice questionnaires in assessing and raising public awareness of genetic disorders. The limitations are not only of language and meaning, not even just of culture, but of the incommensurate nature of different discourses. Giving a voice to communities of interest has been said to constitute a challenge to the hierarchy of credibility (Becker, 1967) such knowledge assessment questionnaires impose.

Yet merely giving a voice to communities of interest is problematic within critical realism for it raises the potential of an extension of social control. In such cases to challenge the basis of oppressions requires clarity about how an oppression is constituted. Such decisions are not always easy to make. In gender terms it may be that genetic knowledge simultaneously increases the choices of (some) individual women whilst simultaneously foreclosing debates about the values and nature of children and child rearing. In terms of ethnicity it is not clear whether prenatal diagnosis and selective termination is an unwelcome curtailment of fertility or whether such techniques are disproportionately denied to black families by health workers who make racist assumptions that Moslem groups, for example, will find such techniques unacceptable (Petrou et al, 1990). Indeed an anti-essentialist view of ethnicities renders the very link between haemoglobin disorders and particular ethnic groups problematic (Dyson, 1998). Moreover, the blurring of genetic information by community members (Hill, 1994) suggests a problem for disability rights authors (Oliver, 1992) for an invitation to change the social relations of research may itself be experienced as an imposition (Hammersley, 1995) in that what may be desired is no research at all.

Postmodernist commentators are variously concerned to question the essentialist fallacies of genetic discourses. In Foucauldian terms the ante-natal clinic creates not only the genetic subject but also simultaneously the ethical affronts to that subject. If professions are key resources in the process of governmentality, in which populations learn to exercise control over themselves (Johnson, 1995), then public awareness through genetic information is one medium through which governmentality is enacted.

Other authors have emphasised the previously underestimated centrality of the body to social relations. Genetic information is not passively learned but becomes a resource in the active construction of social identities. Such postmodern sociologies can offer the promise of channelling communication between competing discourses in the hope of promoting the democratisation of science (Kerr et al, 1997).

In summary, subjectivisms can teach us how to listen. Critical theories may alert us, albeit sometimes too programmatically, to what communities of interest experience as 'wrong'. Postmodernisms may help us mutually reveal our standpoints. For all the contradictions in their respective resolutions to the dilemmas of knowing, these sociologies do, collectively, caution against certainty. They caution against certainty in the conception of genetic knowledge, in the action of health professionals and in the implementation of genetic policy.

REFERENCES

Ahmad, W.I.U. (1994), 'Reflections on the consanguinity and birth outcome debate', *Journal of Public Health Medicine,* Vol. 16 No. 4 pp. 423–428.

Ahmad, W.I.U. (1996), 'The trouble with culture', in Kelleher, D. and Hillier, S. (eds), *Researching Cultural Differences in Health,* Routledge, London, pp. 190–219.

Andrews, L.B.; Fullarton, J.E.; Holtzman, N.A. and Motulsky, A.G. (eds), (1994), *Assessing Genetic Risks: Implications for Health and Social Policy,* National Academy Press, Washington D.C.

Angastiniotis, M.; Kyriakidou, S. and Hadjiminas, M. (1986), 'How thalassaemia was controlled in Cyprus', *World Health Forum,* Vol. 7 pp. 291–297.

Armstrong, D. (1993), 'Public health spaces and the fabrication of identity', *Sociology,* Vol. 27 No. 3 pp. 393–410.

Armstrong, D. (1983), *The Political Anatomy of the Body,* Cambridge University Press, Cambridge.

Bauman, Z. (1992), 'Survival as a social construct', *Theory, Culture and Society,* Vol. 9 pp. 1–36.

Becker, H. (1967), 'Whose side are we on?', *Social Problems,* Vol. 14 pp. 239–247.

Bijker, W.E.; Hughes, T.P. and Pinch, T. (1989), *The Social Construction of Technological Systems,* Massachusetts Institute of Technology Press, Massachusetts.

Bijker, W.E. (1989), 'The social construction of bakelite: towards a theory of invention', in Bijker, W.E.; Hughes, T.P. and Pinch, T. (eds), *The Social Construction of Technological Systems,* Massachusetts Institute of Technology Press, Massachusetts, pp. 159–187.

Bourdieu, P. (1984), *Distinction: A Social Critique of the Judgement of Taste,* Routledge and Kegan Paul, London.

British Medical Association (1992), *Our Genetic Future: the science and ethics of genetic technology,* Oxford University Press, Oxford.

Burn, J. (1993), 'Screening for cystic fibrosis in primary care', *British Medical Journal,* Vol. 306 pp. 1558–9.

Cicourel, A.V. (1964), *Method and Measurement in Sociology,* Collier-MacMillan, London.

Clarke, A.; Wallgren-Petersson, C; and Hughes, H. (1992), 'Children with genetic disease: who should pay?', *Lancet,* Vol. 339 pp. 1614–5.

Crawford, R. (1980), 'Healthism and the medicalization of everyday life', *International Journal of Health Services,* Vol. 10 pp. 365–388.

Darr, A. (1990), *The Social Implications of Thalassaemia Among Muslims of Pakistani Origin in England: family experience and service delivery,* Ph.D Thesis, University of London.

Davis, A. (1982), *Women, Race and Class,* The Women's Press, London.

Decruyenaere, M.; Evers-Kiebooms, G. and Van den Berghe, H. (1992), 'Community knowledge about human genetics', in Evers-Kieboom, G.; Fryns, J.P.; Cassiman, J.J. and Van den Berghe, H. (eds), *Psychosocial Aspects of Genetic Counselling,* Wiley-Liss, New York, pp. 167–184.

Department of Health (1993), *Report of a Standing Medical Advisory Committee on Sickle Cell, Thalassaemia and Other Haemoglobinopathies,* HMSO, London.

Donzelot, J. (1979), *The Policing of Families,* Hutchinson, London.

Dyson, S. Davis, V. and Rahman, R. (1993a), 'Thalassaemia: Establishing Basic Awareness', *Health Visitor,* Vol. 66 No. 10 pp. 360–1.

Dyson, S. Davis, V. and Rahman, R. (1993b), 'Thalassaemia: Current Community Knowledge in Manchester', *Health Visitor*, Vol. 66 No.12 pp. 447–8.

Dyson, S. (1995), 'Clients as researchers: issues in haemoglobinopathy research', *Social Action*, Vol. 2 No. 4 pp. 4–10.

Dyson, S.; Fielder, A. and Kirkham, M. (1996), 'Midwives' and Senior Student Midwives' Knowledge of Haemoglobinopathies in England', *Midwifery* Vol. 12 pp. 23–30.

Dyson, S. (1997), 'Knowledge of sickle cell in a screened population', *Health and Social Care in the Community*, Vol. 5 No. 2 pp. 84–93.

Dyson, S. (1998), ''Race', ethnicity and haemoglobin disorders', *Social Science and Medicine*, (forthcoming).

Ebomoyi, E. (1988), 'Knowledge and attitude towards genetic screening for sickle cell disease', *Hygie: International Journal of Health Education*, Vol. 7 No.1 pp. 33–38.

Elias, N. (1978 [1939]), *The Civilizing Process Volume I: The History of Manners*, Basil Blackwell, Oxford.

Elias, N. (1982 [1939]), *The Civilizing Process Volume II: State Formation and Civilization*, Basil Blackwell, Oxford

Evans-Pritchard, E.E. (1950), *Witchcraft, Magic and Oracles among the Azande*, Oxford University Press, Oxford.

Farrant, W. (1985), 'Who's for amniocentesis?', in Homans, H. (1985) (ed), *The Politics of Reproduction*, Gower, Aldershot, pp. 96–122.

Foucault, M. (1971), *Madness and Civilization: a history of insanity in the age of reason*, Tavistock, London.

Foucault, M. (1973), *The Birth of the Clinic: an archaeology of medical perception*, Tavistock, London.

France-Dawson, M. (1990), 'Sickle cell conditions and health knowledge', *Nursing Standard*, Vol. 4 No.35 pp. 30–34.

Ghanei, M.; Adibi, P.; Movahedi, M.; Khami, M.A.; Ghasemi, R.L.; Azarm, T.; Zolfaghari, B.; Jamshidi, H.R. and Sadri, R. (1997), 'Pre-marriage prevention of thalassaemia: report of a 100,000 case experience in Isfahan', *Public Health*, Vol. 111 pp. 153–156.

Giddens, A. (1992), *The Transformation of Intimacy: Love, Sexuality and Eroticism in Modern Society*, Polity Press, Cambridge

Gilbert, G.N. and Mulkay, M. (1984), *Opening Pandora's Box*, Cambridge University Press, Cambridge.

Hammersley, M. (1995), *The Politics of Social Research*, Sage, London.

Hill, S.A. (1994), *Sickle Cell Disease in Low Income Families*, Temple University Press, Philadelphia.

Hindess, B. (1973), *The Use of Official Statistics in Sociology*, MacMillan, London.

Jennings, P. (1990), 'Thalassaemia. Levels of knowledge', *Paediatric Nursing*, Vol. 2 No. 7 pp. 22–3.

Johnson, T. (1995), 'Governmentality and the institutionalization of expertise', in Johnson, T., Larkin, G. and Saks, M. (eds), *Health Professions, The State and Europe*, Routledge, London, pp. 7–24.

Kerr, A.; Cunningham-Burley, S.; Amos, A. (1997), 'The new genetics: professionals' discursive boundaries', *Sociological Review*, Vol. 45 No. 2 pp. 279–303

Lorenzi, E. (1993), 'The effects of comprehensive guidelines for the care of sickle cell patients in crisis on the nurses' knowledge base and job satisfaction for care given', *Journal of Advanced Nursing*, Vol. 18 pp. 1923–1930.

Melrose, D. (1982), *Bitter Pills: medicines and the third world poor*, Oxfam, Oxford.

Merz, J. (1987), 'Matchmaking scheme solves Tay-Sachs problem', *Journal of the American Medical Association*, Vol. 258 pp. 2636–2639.

Midence, K.; Graham, V.; Acheampong, C. and Okuyiga, E. (1994), 'Increasing awareness for higher quality care: measuring knowledge of sickle cell disease in adult patients', *Professional Nurse*, January 1994, pp. 255–258.

Oakley, A. (1980), *Women Confined: towards a sociology of childbirth*, Martin Robertson, Oxford.

Oakley, A. (1984), *The Captured Womb: a history of the medical care of pregnant women*, Blackwell, Oxford.

Oliver, M. (1992), 'Changing the social relations of research production', *Disability, Handicap and Society*, Vol. 7 No. 2 pp. 101–114.

Pearson, M. (1986), 'Racist notions of ethnicity and culture in health education', in Rodmell, S. and Watt, A. (eds), *The politics of health education*, Routledge and Kegan Paul, London, pp. 38–56.

Pearson, M. (1989), 'Sociology of race and health', in Cruikshank, J. and Beevers, D. (eds), *Ethnic Factors in Health and Disease*, Wright, Guilford, pp. 71–83.

Petrou, M.; Modell, B.; Darr, A.; Old, J.; Kin, E.I. and Weatherall, D. (1990), 'Ante-natal diagnosis: how to deliver a comprehensive service in the United Kingdom', *Annals of the New York Academy of Sciences*, Vol. 612 pp. 251–263.

Phoenix, A. (1990), 'Black women and the maternity services', in Garcia, J.; Kilpatrick, R. and Richards, M. (eds), *The Politics of Maternity Care: Services for childbearing women in the twentieth century Britain*, Clarendon Press, Oxford, pp. 274–299.

Pinch, T. and Bijker, W.E. (1989), 'The social construction of facts and artifacts: or how the sociology of science and the sociology of technology might benefit each other', in Bijker, W.E.; Hughes, T.P. and Pinch, T. (eds), *The Social Construction of Technological Systems*, Massachusetts Institute of Technology Press, Massachusetts, pp. 17–50.

Powars, D. (1989), 'Newborn screening for sickle cell disease and other haemoglobinopathies. Diagnosis at birth improves survival of children with sickle cell anaemia', *Paediatrics,* Vol. 83 No.5 pp. 830–833.

Punales-Morejon, D. and Penchaszadeh, V.B. (1992), 'Psychosocial aspects of genetic counselling: cross-cultural issues', in Evers-Kieboom, G.; Fryns, J.P. and Cassiman, J.J. and van den Berghe, H. (eds), *Psychosocial Aspects of Genetic Counselling,* Wiley-Liss, New York, pp. 11–16.

Radley, A. and Billig, M. (1996), 'Accounts of health and illness: dilemmas and representations', *Sociology of Health and Illness,* Vol. 18 No.2 pp. 220–240.

Rapp, R. (1988), 'Chromosomes and communication: the discourse of genetic counselling', *Medical Anthropology Quarterly,* Vol. 2 pp. 143–157.

Rothman, B.K. (1993), 'The tentative pregnancy: then and now', *Fetal Diagnosis and Therapy,* Vol. 8 (Supplement 1) pp. 60–63.

Rothman, B.K. (1994), *The Tentative Pregnancy,* Revised Edition, Pandora, London.

Rowley, P.T.; Fisher, L. and Lipkin, M. Jr (1979), 'Screening and genetic counselling for beta-thalassaemia trait in a population unselected for interest: effects on knowledge and mood', *American Journal of Human Genetics,* Vol. 31 pp. 718–730.

Schilling, C. (1993), *The Body and Social Theory,* Sage, London.

Shakespeare, T. (1995), 'Back to the future? New genetics and disabled people', *Critical Social Policy,* Vol. 15 No. 2/3 pp. 22–35.

Smaje, C. (1996), 'The ethnic patterning of health: new directions for theory and research', *Sociology of Health and Illness,* Vol. 18 No. 2 pp. 139–171.

Smith, J.A.; Epseland, M.; Bellevue, R.; Bonds, D.; Brown, A.K. and Koshy, M. (1996), 'Pregnancy in sickle cell disease: experience of the cooperative study of sickle cell disease', *Obstetrics and Gynaecology* Vol. 87 No. 2 pp. 199–204.

Stacey, M. (1996), 'The new genetics : a feminist view', in Marteau, T. and Richards, M. (eds), *The Troubled Helix: social and psychological implications of the new human genetics,* Cambridge University Press, Cambridge, pp. 331–349.

Stamatoyannopoulos, G. (1974), 'Problems of screening and counselling in the haemoglobinopathies', in Motulsky, A.G. and Ebling, F. (eds) *Proceedings of IVth International Conference on Birth Defects,* Excepta Medica, Amsterdam, pp. 268–276.

Turner, B.S.(1992), *Regulating Bodies: essays in medical sociology,* Routledge, London.

Turner, B.S. (1996), *The Body and Society,* Second Edition Sage, London.

Vichinsky, E.; Hurst, D.; Earles, A. et al (1988), 'Newborn screening for sickle cell disease: effect on mortality', *Paediatrics* Vol. 81 No. 6 pp.749–755

Watson, E.K.; Mayall, E.S.; Lamb, J.; Chapple, J. and Williamson, R. (1992), 'Psychological and social consequences of community carrier screening for cystic fibrosis', *Lancet* Vol. 340 No. 8813 pp. 217–220.

Wood-Harper, J. and Harris, J. (1996), 'Ethics of human genome analysis: some virtues and vices', in Marteau, T. and Richards, M. (eds) *The Troubled Helix: social and psychological implications of the new human genetics,* Cambridge University Press, Cambridge, pp. 274- 294.

World Health Organization (1988), *The Haemoglobinopathies in Europe: combined report of two WHO meetings,* (Document EUR/ICP/MCH 110), WHO Regional Office for Europe, Copenhagen.

World Health Organization (1994), *Guidelines for the control of haemoglobin disorders Geneva: World Health Organization,* Unpublished WHO document of the VIth annual meeting of the WHO working group on haemoglobinopathies, Cagliari, Sardinia 8–9th April 1989.

Zare, N.; Sorenson, J.R.; and Heeren, T.C.(1984), 'Sex of provider as a variable in effective genetic counselling', *Social Science and Medicine,* Vol. 19 pp. 671–675.

Zola, I.K. (1977), 'Healthism and disabling medicalization', in Illich, I.; Zola, I.K.; McKnight, J.; Caplan, J. and Shaiken, H. (eds), *Disabling Professions,* Marion Boyars, London, pp. 41–67.

24

GERM-LINE GENE THERAPY

Is the Existing UK Norm Ethically Valid?

Sheila Dziobon*

University of Exeter
Rennes Drive, Exeter, EX4 4RJ, UK

Just those action norms are valid to which all possibly affected persons could agree as participants in rational discourse
—Habermas, 1996, p.107

1. INTRODUCTION

According to Jürgen Habermas, regulation of a society by "action norms" remains valid for those to whom they are addressed if they are a product of rational discourse within a pre-existing framework of private and public rights granted to an individual as a member of a socially determined association - the society of which the addressee is a member (Habermas, 1996, p.107). This pre-existing framework places emphasis on rights in a broader sense than either the "liberal" tradition's concept of human rights as the expression of moral self-determination or the idea of rights as the legal protection of the individual against institutions of the state. The basic requirement is the "*right to the greatest possible measure of equal individual liberties*" which can be achieved only with the associated rights of "*status of member*" and the political facility to exercise those rights, including "equal opportunities to participate in processes of opinion-and will-formation in which citizens exercise their *political autonomy* and through which they generate legitimate law." (Habermas, 1996, p.122–3).

The purpose of this paper is twofold: first, to look at the validity of "action norms" in relation to germ-line gene therapy in the light of Habermas's theory and consider who has been, and continues to be, involved in the "rational discourse" leading to the current norms; and second, to question the validity of such regulation for persons "possibly af-

* Telephone: ++44 1392 263151, Fax: ++44 1392 263196, email: S.J.Dziobon@ex.ac.uk

Genetic Information, edited by Thompson and Chadwick.
Kluwer Academic / Plenum Publishing, New York, 1999.

fected". The basis for discussion in respect of the first issue will be the Clothier Report (Report, 1992), (and subsequent reports from the Gene Therapy Advisory Committee (GTAC)) together with the Human Fertilisation and Embryology Act 1990. In relation to the second question, some important issues are raised with respect to the consequential effect on women encouraged to consider genetic testing as part of normal antenatal care. As a result of this examination the question will remain - can the existing norm with respect to the recommendation not to attempt germ-line gene therapy retain validity?

2. BACKGROUND

The human genome initiative (Beckwith, 1991) has been compared to scientific projects on the scale of "the building of the first atomic bomb or sending astronauts to the moon" (Davis et al, 1990). The genetic research and treatment under discussion takes two forms: somatic cell gene therapy (that which is administered or aimed at the living being, including the foetus);[1] and germ-line gene therapy (that which is aimed at altering the genetic make-up of the gamete or germ-line).

Somatic cell gene therapy involves the insertion of a normal gene into the body cells of a patient with a genetic disorder and has the potential for reducing or eliminating the effects of that genetic disorder.[2] Somatic cell gene therapy does not effect an individual's germ-line and, therefore, does not (as far as scientific knowledge is currently aware) effect the genetic make-up of their offspring and future generations. Germ-line gene therapy has the potential to affect subsequent generations (Anderson, 1989).

Private autonomy in the area of health care and reproductive choice is no stranger to conflict with the promotion of public values through legal constraint.[3] Apparently accepted ethical values underpin the decisions to curtail an individual's freedom in respect of their bodies, or body parts, for the common good even when the decision to allow individual action results in no obvious 'harm' to the rest of society.[4] With respect to germ-line gene therapy there is no specific legislation only influential guidelines, but these will operate with existing legislation to regulate the future of research in this area.

3. THE REPORT OF THE COMMITTEE ON THE ETHICS OF GENE THERAPY (REPORT, 1992)

3.1. The Composition of the Committee

The Committee was formed in 1989 and presented its report to Parliament in January 1992. Its terms of reference were *inter alia*:

> To draw up ethical guidance for the medical profession on treatment of genetic disorders in adults and children by genetic modification of human body cells

Having deliberated and drawn up these guidelines, the Committee (and more recently the GTAC) use these as the basis for deciding whether or not to allow individual research projects and pilot clinical studies. These guidelines, therefore, form the basis of the United Kingdom view on genetic modification of human body cells and, although, not in the form of legislation, have the effect of allowing or prohibiting various forms of research and development in this area. It would be thought, therefore, that the ethical values underpinning these guidelines are representative of values existing throughout the community and for those whose values are not reflected in the guidelines, equality of opportunity has been given to enable them to enter into the "opinion- and will-formation" preceding their adoption.

The Committee, chaired by Sir Cecil Clothier KCB QC,[5] comprised nine members, seven of which are Members or Fellows of the Royal College of Physicians, one a female lay member of a Regional Health Authority and one a male broadcaster and journalist.[6] The Committee approached organisations and individuals for input into the discussion. Out of the eighty-two submissions received, 26 were from organisations representing those working in the medical or scientific field, 12 from religious organisations, 2 from women's organisations and 52 from Health Authority Research Ethics Committees (Report, 1992, p.31). Whilst obviously eminent in their fields those participating in the rational discourse leading to the ethical guidelines and subsequent recommendations are overwhelmingly those working in, academically qualified in, or participating in, the application of health services to the rest of the community. The large number of health service providers involved in the deliberations implies a belief that their values represent the values of all members of society as the private autonomy of all possible addressees may be affected by the norms emanating from the Committee. Can this belief be ethically justified because their particular qualities and expertise indicate that their values are the right ones or is it that their values are the ones they believe ought to prevail? If it is the latter does the resulting action retain a quality of paternalistic imposition overriding alternative values of some addressees within our society? If this is the case then the resulting norm fails to meet the indicated criteria for validity as it fails to respect the most basic rights attached to membership of our society.

3.2. The Ethical Basis

The Report begins by giving some background to the issues under discussion. This rhetoric paves the way for their later recommendation that germ-line gene therapy should not be attempted by appealing to "public concerns" and acknowledges the potential existence of "irrational fears which derive from misunderstandings of biology, and are compounded by effects of popular creations of fiction, such as Frankenstein's monster." (Report, 1992, p.2).[7] This effectively, but subtly, reinforces the premise that it is those with knowledge, those "in the business", who are the best equipped to discuss and draw up guidelines which have the potential to affect every member of our society.

The Committee states the ethical basis of their deliberations. The questions mentioned above concerning the representative constitution of the decision-making group is compounded:

> We begin from the basis that ethics are the moral convictions of *thoughtful, conscientious and informed people*, to be rejected only when they are in conflict with other convictions which stand better the test of reflection. (1992, p.10).[8]

The Committee is, in effect, making a judgement as to whose values are to be reflected, presenting their subjective conclusion as if undisputed fact. The inclusion of the word "informed" again focuses on the qualifications for membership of this exclusive discourse: health providers, academics or church leaders are considered sufficiently "informed" to enter the discussion.

In an increasingly pluralist society the temptation to promote a conception of the good which reflects temporally and historically a society which may or may not exist now, or even in the future, is a route to be guarded against in order to ensure validity of the resultant norm. This promotion of a concept of the good manifests itself in the two principles which the Committee applies to the task in hand:

> The first ethical principle which has commanded our attention is the obligation inherent in human nature to enquire, to study, to pursue and apply research by ethical means. (1992, p.10).

The Committee here is attracted by theories of natural law, referring to "convictions derived from a compound of natural philosophy and religion", and this discourse has the effect of excluding those who hold alternative values. They continue by reiterating ethical values in a circular-type argument which reinforces their own beliefs. The principle states no positive norm, but has the effect of marginalising members of society who do not follow their ethical guidelines by partially dehumanising them ("inherent in human nature"). The Committee suggest that these statements of principle are "intuitive" (1992, p.10). This is another marginalising phrase alienating those who may not "feel" this way at all, and the Report continues by outlining an utilitarian approach with obvious references to the balancing of risks and benefits for individual patients. These phrases return to play a prominent part in their later recommendations.

> The second ethical principle which has governed the deliberations of the Committee is that in the sometimes inescapable tensions between the pursuit of knowledge and the protection of patient's interests, the latter must prevail. (1992, p.10).

Again, in the guise of a statement of fact, the Committee promotes a particular definition of "patient's interests" which includes only current and existing patients. This could be argued in the alternative suggestion that any risks associated with creativity and development at the boundaries of scientific knowledge and research are equally, if not more, relevant to future generations of patients. A shifting of the boundaries of a risk/benefit analysis could incorporate with greater weight potential benefits to future generations; and equally, a principle intending to deny future generations possible relief from the very pain and suffering offered to existing patients could be emphasised.

It is not suggested here that the ethical principles as stated by the Committee would not be those reached if there had been wider consultation and deliberation, as this may be so. Without a greater and more varied input into the discussion, however, this is not known and the guidelines which result may alienate, or fail to address important concerns and beliefs honestly held by members of our society whose values differ from these principles out of which the guidelines emerge.

If all members of society have the right to political participation[9] it is plausible that a great many of these may have different ideas of which ethical values should apply in respect to gene therapy and in particular germ-line gene therapy. The Report may be a reflection of the ethical values of the Committee (and those who responded to the consultation documents as indicated above) but it may marginalise and exclude, without consideration, many other equally valid values. The composition of the Committee does not seem to reflect the membership of our society and the ethical position adopted may exclude valid and rational alternatives of those not heard, many which may be "thoughtful, conscientious", some which may not. Many will undoubtedly not be "informed", but all should have the equal opportunity to participate in the decision-making process. Does the resulting Report, therefore, remain valid for those invisible members of our society? For those potentially affected it is suggested that the majority will remain ill-informed unless and until faced with an individual decision involving gene therapy. At that juncture, however, the existing and adopted norms will be applied and the curtailment of their private autonomy will become a factual reality.

3.3. Equality of Opportunity in Practice

The input into the Report demonstrates the narrow field of enquiry prior to the adoption of the guidelines. Criticism may be levelled in respect to this limitation but the question will be raised as to how it would be possible to realistically give everyone a voice in such a discourse. There is undoubtedly a difference between theory and reality in the application of an ideal. A more representative group is certainly a realistic possibility, but how would rational discourse prevent the exclusion of minorities and the ill-informed? It is self-evident that in a discursive situation individual actors, or actors who find themselves as a part of a like-minded group, will seek to promote their own values in an attempt to persuade others to adopt them as their own. Such individuals will have varying bargaining power and this form of argumentation has the potential to destroy or distort the fundamental right of equal opportunity for all. Presupposing and promoting an alternative form of discourse, "communicative action discourse," Habermas suggests that participants in the discussion have an obligation to all other particpants to allow others to be heard, as only by respecting the rights of others will their own rights remain valid within their society (Habermas, 1981, p.273–337). By entering into a discussion empathising and considering the views of others, each must reflect on their own values in the light of the values of the other participants and reach a reflective equilibrium which will represent the basis for any resulting norm. This is recognised by the Committee within their introduction to the ethical basis:

> [E]thics are...to be rejected only when they are in conflict with other convictions which stand better the test of reflection. (Report, 1992, p.10).

In any discourse there will be those whose values remain unrepresented, who have not been persuasive, or who adopt an extreme or different opinion with respect to the issue under consideration. Ideally, therefore, there will be two "groups" of people. Firstly, representatives from all sections of society who will consult and consider the issue under discussion. Following a process of reflection, values will be adopted which reflect only the values of a finite portion of those deliberators. However, those whose values have not been adopted as influential for any subsequent action norms will continue to want their values to be reflected. When the same, or related issues are presented again for discussion they will have the opportunity to persuade others of the rationality of their particular moral, ethical or religious values and retain the potential to move into the core group. This produces a constantly shifting,[10] but free and equal opportunity for meaningful discourse and avoids the alienation and marginalisation of any member of society. As participation can, however, at a practical level only be representative, it is submitted that those involved in deciding the value base of the Report failed to meet this need. Before looking at the application of the ethical basis to potential addressees it is necessary to look at the Report's recommendations and how it will interact with existing legislation in the form of the Human Fertilisation and Embryology Act 1990.

3.4. The Report's Recommendations

3.4.1. Somatic Cell Gene Therapy. Bearing in mind the purpose of somatic cell gene therapy the Committee concludes that this is "a proper goal for medical science"(Report, 1992, p.13). The Committee, however, suggests strict controls over such therapy by classifying it as "research", subjecting it, therefore, to the more rigorous guidelines as laid down by The Royal College of Physicians of London (RCP, 1990a & b). The Committee includes foetuses among those suitable for consideration for somatic cell gene therapy research, but only after the differentiation of the germ-line cells estimated at approximately five weeks after fertilisation (Report, 1992, p.13).[11] Any decision involving the use of such research will follow extensive prior assessment of the balance of potential benefits and risks for the individual patient (1992, p.14). The Committee applies their own ethical guidelines: the first principle manifests itself by reference to existing guidelines in respect to medical research generally, and the second principle reiterates the emphasis on the individual, existing patient to the exclusion of the future patient. The Report constantly reiterates the inherent dangers, the potential harm and the need to ensure non-interference with the germ-line in advance of its later recommendation against any attempt at modification of the germ-line.

3.4.2. Germ-Line Gene Therapy. On the basis of "insufficient knowledge" the Committee recommends no attempt should yet be made for modification of the germ-line (1992, p.18). This conclusion seems inevitable. The Committee sets the scene early in the Report when it states:

> There is a belief, which was strongly affirmed in the responses to our consultation, that genetic treatments of human beings should not be undertaken if they are designed to, or may inadvertently, affect future generations. (1992, p.1).

Where this "belief" is founded is not clear, but as it is stated prior to consultation the impression is that it is a pre-existing value judgment of the Committee itself. Having stated its position, the Report continues its rational response by proposing an examination "of this subject too". The illustration of the potentially damaging consequences of gene therapy in general are graphically described in the Report in the following paragraphs: the "fear of the unknown"; the potential "to influence life and health more fundamentally than could any treatment available hitherto" and the recognition that therapy could alleviate suffering "but that [it] should be used prudently." (1992, p.1–2). In its introductory guide the Report, in an otherwise descriptive passage, mentions biodiversity "[T]he great variety of living things is made possible by their rich genetic diversity, which provides an equally rich diversity of proteins." (1992, p.3). The Committee appears unsure in respect to the effect on the germ-line by somatic cell gene therapy (1992, p.6), but, clearly, the aim of treating the germ-line, as opposed to the somatic cell, is to directly affect the genetic make-up of future generations. The Report continues by separating their consideration of germ-line gene therapy from that of somatic cell gene therapy, a dislocation which affirms the belief that the underlying ethical issues are in major ways different from each other. The Report makes this distinction "because little is known about the possible consequences and hazards, and any harm to future generations would take a long time to discover and deal with" (1992, p.9). Scientific development is certainly in its infancy in relation to germ-line gene therapy, but it is difficult to understand why this necessitates separate consideration. From the utilitarian perspective (as adopted as part of the ethical basis of the Report) such issues, whether in respect of somatic cell or germ-line gene therapy, would surely be integral to any risks/benefits analysis. The dénouement appears, therefore, rational and unsurprising—the recommendation at paragraph 5.2 that "gene modification of the germ-line should not yet be attempted."[12] There is no explanation of when, if ever, it might be, or what criteria or ethical principles will be applied at that time should they be different (as implied by paragraph 2.26) from those concerning somatic cell gene therapy. The recommendation at paragraph 5.2 recognises the dilemma for couples wanting to have a child but who are identified at risk from an inheritable genetic disorder. With the option of genetic therapy aimed at eradicating such a genetic disorder in the germ-line prohibited, the possible alternatives for such parents are "embryonic diagnosis and selective implantation". Such alternatives are possibilities when fertilisation takes place *in-vitro* but for other couples, testing during pregnancy may reveal a genetic disorder and somatic cell gene therapy might be available for the foetus (after differentiation of the germ-line cells) or the new-born infant, or they may decide to terminate the pregnancy. Somatic cell gene therapy may, or may not, be effective but the future reproductive dilemma for the new-born child may remain the same as that of her/his parents.

4. THE HUMAN FERTILISATION AND EMBRYOLOGY ACT 1990 (HFEA)

The HFEA details the setting up of a Licensing Authority (ss. 5–10) to deal with, *inter alia*, the issuing of licences for treatment and research in relation to human genetic therapy (ss.10–22). This is the legal regulation which claims validity in the real world of all addressees. The Act defines the meaning of "embryo" and "gamete" (s.1) and, in Schedule 2, lists activities for which licences may be granted. S1.(4) of Schedule 2 Licences for treatment states that:

> A license under this paragraph cannot authorise altering the genetic structure of any cell while it forms part of an embryo.

S.3(4) Licences for research states that:

> A licence under this paragraph cannot authorise altering the genetic structure of any cell while it forms part of an embryo, except in such circumstances (if any) as may be specified in or determined in pursuance of regulations.

Such "circumstances" may be found, for example, under s.3(2)(e) "developing methods for detecting the presence of gene chromosome abnormalities in embryos before implantation", but in such circumstances a licence can only be authorised if it appears to the Authority to be *necessary or desirable* for this purpose. There is, therefore, no specific banning in the Act on treatment or research in relation to gametes, but it seems likely that any such research or treatment would need to meet the requirements of necessary or desirable as interpreted by the Licensing Authority. In its deliberations the Authority will, no doubt, take into consideration the recommendations of the Report and refuse any licence which asks to conduct germ-line gene therapy, whether on the embryo, which seems to be specifically excluded by the Act, or on gametes prior to fertilisation.

The question remains, therefore, that if the Committee supports its findings that germ-line gene therapy should be banned as there is the alternative and less risky somatic cell gene therapy which can be used to treat genetic disease in-utero or post-natal, should the parents elect for this rather than termination? The HFEA bans genetic manipulation of embryos *in-vitro* (s.3(4))[13] allowing only genetic testing and selective implantation. For those whose future children are at high risk of being born with a genetic disorder there is little current regulation can offer. Science, on the other hand, may be able to and it must be asked why these directly affected members of society were not represented at the initial discussion of the ethical values from which the subsequent regulation was to emerge. There were no submissions to the Committee from disabled organisations or carers out of the professional arena who might be thought to be some of the most affected and certainly well-informed members of society. This dismissal of germ-line gene therapy as a route to the eradication of genetic disorders for future generations leads into the final issue for consideration in this paper. Will potential parents consider these proposals and legislation valid for them?

5. SOME CONSEQUENCES OF THE COMMITTEE'S FINDINGS

Although in its infancy, applications for gene therapy research using somatic cell gene therapy are presented for approval before the Gene Therapy Advisory Committee which was set up following the Committee's Report.[14] In its initial year, 1993/4, a total of ten research protocols were approved, in 1994/5 an additional three, and in 1995/6 an additional five. Cystic fibrosis is the most investigated single gene disorder. Genetic testing kits for cystic fibrosis are now available by mail-order placing further pressure on researchers and individuals.[15] Detection alone of a single gene disorder can provide information for prospective parents but adds little to existing reproductive choices (Lippman, 1991, p.20).[16] The increase in trials in somatic cell gene therapy must always be considered in the light of the existing moratorium on germ-line gene therapy.

5.1. Potential Implications on Birth Choices

The paper so far has attempted to analyse the discourse leading to regulation of genetic therapy and particularly the proposed ban on germ-line gene therapy. As mentioned earlier, all members of society must be enabled to freely and equally participate in this discourse, whether they chose to do so or not, in order to give regulation validity as a norm operating on members of that society.

When considering the interaction between the Report and the HFEA it was noted that, on the ethical basis of discussion, germ-line gene therapy was too risky to consider and that the same result (the eradication of genetic disorders) could be achieved by somatic cell gene therapy or termination. It is interesting to note that the ethical basis of the Committee's Report raised no issue in respect of this assumption.

As genetic manipulation of the *in-vitro* embryo or pre-fertilisation gamete is banned, those parents at high risk of conceiving a 'diseased' child, but who want children, will have the option (or their medical supervisors will advise the options) to discard embryos tested as genetically unsound, or to go ahead with a pregnancy in the hope that somatic cell gene therapy will be able to treat the foetus or future infant. The parents themselves may have benefited from somatic cell gene therapy but live with the knowledge that their germ-line will carry a defect.

5.2. Prenatal Genetic Testing

The use of pre-natal testing has become accepted over the past twenty years. Routine use is made of ultrasound scans to date and check the development of the foetus and are accepted by parents as a normal part of the ante-natal process. In addition to blood testing (and in some cases this includes HIV), many of these tests are accepted to the extent that they barely need consent (Lippman, 1991, p.21).[17] The acceptance, or otherwise, of such prenatal testing is regarded as responsible parenting and failure to conform implies an increased risk of harm to the foetus and may incur the alienation of the health service provider or members of society generally.[18] It is considered the norm.

> [T]hrough the use of prenatal diagnosis women can avoid the family distress and suffering associated with the unpredicted birth of babies with genetic disorders or congenital malformations, thus preventing disability whilst enhancing the experience of pregnancy. (Lippman, p.25).[19]

There already, therefore, exists a precedent in our culture which will lay the path for prenatal genetic testing to become the norm. As the human genome project enables the identification of more single and multiple gene abnormalities, whether directly or indirectly connected with disease or disability, it is feasible to assume that without too much pressure pregnant women will be 'persuaded' by society and health care providers to have these tests. They, too, will become the norm.

5.3. The Human Genome Project and Humanness

It is suggested that the human genome project will affect our view of ourselves and our relationships with others (Kennedy & Grubb, 1994, ch. 1). Reductionism of the human being to a collection of identifiable genes, neatly labelled and stored, will result in a more mechanistic view of ourselves and others. Reproductive choices are currently "engineered" by social conditions, education and economics. If we are no more than a catalogue of genes it will become an increasingly large part of parenting choices to consider the genetic make-up of your selected partner (1994, ch.1).[20]

Prenatal genetic testing may become a regular and automatic occurrence and the resultant pressure will be on women to abort for genetic disorders, some of which may be largely dependent on environmental circumstances and some of which will carry a small and indeterminate risk of development. If somatic cell gene therapy develops successfully, the incidents of disability and disease within society will decline correspondingly, and it will become increasingly anti-social to bring into the world a new life with a genetic disorder which could be tested for prior to birth. The Committee's recommendation and the HFEA currently prevent the possibility of human research of germ-line gene therapy. The ethical dilemma for couples at risk has increased.

6. CONCLUSIONS

The recommendations of the Report of the Committee on the Ethics of Gene Therapy have the potential to affect the lives of all members of UK society, directly or indirectly. As detailed above, the consultation and deliberation prior to the Committee's Report was almost exclusively among an elite group of specialist persons. The resultant norms may have the effect of marginalising the values of many individuals and groups within our society who have not had the equal opportunity of participating in these deliberations. Changes to the personnel of the current Genetic Therapy Advisory Committee are to be welcomed as being more widely representative. For all those "possibly affected" the resultant norms would appear to be "merely paternalistic imposition" and when applied may lack validity in the real world. As scientific knowledge and discovery moves inexorably into the 21st century such issues will inevitably give rise to wider public debate.

NOTES

1. The UK legal definition of a "life in being", however, excludes the foetus for purposes of criminal law sanctions in this context.
2. For example, in November, 1996 the UK Gene Therapy Advisory Committee (GTAC) gave conditional approval to gene therapy trials for cystic fibrosis, (GTAC *Third Annual Report*, Health Department of the United Kingdom, 1997).
3. There are many examples attracting criminal sanction in UK legislation; the regulation of the termination of pregnancies, commercial contracts for surrogacy, the acquisition of organs for transplant etc.
4. For example *R v HEFA, ex-parte Blood*, The Times Law Reports February 7, 1997. The underlying ethical values manifested in the Human Fertilisation and Embryology Act were possibly compromised by application of an "amalgam of domestic and European Community law", and permission given to Mrs. Blood to travel with the sperm of her deceased husband for treatment outside the UK. This outcome has the effect of shifting the values which gave rise to this particular "action norm".
5. Sir Cecil Clothier KCB QB, Chair, Council on Tribunals: formerly Parliamentary Commissioner for Administration and Health Service Commissioner for England, Scotland and Wales, and Chair, Police Complaints Authority.
6. For full details of the members, their qualifications and status, see page iv of the Report. *The Third Annual Report of GTAC* (Department of Health, 1997) lists current membership, p.16. These include representatives from The Cystic Fibrosis Trust, The Tuberous Sclerosis Association, The Peptide Therapeutics Group, The Church in Wales and a Barrister, in addition to health care workers and scientists.
7. The Report draws on an editorial comment, Lancet, 1989, I: p.193–4, but compare Kerr, A, Cunningham-Burley, S, Amos, A., *The New Genetics: Public Awareness and Public Involvement*, Department of Public Health Sciences, Medical School, University of Edinburgh, presented Preston, 1997.
8. Author's italics.
9. Providing they have membership rights within that society as agreed. (Habermas, 1996, p.122).
10. This is not to suggest that no decision is made, but is more a reflection of shifts in societal values and changes in political power within a society.
11. Report, Ref. 23. Moore, K.L. *The Developing Human—Clinically Orientated Embryology*, Philadelphia: Saunders, W.B., 1977.
12. In November, 1996 the Council of Europe adopted the *Convention on Human Rights and Biomedicine* and this includes a moratorium, for a minimum of five years, on germ-line gene therapy in humans.
13. HEFA s.3(4) restricts such testing and selection of embryos up to a maximum of 14 days beginning with the day when the gametes are mixed, not counting any time during which the embryo is stored.
14. GTAC Annual Reports November, 1993—December 1996 show a total of 18 trials involving 145 patients. Of these trials 6 are for cystic fibrosis (66 patients), 10 for various forms of carcinoma (78 patients), one for ADA (1 patient) and one for Huler's Syndrome (0 patients specified).
15. Unsubstantiated comment suggests that approximately 1,000 kits have been ordered by mail since advertising appeared in the UK (there are no official figures).
16. At the beginning of this decade over 150 "single gene" disorders were detectable and this is continually increasing.
17. As Lippman comments, "It is worth noting that ultrasound is no longer the only genetic technology applied without prior consent. Screening for carriers of haemoglobin disorders, for example, is also done unbeknownst to the individuals being tested in certain jurisdictions."
18. It is possible to imagine a situation where it is required by law that a pregnant women at risk should have genetic testing. The court has previously intervened in the conflict between a pregnant women's rights and

those perceived as being in the "best interests of the child" with an order, for example, for a Caesarian section. See *Re S's Application for Judicial Review*; Court of Appeal, Independent LR 10 July 1997.

19. Lippman, fn. 35, refers to comment on the test of amniocentesis. "The procedure offered those at risk the possibility of 'health' ... [it] provided parents with reassurance and avoided abortion ... [and it] prevent[ed] desease and disability." McDonagh, J. (1990), *Congentital Disability and Medical Research: The Development of Amniocentesis*, 16 Women & Health 137, 143–44.
20. Kennedy & Grubb, chapter 1, quoting from the US Congress's Office of Technology Assessment (OTA) in 1988. The Office as discussing the 'Social and Ethical Considerations' raised by the Human Genome Project: "Human mating that proceeds without the use of genetic data about the risks of transmitting diseases will produce greater mortality and medical costs than if carriers of potentially deleterious genes are alerted to their status and encouraged to mate with noncarriers or to use artificial insemination or other reproductive strategies." There is clearly a tension between this deliberation and the need to maintain "rich genetic diversity" as detailed by the Committee's Report and noted earlier in this paper at paragraph 3.4.2.

REFERENCES

Anderson, W.F., (1989) 'Human Gene Therapy: Why draw a line?, *The Journal of Medicine and Philosophy* 14, 681–693.

Beckwith, J. (1991), 'Foreword: The Human Genome Initiative: Genetics' Lighning Rod', *American Journal of Law and Medicine*, Vol. XVII. No. 1–2.

Council of Europe: *Convention for the Protection of Human Rights and Dignity of the Human Being with regard to the Application of Biology and Medicine: Convention on Human Rights and Biomedicine.* European Treaty Series 164, Oviedo, 4.IV.1997.

Davis et al (1990),'The Human Genome and Other Initiatives', 249 *Science* 342, 342.

Gene Therapy Advisory Committee (GTAC), (1997) *Third Annual Report, Health Department of the UK.*

Habermas, J (1981) *The Theory of Communicative Action* (trans. Thomas McCarthy, 2 vols, Boston: Beacon, 1984, 1987).

Habermas, J. (1996), *Between Facts and Norms. Contributions to a Discourse Theory of Law and Democracy*, Blackwell, Oxford. (Translated by William Rehg).

Human Fertilisation and Embryology Act 1990 (1990 c.37).

Kennedy & Grubb, (1994) *Medical Law: text with materials*, 2ed. Butterworths, London.

Lippman, A (1991), 'Prenatal Genetic Testing and Screening: Constructing Needs and Reinforcing Inequities', *American Journal of Law & Medicine*, Vol. XVII. No. 1 & 2, p. 15–50.

Report of the Committee on the Ethics of Gene Therapy, Cm. 1788, (1992) HMSO, London, (The Clothier Report)

Royal College of Physicians, (1990a) *Research Involving Patients. A Working Party Report*, London.

Royal College of Physicians. (1990b) *Guidelines on the Practice of Ethics Committees in Research Involving Human Subjects*, 2ed London.

25

NEGOTIATING THE DILEMMAS OF PRENATAL TESTING FOR GENETIC DISORDERS

What Is the Virtuous Person to Do?

Ann Gallagher

Faculty of Healthcare Sciences, Kingston University, and
St. George's Hospital Medical School, Kingston Hill Campus
Kingston upon Thames, Surrey KT2 7LB

1. INTRODUCTION

There is little doubt that potential parents have always been concerned about the well-being of the foetus, their invisible baby-to-be. In the past reassurance, if it came, came only with birth. Prenatal testing techniques and genetic knowledge now provide the means to see the foetus and to test it for genetic disorders. It has been argued that 'the availability of foetal surveillance techniques has transformed the pregnancy experience from a developmental process and a miracle of nature to a risk-dominated and technology-guided event'(Raines 1996). Rather than a transformation, however, it seems that awe continues to co-exist with varying degrees of agonizing in relation to calculated risk.

A variety of prenatal tests are available in this country. A blood test is offered (AFP test, the Triple test, the Triple Plus test, the Bart's test or the Biomark test) at about 16 weeks to test initially for genetic disorders, such as Down's syndrome. Several substances, or markers, in the blood are measured. The result and the woman's age is then programmed into a computer and the risk calculated. High risk is generally interpreted as less than one in 250 and low risk is calculated as more than 250 (Barnes & Bryan 1996). An early pregnancy scan or ultrasound, at the end of the first trimester can date the pregnancy, diagnose multiple pregnancy, and indicate the risk of chromosomal abnormality. If the odds are considered high, as with the blood test, other tests are offered, for example chorionic villus sampling and amniocentesis. The risk of miscarriage with the latter two tests is approximately 2% and 1% respectively. Other risks of these tests include bacterial infection and foetal injury (Minnuti 1996).

Prenatal testing for genetic disorders generates a number of challenging questions. What, for example, is a genetic disorder? For which genetic disorders and/or predisposi-

Genetic Information, edited by Thompson and Chadwick.
Kluwer Academic / Plenum Publishing, New York, 1999.

tions should prenatal testing be available? Which theory might best throw light on ethical questions which arise. In negotiating the dilemmas, how might the virtues guide the professional and the patient/pregnant? Finally, what guidance might we glean from virtue theory as we tentatively tread into the next millennium?

2. GENETIC DISORDER

There are, according to The Progress Guide to Genetics (Pembrey 1997), approximately 4,000 known, simply-inherited genetic disorders and together they are responsible for suffering in 1–2% of the population. Common examples given include cystic fibrosis, sickle cell disease, the thalassaemias, fragile X syndrome and Huntington's disease. Other genetic disorders include Down's Syndrome and spina bifida.

The question as to what constitutes a genetic disorder raises questions also about normality and abnormality. This is discussed in the Euroscreen Report on the ethics of genetic screening (Chadwick 1997 p. 6–7). The consensus was that 'the purpose of prenatal screening is to detect serious disease and that....there was no right to test for everything.' There is no consensus on what constitutes a genetic disorder, or disease or disability. It has been pointed out by Rothblatt (1997 p.7), for example, that 'One man's disease may be another man's inspiration. One woman's disability may be another woman's different ability. One person's disorder may simply be another person's personality.' Even if there was a consensus on what constituted genetic disorder, it seems inevitable that controversies would continue about prenatal testing and termination (Penticuff 1996 p.786). As the Human Genome is unraveled and additional diseases, disorders and predispositions are detected, questions as to which prenatal test offers to make will become more onerous. Indeed, the possibility of parents being offered a genetic profile of their foetus has become imaginable.

Genetic disorder implies that there is some genetic order and in this a quotation from Michael Kaback is sobering and, I think, helpful 'We're all mutants. Everybody is genetically defective'(Rothblatt 1997 p.145).

3. THEORETICAL POSSIBILITIES

In the quest for a quick fix and a simple explanation some people may look to genetics. It is suggested that 'humans are suckers for easy solutions'(Rothblatt 1997 p.165). The quest for an easy ethical solution to the agonising challenges presented by prenatal testing might lead us to believe that mastery of some ethical principles or theories would resolve our dilemmas. There is no doubt that an understanding of theories such as deontology, utilitarianism and rights approaches can facilitate deliberation and provided some justification for actions and omissions in relation to prenatal testing. They have, however, been criticised for being impersonal, abstract, overly rational and for ignoring emotions (Alderson 1991). The ethical issues surrounding prenatal testing are, most certainly, personal, contextual and emotional. Decisions made by parents, particularly women, can enhance or detract from the rest of their lives; professionals may find themselves counselling women about the most fundamental issues in relation to individual flourishing; and policy-makers may flounder when confronted with societal genetic possibilities. There is much to prenatal testing that falls outside an abstract rational analysis and much which seems to be illuminated by virtue ethics. Beauchamp and Childress (1994 p. 462) state that 'morality would be a cold and uninspiring place without various traits of character, emotional responses, and ideals that reach beyond principles and rules.'

Aristotle, one of the earliest and most significant virtue theorists, is quoted (in his *Politics*) as stating: 'As to the exposure and rearing of children, let there be a law that no deformed child shall live'(Carey 1991) and to hold the view that 'the source of male virtue is thought to be the rational element of the soul which, while present in the female, is ineffective' (Johnstone 1994 p. 109). He may, then, be thought an inappropriate theorist to provide guidance on an issues relating (for the most part) to women and children. Despite these views, virtue ethics seems to offer a more complete picture of the ethical aspects of prenatal testing. Rather than focusing on acts or omissions, a virtue approach focuses on the character of the person involved.

Aristotle believed, as do his philosophical descendants (Geach 1977, Foot 1978, Hursthouse 1987) that living a good life or flourishing was what human beings aspire to and to do this, they must acquire and practice the virtues. According to Hursthouse, 'the fully virtuous person, with full wisdom, would be the person who *knew* what to do or think in exactly those circumstance that many of us find so deeply puzzling....it becomes clear that she is an ideal. We go for guidance to, or try to model ourselves on, the people who approximate to that ideal'(1987 p.247). This suggests the importance of role models and highlights the circularity of the title question—the virtuous person becomes virtuous by practicing the virtues and will just know what the right thing to do is in the particular situation. Virtue ethics, unlike mainstream ethical theories previously mentioned, accepts the place of the emotions or passions in the moral life. Virtues could be said to be the right expression or moderation of the passions 'in the right place and at the right time'(Carr 1991 p.50) or as Aristotle states 'virtue aims to hit the mean...to have those feelings at the right times on the right grounds towards the right people for the right motive and in the right way is the mark of virtue'(Aristotle 1953 ed. p.101).

Accounts of the significant virtues for human flourishing vary. There is general agreement that the 'cardinal virtues' are prudence, justice, temperance and courage. The theological virtues are understood to be faith, hope and charity. However, there are specific virtues which are thought to relate to specific roles and practices. In considering the role of virtue theory in relation to prenatal testing, the virtues of the professional and the patient (the woman who is pregnant) will be considered.

4. THE PROFESSIONAL

Health care professionals are involved in raising the issues of the possibility of genetic disorder prenatally; they provide information about the prenatal tests available; they give potential parents the results of the tests and this, sometimes, means breaking bad news. They are also involved in supporting parents, during and following, agonising decision-making about the termination, or continuation, of the life of a foetus (or child) with a genetic disorder.

Edmund Pelligrino (Shelp 1985 p.246) has written that the 'end of medicine is not simply a technically proficient performance but the use of that performance to attain a good end—the good of the patient—his medical or biomedical good to the extent possible but also his good as he, the patient, perceives it...It is the sensitive balancing of these senses of the patient's good which the virtuous physician pursues to perfection.'

There seems little doubt that the professional's performance in relation to prenatal testing calls for more than technical expertise. Achieving the patient's good as the end of medicine requires, according to a virtue ethics approach, that the professional practices the virtues. There is some variation in the virtues recommended for health care practice and these generally relate to the nature of the practice and the role within that practice.

Beauchamp and Childress (1994) emphasise compassion, discernment, trustworthiness and integrity as core virtues in biomedicine. MacIntyre (Benjamin & Curtis in Shelp 1985 p. 263) points to honesty, courage and justice as virtues which are a necessary part of any practice. Pelligrino (Shelp 1985 p. 246) advocates 'the whole list of virtues spelled out by Aristotle'. From the practice of nursing emerges a list of Nightingale virtues which are out of step with nursing as we now know it. These include obedience, punctuality and observation (Sellman 1997). It is not possible here to consider the implications of each of the virtues in relation to prenatal testing, so I will limit my discussion to three of those most commonly referred to: the virtues are compassion, courage and discernment.

Compassion is defined as a 'trait combining an attitude of active regard for another's welfare with and an imaginative awareness and emotional response of deep sympathy, tenderness and discomfort at the other person's...misfortune or suffering'(Beauchamp & Childress 1994 p.466). The 'emotional tone' of the professional can enhance or detract from the support offered to women in the pre- and post-test period. The anxiety and anguish which is likely to be experienced is unlikely to be alleviated if the professional bringer of bad news is emotionally detached and unsympathetic. The point needs to be made again, however, that virtues are dispositions displayed at the right time, to the right degree and in the right way. If the professional becomes so distraught and bereft at the patient's suffering that s/he becomes ineffectual, then this exceeds the mean of the virtue of compassion.

Courage was identified as one of the cardinal virtues by Geach (1977) and as a necessary practice virtue by MacIntrye. Examples of actions performed by courageous professionals include: caring for those with infectious diseases; handling potentially dangerous drugs and materials; and speaking out when practice conditions are inadequate or unsafe (Benjamin & Curtis in Shelp 1985 p.270). Other senses of courage have particular pertinence to prenatal testing. The fear of litigation may cause professionals to influence patients to opt for every test possible. The fear of disability may cause the professional to present a biased picture of the potential child's future. Knowing what the mean of moral courage entails, in relation to prenatal testing, and distinguishing it from foolhardiness and cowardice necessitates consideration of a third virtue, the virtue of discernment.

The virtue of discernment is associated with practical wisdom (Beauchamp & Childress 1994 p.468) and is described as including 'acute judgement and understanding' and the 'ability to make judgements without being unduly influenced by extraneous considerations, fears, or personal attachments'. There is, at least, anecdotal evidence that the virtue of discernment is sometimes lacking in prenatal counselling. Damon Hill, the Formula One racing driver, who has a son with Down's Syndrome is quoted as saying 'Hospital staff prepared us for the worst—a child with learning difficulties...Down's Syndrome was portrayed as a family tragedy...I'd love those doctors who painted such a gloomy picture to see him now'(Williams 1995 p,47). The point was also made at a learning disabilities conference in Canada in 1992 that 'professionals involved in counselling...know frighteningly little about adult lives of people with Down's Syndrome...and the positive outlook'(Williams 1995 p.47). To practice the virtue of discernment, professionals need to have a fuller understanding of what they are testing for, what the implications of living with such a genetic disorder are for the individual and his/her family and also what the implications are of terminating a pregnancy where a genetic disorder has been diagnosed.

Genetic counselling is essentially non-directive and is described as 'the cornerstone of genetic services'(Pembrey 1996 p.9). It involves a dialogue between the professional and the potential parents, exploring feelings about a positive result and how the latter will cope in response to this (ibid). Resource constraints mean that counselling will generally only be offered following a positive test result (Chadwick 1997 p.6). Clearly, it is not the professional counsellor's role to make judgements as to what the potential parents should do but rather to provide information in an objective and balanced manner, to facilitate their thinking about the pertinent issues and to explore feelings. The virtues of compassion, courage and discernment are necessary (but probably not sufficient) virtues which will enhance the work of the professional involved in prenatal testing.

5. THE VIRTUOUS PATIENT/WOMAN

It is often said, and rightly so, that pregnancy is not an illness. Women who are pregnant, however, have reason to visit what Susan Sontag calls the 'Kingdom of the Sick'(Labacqz in Shelp 1985 p.275) so that their health and the health of the foetus, can be monitored. With monitoring comes a range of onerous offers which women must consider: the offer of testing; the offer of results in the form of odds and ratios; the offer of certainty (that is, a definite genetic diagnosis by opting for amniocentesis or chorionic villus sampling) with the risk of losing the foetus or baby; and the offer of termination if the tests detect serious genetic disorders.

It might be argued that the virtues of a virtuous patient are inappropriate for a pregnant woman. Some discussion of the virtues offered by Karen Labacq, however, suggest that they have relevance to the issue of prenatal testing. The virtues of the virtuous patient, presented by Labacqz (Shelp 1985) are fortitude, prudence and hope.

Both Peter Geach and Rosalind Hursthouse present pregnancy as necessitating courage. Geach (1977 p.xxix) states that courage is what we all need in the end, and it is constantly needed in the ordinary course of life: by (for example) women who are with child'. Hursthouse (1987) states that 'most pregnancies and labours call for courage, fortitude and endurance, though most women make light of them—so why are women not praised and admired for going through with them?'.

Courage, in relation to prenatal testing for genetic disorders means facing the fears which emerge. Two aspects of courage or fortitude are identified as *endurance* and *attack* (Labacqz). Endurance in relation to testing could be related to the decision made, for example, to continue or discontinue the pregnancy as endurance has to do with 'keeping one's spirit from being broken by fear, grief or sadness'(ibid). Attack could represent a woman's efforts to demand support for the child born with a genetic disorder.

The second virtue of prudence necessitates, according to Labacqz (in Shelp 1985 p.282), our ability to perceive accurately, to learn 'the truth of things' and to be able to act in a fitting manner. Contemplation, docility (in the 'sense of teachableness, or openness) and trust are, Labacqz states, necessary for 'receptive silence before the truth of real things'(p.283). There are many issues to be prudent about in relation to prenatal testing—how, for example, to interpret the odds presented and what to do in the light of the odds. One woman may, for example, consider odds of one in seven positively (there are six chances out of seven of having a baby without a genetic disorder) and another may interpret a 1 in 120 chance negatively (her baby could be the one with the genetic disorder). Other questions relate to further tests and the risk factors relating to these tests and also relate to how to respond to positive results in the light of the possibility of false positives and negatives. What prudence seems to require is that information, and perhaps other women's experience of continuing or discontinuing pregnancies where genetic disorders have been diagnosed, are taken into account when contemplating actions. Prudence is unlikely to protect against all negative or painful emotions but prudent decisions should be easier to live with than imprudent decisions. The point should be made also that, as there are time constraints with many of these decisions, the knowledge, support and skills of a professional could be most helpful as a great deal will have to be considered in a relatively short space of time. The third, and final, virtue of the patient on this account is hope which includes 'hope for improvement, hope for the courage to withstand it all, hope for meaning to emerge out of the chaos, pain and sense of injustice'(Labacqz in Shelp 1985 p. 284). Whilst hope can be interpreted in different ways in relation to prenatal testing this would seem to be a virtue which will benefit the woman, making the difficult decisions more bearable.

6. CONCLUSION

In this paper, I have briefly outlined some of the implications of adopting a virtue ethics approach to the area of prenatal testing for genetic disorders in relation to professionals and patients.

In considering the implications of prenatal testing for society Mary Warnock's (1985 p.3) questions in relation to reproductive technologies are helpful:

> Barriers, it is generally agreed, must be set up; but there will not be universal agreement about where. The question must ultimately be what kind of society can we praise and admire? In what sort of society can we live with our conscience clear?

There is, indeed, no quick fix to be had from either ethics or genetics. A rational, purely abstract analysis of the difficult issues involved in prenatal testing is unlikely to capture the emotional, context of a life, circumstantial concerns which are inevitable in prenatal testing. It is pointed out that being virtuous, and practicing the virtues, does not guarantee that we will flourish but rather it is probabilistic (Hursthouse 1987 p.230). So too, with genetics. There is much that is probabilistic. While undoubtedly some genes are responsible for recognised and accepted genetic disorders, they do not govern our responses to the disorders. Also, genetic predispositions are moderated by lifestyles and the environment so certainty (about the development of certain disorders) will not come from a prenatal test. The genetic information we can have access to can be empowering but it may also cause individuals to 'spend a lifetime of PPDs (possibly potentially diseased) or DIWs (diseased in waiting)', according to Kenen (1994).

We must, then, tread warily toward the new millennium. Offers of genetic testing will continue to be made and we need to consider the implications of these offers for the individual and for current, and future, societies. Ethics can, I am hopeful, prevent us from being shipwrecked as a result of the call of sirens offering sometimes dubious tests.

I end with a predictable recommendation: education, education, education. It can, in my view, never be too early to commence moral and genetic education. Our children will be the parents of tomorrow who will contemplate prenatal testing. It seems preferable to explore issues in relation to prenatal testing *before* pregnancy, perhaps in the context of school relationship or social education. A focus also on, what Rothblatt (1997) calls "euthenics", deserves more attention. She defines euthenics as 'the science of improving the physical and intellectual capacities of humanity by control and improvement of living conditions'.

Finally, Thomas Paine (1776) is quoted as saying (Rothblatt 1997 p.165):

> When we are planning for posterity, we ought to remember that virtue is not hereditary.

Thank goodness for that.

REFERENCES

Alderson P. (1991) 'Abstract bioethics ignores human emotions' *Bulletin of Medical Ethics* May pp. 13–21

Aristotle (1953) *The Nicomachean Ethics*, Penguin Books, London.

Barnes G. & Bryan J. (1996) *The Decision—Testing for Down's Syndrome*, Channel 4 Television, London.

Carey J.S. (1991) 'I am not born yet- but that won't stop me suing you' *New Statesman & Society* No. 15, pp. 28–29

Carr D. (1991) *Educating the Virtues—An essay on the philosophical psychology of moral development and education*, Routledge, London.

Chadwick R. (Co-ordinator) (1997) *Genetic Screening: Ethical and Philosophical Perspectives* Euroscreen Final Report

Foot P. (1978) *Virtues and Vices and other essays in Moral Philosophy* Blackwell, Oxford.

Geach P. (1977) *The Virtues—The Stanton Lectures 1973–4* Cambridge University Press.

Hursthouse R. (1987) *Beginning Lives* Open University Press, Milton Keynes.

Johnstone M-J (1994) *Bioethics—A Nursing Perspective* Second Edition, W.B. Saunders/Bailliere Tindall, Sydney.

Jones S.L. (1996) 'The At-Risk Health Status and Technology: a Diagnostic Invitation and the Gift of Knowing' *Social Science and Medicine* Vol. 42, No. 11, pp. 1545–1553

Mennuti M. (1996) 'A 35-Year Old Pregnant Woman Considering Maternal Screening and Amniocentesis' *JAMA* Vol. 275, No. 18pp. 1440–1446.

Pembrey M. (1997) *The Progress Guide to Genetics* The Progress Educational Trust

Penticuff J.H. (1996) 'Ethical Dimensions in Genetic Screening: A Look into the Future' *JOGNN* Vol. 25, No. 9, pp. 785–789

Rothblatt M. (1997) *Unzipped Genes* Temple University Press, Philadelphia.

Raines D.A. (1996) 'Fetal Surveillance: Issues and Implications' *JOGNN* Vol. 25, No. 7, pp. 559–564

Scott P.A. (1995) 'Aristotle, Nursing and Health Care Ethics' *Nursing Ethics* Vol. 2, No. 4, pp. 279–285.

Sellman D. (1997) 'The Virtues in the Moral Education of Nurses: Florence Nightingale Revisited' *Nursing Ethics* Vol. 4, No. 1, pp.2–11.

Shelp E.E. (ed.) (1985) *Virtue & Medicine* D. Reidel Publishing Company

Warnock M. (1985) *A Question of Life—The Warnock Report on Human Fertilisation and Embryology* Basil Blackwell Ltd.

Williams P. (1995) 'Should We Prevent Down's Syndrome?' *British Journal of Learning Disabilities* Vol. 3, pp. 46–49.

26

GENETIC INFORMATION

Questions and Worries from an African Background

Godfrey B. Tangwa*

University of Yaounde 1, Department of Philosophy
Faculty of Arts, Letters and Social Sciences
P.O. Box 755, Yaounde, Cameroon

1. INTRODUCTION

The genes, we are told, are responsible for our hereditary traits (nature). These hereditary traits combine with environmental and socio-cultural factors and influence (nurture) to make us what we are. It is, of course, not easy, in practice, to separate nature from nurture or purely biological from environmental and socio-cultural factors.[1] In African idiom, what the genes are responsible for in the make-up of an individual is said to be "in the blood" (*kìdzë̀ metzë̈*, as we say in *Lamnso'*) and is considered ineradicable by any ordinary means. What is not "in the blood" can easily be learned or unlearned through upbringing and education (*liy wuna yeey*) but, of course, in practice, it may be as difficult to distinguish what is learned from what is genetic as it is to distinguish the black from the white chicken in the soup.

The genes are like the inner core, or the building blocks of living organisms, containing information essential to the organic functioning/dysfunctioning of living things. The genes provide the key to inherited characteristics. Because of this, one area in which genetics and genetic information are of obvious and indisputable importance is in the domain of hereditary. (Lemmens, T. and Bahamin, P., 1996, pp. 10–17). There is a wide range of hereditary genetic diseases whose understanding, early detection and prevention or treatment would be greatly facilitated by genetic information. Within the context of medicine, whose primary objective is health, the non-abusive use of genetic information poses no particular ethical problems and would not only be supported but encouraged by peoples of all cultures and cultural backgrounds, irrespective of their level of technological development. The only possible danger here may be that of *reductionism to genetics*

* E-mail : GBtangwa@cam.healthnet.org

Genetic Information, edited by Thompson and Chadwick.
Kluwer Academic / Plenum Publishing, New York, 1999.

whereby personal behaviour and particular orientations or habits such as alcoholism, smoking, homosexuality, paedophilia etc., may be attributed to or blamed on the genes, with or without any leading evidence.

But, outside of preventive or therapeutic medicine, genetic information and all types of genetic manipulation and experimentation pose serious ethical problems, some of which might more easily be swept under the carpet within certain cultures, such as Western culture, because of certain dominant philosophies, ideas, attitudes and practices: for example, the idea that knowledge is power and that all knowledge is unqualifiedly good. Other examples are the impulse to know everything knowable, the spirit of omnivorous discovery and the practice of patenting, monopolising and then commercialising discoveries and inventions. Since the Industrial Revolution, the Western world has increasingly looked upon knowledge as power convertible into commercial value, on nature as, in principle, completely knowable and controllable and on the entire universe as something that should be explored, dominated and commercially exploited. This outlook has led to secularisation, desacralisation and profanation of nearly everything in the universe. The Western world today is one in which there are no taboos but one in which a highly efficient technology is propelled and sustained by the profit motive and so-called open market forces. Because of the power and influence of the Western world, the entire world is today reaping the fruits (both beneficial and harmful) of this state of affairs.

2. THE AFRICAN OUTLOOK

By contrast with the Western outlook, the traditional African outlook stands in respectful awe of both knowledge and nature and considers human beings as mere partners or share-holders, with all other creatures, inanimate objects and intangible/invisible forces, of the resources of the earth. It is an outlook with a strong religious dimension and one in which the sacred and the profane are still inseparable. This metaphysical outlook influences and is harmonised with ethical imperatives, customs, laws and taboos, which form a single continuum in such a way that distinctions between them are not usually made. (Tangwa, G.B., 1996a). Taboos are different from positive laws in that their violation, whether conscious or unconscious, whether in public or private, attracts sanctions, not only on their violator but sometimes also on his/her entire community as a whole, whereas the consequential effects of violating a positive law concern only the violator, if and when s/he happens to be caught. In like manner, moral imperatives have the great advantage over both positive laws and taboos in that no external threats or sanctions are required for their observance. The moral agent requires no undue persuasion or encouragement for doing what is morally right or avoiding what is morally wrong because the determinant of action in this case is autonomous conviction. Sanctions are external to morality although that is not to say that they do not play a role in the explanation of human moral conduct. But morality is on the verge of degenerating into law when sanctions attain a commanding status in its domain (Wiredu, K., 1983).

2.1. Traditional African Societies

In traditional African societies, there was great emphasis on using specialised or extraordinary knowledge benevolently (Tangwa, G.B., 1996b, pp. 191–192). This obsession with the moral implications of knowledge was responsible for the fact that, within traditional African societies, people did not usually earn a living from their specialised knowledge but rather from the same sort of occupations, such as farming, fishing, hunting, animal husbandry, etc. by which every other person earned their living. Medical practitioners, carvers, entertainers and other artists, for example, usually never charged any fees for their professional services, for fear of losing their specialised skills and natural endowments. They could ask to be supplied the material ingredients, instruments or implements necessary to enable them accomplish their tasks, but, for the rest, they could only be offered appreciative presents after their tasks had been accomplished. A medical practitioner, for instance, would usually ask the patient for all the common ingredients of his intended preparation such as salt, oil, honey, *egusi* (melon seeds), palm-wine etc., but s/he herself/himself went in search of the uncommon ones such as herbs, tree barks, roots, chemical substances etc. Treatment itself was completely free of charge but the treated patient usually came back, as a matter of custom, at his/her own convenience, to "thank" the medical expert with appropriate appreciative gifts and presents. The level of egalitarianism presupposed in such a context and especially the implicit rejection of any unqualified labour theory of value are, perhaps, no longer attainable today, but some of the spirit and the moral imperatives underlying such an outlook can certainly still be salvaged. The same moral obsession and levelling tendency is responsible for the ubiquity of something like witchcraft discourses all over Africa, being, basically, an attempt to come to grips with the uses and abuses of esoteric knowledges and access to inegalitarian powers. (Geschiere, P., 1997).

2.2. Human Handicaps and Deformities

Another relevant aspect of the traditional African outlook is that human handicaps and deformities, such as some of those which might result from genetic defects, were usually considered as a deliberate doing of God and those suffering from such handicaps or deformities were usually not looked down upon or discriminated against in any way. Moreover, they were often considered as possessing extraordinary "depth" (*sëm*) as well as being liable to be used as disguises by God and other spirits. (Tangwa, G.B., 1996b, p. 190). Within such a framework, people could live with their handicaps, whether of genetic origin or otherwise, more easily.

2.3. Patenting and African Communitarianism

The traditional African outlook is also communitarian. This communitarianism makes a Western practice such as patenting problematic within the African context where knowledge generally and discoveries in particular are not credited to individuals or to a select group of individuals but considered rather as communal property. So there is here a worry that the technologically more advanced countries might appropriate, for free, the genetic heritage and accumulated knowledges of African communities, patent these acquisitions, and go ahead to commercialise them in a manner that would put their benefits out of the reach of those very communities, in a manner similar to what has sometimes happened in the case of other industrial raw materials. To avoid the prospect of bio-colonialism and exploitation, it would perhaps not be inappropriate if African and other third world countries were to consider imposing, for the time being, a moratorium on all bioprospecting within their borders until such a time that the ethical issues and wider consequences involved would have been thoroughly considered from their own point of view and satisfactory legislation enacted, although such a procedure, given the present global excitement over genetic information, might tend, in the meantime, to encourage bio-piracy. For bio-prospecting, as Bell (1997, p. 3) remarks,

> ...helps to cajole or bully gene-rich countries of the South into accepting and adopting Northern models of intellectual property rights (IPRs) and the privatisation of genetic resources, to which... [they]... are opposed.

And, as Zamora (1997, p. 13) also points out, with regard to the Philippines, "Privatisation of biodiversity is disenfranchising local communities." Furthermore, within many traditional African communities, it was strictly taboo to sell or otherwise commercialise certain things, such as water, housing, fuel wood, staple foods, etc. The idea behind such taboos would seem to be that that which is in no way a luxury but is essential for mere survival, and which has, moreover, been provided in sufficient abundance by God and nature, should be at the disposal of and within the reach of all and sundry. A taboo against the sale or commercialisation of, say, genes and gametes would be seen very much in the same light within this cultural outlook. So would a taboo against any manipulation of the genes, such as in cloning, which, in the domain of humans especially, can be viewed as an attempt by human beings to produce human beings for human beings. Despite the scientific excitement aroused by such a possibility, cloning of humans, more than anything else, evokes the dangers of epistemological greed resulting in a catastrophic fall over a fatal threshold. In metaphoric language, it is an attempt to challenge God to a race, but no human being can run faster than God. African mythologies are replete with stories, similar to the biblical story of Adam and Eve in the Garden of Eden, about the irreversible catastrophic consequences of such human greed and foolhardiness when it recklessly breaks the bounds of human competence and propriety. There is a myth, for instance, told with varying details among many African peoples, about how, originally, the sky, which provided all human needs, was lower down, close to the earth, and within reach of everybody, but now, owing to the greed of a single individual, it receded out of the reach of all humans.[2]

2.4. A Piecemeal Approach

As human beings, we emerged into the world after it had been in existence for billions of years and, if we happen to disappear out of the world again, it would continue to exist as if we had never existed. To stay on the safe side, ethically and otherwise, human beings would do very well to adopt an attitude of cautious and respectful piecemeal tinkering with the world, always first seeking to know what is wrong with the state of things as they are before introducing changes aimed at modifying or correcting what is already there. From such a perspective, the use of genetic information for therapeutic purposes is clearly justifiable while other possible uses such as in cloning are clearly unjustifiable.

3. GENETIC INFORMATION AND INSURANCE

One area in which the use of genetic information is highly questionable is insurance. Although genetic information and insurance business, as conceived and practised in the Western world, might appear "made for each other" and although it might be true that genetic information is not really different from other health information already used by insurers, this in no way proves that the use of genetic information in insurance is ethical.

The morality of a practice cannot be decided by considerations of its suitability or acceptability as "normal business practice".

The use of genetic information in insurance calls into question the morality of insurance itself as it has developed and is practised in the Western world today, difficult as it is now to imagine that it could be otherwise. The definition of insurance as "a private contract in which one party, the insured, transfers certain risks of loss to another, the insurer, for monetary considerations"[3] already clearly exhibits the unmistakable trademarks of the Western ideologies of possessive individualism and open free market. Against such a definition, can be proposed the following prescriptive and more general definition which might be more acceptable in all cultures and ideological backgrounds: "Insurance is the practice of putting something aside today (in good times) for the unpredictable rainy day tomorrow (in bad times)." In Africa, the justification of insurance so conceived would be expressed in proverbs such as the following: *Kila kibvëshi* ? (Nobody knows tomorrow); *Dzë mo lan a dzë wo kibvëshi.* (It is my turn today, yours tomorrow); *Ngeh yoh yi nyah wir.* (Adversity and misfortune have no favourites).

3.1. Egalitarian Impulse of African Culture

In traditional African culture, there is both a highly egalitarian impulse and a strong levelling tendency which some scholars have identified as the causative explanation for a practice such as witchcraft (Geschiere, 1997) which I would basically consider a metaphor[4] for practices and discourses fundamentally aimed at reining in super-human power and extraordinary intelligence so that these might not be abused or used selfishly and immorally but rather altruistically and benevolently. A society in which such control exists may not develop or progress very quickly but it may develop and progress more surely and safely. Western culture, of course, is not without an egalitarian component; but this is mainly egalitarianism in the sense of *equality of opportunity* which, in practice, always ends up with unequal results which are then accepted and protected, thereby undermining empathy and solidarity as well as making further equality of opportunity impossible. Because human beings are very different in their capacities and also because of the element of chance, an equal starting point, though fair in itself, does not guarantee equal results and, if such unequal results are to be the basis of further competition, then the competition from that point is no longer quite equal. So there will always be room for a bit of deliberate levelling and equalising on moral grounds.

3.2. Insurance as a Gamble

One problem with insurance as it has developed and as it is practised in the Western world and its dominions, dependencies and ever-expanding spheres of influence, is that, even though its necessity, origin and *raison d'etre* are based on uncertainty and unpredictability of the future, it has tended, in practice, to operate more and more on assumptions of certainty and predictability. Even then, such assumptions raise further questions. For example, an individual's genes may hold the key, as it were, to his/her future as a biological organism, but the predictive value of such genetic information will always be less than absolute. In other words, the information that the genes supply can be considered to reveal *dispositional properties* which, in themselves, do not guarantee their actualisation with any absolute degree of certainty. There will always be an element of uncertainty and therefore a gamble in basing particular predictions on such dispositional properties, no matter how certain they may be according to statistical indices. If some rich people in the society want to gamble around these elements of partial certainty and partial uncertainty in the manner they do in, say, casinos, there is nothing particularly wrong with that, but it is not something that ought to be generalised to cover all members of society. For the ordinary majority of people, common thrift will always be more sensible and more profitable than gambling. Life insurance as a type of gamble could be considered appropriate for only a small percentage of people in any society.

However, the main ethical objection to the use of genetic information in insurance is that it does not seem morally right or even ordinarily fair to discriminate for or against, to reward or punish, people on the basis of involuntary and ineradicable biodata. For this reason alone, I believe that the Western concept and practice of insurance needs fundamental rethinking even though this might seem impossible, given the fact that Western insurance businesses have developed into giant multi-national companies which shape the pace and tempo of economic life all over the world. In any case, other cultures should not be prevented but rather encouraged and even aided to develop different forms of insurance along different trajectories based on different conceptions of life and the world. If biodiversity in nature is such a good thing, socio-cultural-economic diversity may not be such a bad thing.

3.3. An African Idea of Insurance

On a small scale, some Africans, especially those now living in urban areas where traditional kinship ties are no longer so important, have been practising such different forms of insurance. In Cameroon, for instance, various groups such as alumni of the same educational institution, co-workers in the same establishment, professional colleagues or people originating from the same village or region etc., usually have a rudimentary insurance arrangement as part of their regular social structure (*Angwa'* or *Tontins*). In such an arrangement, a special fund is usually created into which all members contribute regularly and compulsorily and for which money may also be raised through various group fund-raising activities. Such a fund may be called "trouble fund", "adversity bag" (*Kibam ke ngeh*) etc. Money from such a fund is used to help any of the members of the group in case of, say, serious illness, birth of a child, loss of job, death or other such calamity etc. It is significant here that those who contribute to such a fund, but who have not yet benefited from it, would not think of complaining that the arrangement is unfair or that they are in any way being cheated. Since the general belief is that illness or adversity does not discriminate, that trouble does not sound a trumpet, and that death does not spare anyone, those who have not yet benefited from such an "insurance policy" in fact consider themselves fortunate, for which reason they feel obliged to contribute even more to the fund while awaiting, without any enthusiasm, their inscrutable but inevitable turn at the receiving end.

Radical egalitarianism underlies such practices which, if properly modernised and/or enlarged, can yield novel forms of insurance which differ from the Western idea where those considered to be more at risk (through no fault of theirs) are required to pay higher insurance premiums. Such an alternative type of insurance may be more suitable for the more ordinary citizens of communities the world over, including the Western world. This, of course, would not prevent richer members of any society or people of generally more affluent societies from practising other forms of insurance based on predictability of inequalities. We would then have a wide variety of insurance ideas and practices, some of which would have no use for genetic information, important as it may be in other domains. From this perspective, some aspects of current structural adjustment programmes

(SAPs), being imposed by the IMF and World Bank on African and other third world countries, which seek to change, streamline and model some local socio-economic practices on Western paradigms are questionable and need fundamental rethinking.

4. CONCLUSION

Genetic information is evidently very important, and some of its possible uses, such as in preventive and curative medicine or in agriculture, relatively unproblematic. But some other possible uses of genetic information, such as in commerce, cloning or underwriting in insurance, are highly controversial. From an African point of view, and against the general background of traditional African communalism, egalitarianism, epistemological humility and respect for nature, such uses are clearly unjustifiable and can be considered as undesirable developments of certain corollaries of Western culture and outlook on life.

NOTES

1. I am highly indebted to a recent work of Trudo Lemmens' and Poupak Bahamin's: *Genetics in Life, Disability and Additional Health Insurance in Canada: A Comparative Legal and Ethical Analysis*, for much of my understanding of genetics and its actual potential implications today.
2. For a similar myth among the Maka of Cameroon, see Peter Geschiere, op. cit., pp. 38–39.
3. *Black's Law Dictionary*, quoted in Trudo Lemmens and Poupak Bahamin, op. cit., p. 35.
4. Of course, a good metaphor might and usually is taken quite literally by the vast majority of ordinary people, without much harm. A good metaphor like that of Jesus Christ being the son of God is taken quite literally by most christians without any harmful consequences.

REFERENCES

Bell, J. (1997), 'Biopiracy's Latest Disguises', *SEEDLING : The Quarterly Newsletter of Genetic Resources Action International*, Vol. 14, N. 2, pp. 2–10.

Geschiere, P. (1997), *The Modernity of Witchcraft : Politics and the Occult in Postcolonial Africa*, University Press of Virginia, Charlottesville and London.

Lemmens, T. and Bahamin, P. (1996), *Genetics in Life, Disability and Additional Health Insurance in Canada : A Comparative Legal and Ethical Analysis*. [Pre-print version of Report Submitted to the Medical, Ethical, Legal and Social Issues (MELSI) Advisory Committee of the Canadian Genome Analysis and Technology Programme].

Tangwa, G.B. (1996a), 'Bioethics and Biolaw in the Present Situation of Africa'. Unpublished paper presented at the International Conference on Bioethics and Biolaw, Copenhagen, Denmark.

Tangwa, G.B. (1996b), 'Bioethics : An African Perspective', *Bioethics*, Vol. 10, N. 3, pp 183–200.

Wiredu, K. (1983), 'Morality and Religion in Akan Thought', Oruka, O. and Masolo, D.A. (eds.), *Philosophy and Cultures*, Bookwise Limited, Nairobi, Kenya, pp 6–14.

Zamora, O.B. (1997), 'The Philippines : A Bridle on Bioprospecting?', in *SEEDLING: The Quarterly Newsletter of Genetic Resources Action International*, Vol. 14, N. 2, pp. 11–19.

27

GENETIC KNOWLEDGE IN A JUST SOCIETY

Nils Holtug*

Department of Education, Philosophy, and Rhetoric
University of Copenhagen
Njalsgade 80, DK-2300 Copenhagen S, Denmark

1. INTRODUCTION

Your genes are causally responsible for much of what happens in your life. They may have caused you to have a painful disability, a bad temper, a fair skin colour, a musical talent, or a beautiful appearance. These properties may in turn bestow certain benefits or burdens on you. In short, the effects of your genes on your life are pervasive, and your life may go better or worse because of them.

These are just facts. What matters, from a moral point of view, is the attitudes we adopt regarding these facts. We may believe it unfair that nature distributes benefits and burdens unequally. It would have been better if a more equal distribution had been the case. We may then be convinced that, in organizing our society, we should aim at compensating the people who have been disadvantaged in nature's distribution. For instance, we may think that our society should provide health care and other necessities to people with disabilities. Alternatively, we may believe that people are entitled to their personal assets and to the benefits they can derive from them. If this is our considered opinion, we may be convinced that it would be wrong if the state were to redistribute the benefits and burdens that arise from differences in people's natural assets.

In the following, I shall argue that the former view is largely right. When people are disadvantaged because of their misfortune in nature's distribution, it is appropriate to compensate them. In order to make this argument, I shall need to discuss some rather complex issues in both metaphysics and political philosophy.

Now, from the view that people should be compensated for their misfortune in nature's distribution, it more or less directly follows that the more we attribute those of our properties that are relevant for our shares of benefits and burdens to the causal influence of our genes, the more we should be willing to compensate people who have been disad-

* e-mail: nhol@coco.ihi.ku.dk

Genetic Information, edited by Thompson and Chadwick.
Kluwer Academic / Plenum Publishing, New York, 1999.

vantaged in the distribution of these benefits and burdens. So if, as is probable, future developments in the human genome project will cause us to believe that more and more of our properties are genetically determined or have a genetic component, this should increase our willingness to redistribute resources to accommodate the interests of the worse off.

Having made this argument, I shall consider what it implies regarding different systems of health care. In particular, I shall discuss whether a system of private health insurances, in which insurers and people seeking insurance have access to genetic information, is compatible with the requirements of justice.

I shall then go on to consider another question pertaining to justice and genetic information. As we gain more knowledge about the genome, it is likely that we will be able to medically alter more of the properties people have in virtue of their genetic make-ups. What kinds of medical interventions, then, may justice require us to perform in order to compensate those who have been disadvantaged in the distribution of genes? In this context, the correction versus enhancement distinction is of particular interest.

2. THE GENETIC LOTTERY

Each and every one of us, whether we want to or not, participate in the genetic lottery. In the genetic lottery, different sets of genes are distributed upon us. Thus, we come into being with a set of genes that we have not chosen. Furthermore, the mechanism by which genes are distributed can be described as random in the sense that during meiosis, genes are arbitrarily assigned to the germ cells from which we originate. Regarding the facts that we do not choose the outcome and that the outcome is selected randomly, the genetic lottery works like an ordinary lottery. However, unlike in ordinary lotteries, we do not voluntarily participate in the genetic lottery.

As we have seen, the consequences of the lot one draws in the genetic lottery are pervasive and of crucial importance. Some people are born with diseases that cause them to suffer much and live for only a few years, whereas others are born healthy and with much potential, which enables them to live long, rich and fulfilling lives. But exactly how deep does the genetic lottery run? How much of my life is determined by my lot in this lottery?

Future developments in the human genome project will enable us to give a more precise answer to these questions than we can today. However, some of those working on this project seem to suggest that ultimately we will be able to account for all of our (important) properties by referring to our genetic make-ups. For instance, the Nobel-Prize winner Walter Gilbert is reported as saying "The total human sequence is the grail of human genetics," it is "the ultimate answer to the commandment, 'Know thyself'" (Bishop and Waldholz, 1990, p. 218). The underlying idea seems to be that in some sense we are reducible to our genes. Now clearly, each of us consists of more than just genes, so we cannot be identical to our genes. Rather, the idea would seem to be that our genes *causally determine* which (important) personal properties we develop.

I have criticised this idea elsewhere and shall not repeat my criticism here (Holtug, 1995). However, it is worth noting that, even if it is false that our genes causally determine all our (important) personal properties, determinism as such may be true nonetheless. Every event may be causally determined by some of the events and states of the world that precede it. So all of our properties and all of what happens in our lives may be determined by our physical constitution and the environment in which we live. This is relevant to the question of how much of what happens in a person's life we can hold that person responsible for.

Besides the genetic lottery, there are other similar lotteries that affect our lives. The genetic lottery is itself part of a larger lottery, namely the natural lottery. Apart from our genes, other factors affect the sort of physical constitution with which we are born. For instance, our mothers' physical constitution (e.g. the environment provided by the womb) and behaviour (e.g. drinking and smoking habits) affected our fetal development.

Furthermore, each of us participates in what has been called the social lottery. By way of illustration, some are born to parents who have ample financial and emotional resources, others to parents who are poor, have drinking problems and are violent. Like the genetic lottery, the effects of this lottery are pervasive and often crucial as to how well our lives go. But, although each of these lotteries is relevant for the question of how we should organize our society, I shall mainly focus on the genetic lottery.[1]

How well we have fared in the genetic lottery seems to be a matter of brute luck. We have not chosen our genes, rather, they have been selected by a sort of random device. This is important for the following reason. Since we have not chosen our genetic constitutions, we cannot be responsible for them. And since we cannot be responsible for them, we cannot be said to deserve them.

What, then, of the benefits and burdens that befall us because of our genetic constitutions? If you have a severe genetically determined disease, this may mean that you need a lot of expensive medical treatment and that you are not able to hold down an ordinary job. Now, just as you are not responsible for your genetic constitution, neither are you responsible for needing medical treatment and for having a reduced working capacity. This is because, just as you have not chosen your genetic constitution, you have not chosen to need medical treatment or to be unable to hold down an ordinary job. And so, you cannot be said to deserve these burdens.

There are further benefits and burdens that befall us because of our genetic constitutions. Suppose you are born with some special talent, say for athletics, mathematics or music. I, on the other hand, have not been gifted in this way. We pursue similar careers, and we put in the same effort. However, because of your special gift, you do much better than I. Do you deserve to do better, and do I deserve to do worse? I think not. Just as in the case of being born with a severe disease, the reason that I do worse is to be found in my genetic make-up. My doing worse is due to this fact, not to anything I am responsible for myself.

It may be objected that I do deserve to do worse because I could have chosen a career for which I had more talent. But, in general, I do not think that this objection is fair. After all, in many cases, I do not know how far my talents will take me in some particular career unless I actually try it out. Furthermore, we may imagine that I do not have *any* talents that match your special talent, and so my doing worse cannot be said to be due to my choices. Therefore, I do not deserve to do worse.

Imagine instead that you and I have equal talents to begin with, but that you work much harder than me at developing them. As a result, you do better and I do worse. Should we say that this is deserved? That may depend upon what it is that makes you but not me put in the extra effort. Suppose that the ability to make this effort is somehow genetically wired into you but not into me. Then again, we may hold that the reason that I do worse is not, in the relevant sense, due to my choices. With my genetic make-up, I *could* not choose to make the extra effort. So again, it may be tempting to say that I do not deserve to be worse off and you do not deserve to be better off.

Suppose, instead, that the reason why you make the extra effort and I do not is that you were brought up by parents who encouraged such a disposition in you whereas my parents taught me to resign myself to fate. You would then experience social luck rather than genetic luck, but it is luck nonetheless. So again, it may be argued, the differences in how well we do cannot be said to be deserved.

Consider, finally, the following case. You and I have equal talents, and we are equally able to develop these talents. However, you choose to develop them and I choose not to. Because of our choices, you do better and I do worse. Most people will believe that, in this case, the differences in how well we do are deserved. They are due *only* to our choices, not to our luck in the natural and social lotteries. But the philosophical question, of course, is whether there are any such pure choice cases. If determinism is true, then all our choices are determined by our biological inheritance and by the environment in which we live. It may then be difficult to argue that we are responsible for or deserving of anything.[2]

However, my aim here is not to try to settle the limits of human responsibility. Rather, more modestly, it is to argue that as we increasingly come to think of certain of our properties as causally determined by our genes, which no doubt we will as the human genome project progresses, we should give up the belief that people who do less well because of these properties are themselves responsible for or deserving of this. This is compatible with there being other properties we gain or things we achieve that we are wholly or in part responsible for, such that we may be said to deserve these properties and achievements and their effects.

3. THE CLAIMS OF JUSTICE

If we accept this view about our (lack of) responsibility for the effects of the genetic lottery, what moral and political conclusions should we draw? John Rawls has famously argued that when inequalities between people are due to the natural and social lotteries, and so due to facts that people cannot be held responsible for themselves, these inequalities should be eliminated (Rawls, 1971, p. 72).[3] After all, the people who are worse off cannot be said to *deserve* to be worse off. Nor can the people that are better off be said to deserve this. So if a person is worse off because her genes have caused her to suffer from a painful disability, she should be compensated. For instance, health-care services should be provided and a policy to provide special job opportunities for the disabled should be enforced.

Furthermore, as we have seen, there are many other genetically determined features that may cause a person to be worse off than others. Intelligence, physical beauty and athletic talent have a genetic component, and so may musical talent, the ability to make an effort, and much more. So, if inequalities that are due to genetic differences are to be eliminated, clearly there is a whole lot of redistribution to be done.

The principle underlying these political claims is:

> The Principle of Equality: It is in itself bad if some people are worse off than others (through no fault of their own).

According to the Principle of Equality, since people are not responsible for being worse off when it is due to their genetic constitutions, such inequalities are bad and so (everything else being equal) they should be eliminated.

Rawls' argument has important implications for what use we should make of the knowledge we will gain from the genome project. The more we attribute those of our properties that are relevant for our shares of benefits and burdens to the causal influence of our genes, the clearer it becomes that these benefits and burdens fall within the scope of the Principle of Equality. This is because inequalities that are due to genetic differences are paradigm examples of inequalities that we are not responsible for ourselves. So if, as is probable, future developments in the genome project will cause us to believe that more and more of our properties are genetically determined, or have a genetic component, this should increase our willingness to redistribute resources to accommodate the interests of the worse off.

Before I proceed with a discussion of Rawls' argument, I need to mention the unresolved issue of what, exactly, the benefits and burdens with which we are concerned are. We may be concerned with the distribution of resources, or of welfare, or of opportunities for obtaining either resources or welfare. Now, I mention this issue merely to ignore it.[4] I can safely ignore it since, whatever the proper distributive unit, inequalities regarding this unit may arise as a consequence of genetic differences in people. For this reason, whatever the proper unit, we have an issue of whether we should redistribute in order to ensure equality.

I should like to say, however, that money should not be considered a serious contender for being the relevant unit. You and I may have the same amount of money but, if I have a serious disability, I may have to spend all mine on an expensive medical treatment, leaving almost nothing left for food, housing etc. Anyone with egalitarian inclinations will think that, even though we have the same amount of money, a serious and regrettable inequality exists between us. So even if we believe that the proper distributive unit is resources, we need a notion of 'resources' that is robust enough to capture this inequality.

Should we accept Rawls' argument and conclude that, in so far as inequalities are due to differences in people's genetic constitution, they should be eliminated? That would be too hasty. All that Rawls establishes is that those who are worse off do not deserve to be worse off (and those who are better off do not deserve to be better off), not that those who are worse off deserve *not* to be worse off. In other words, we cannnot go from the claim that there is a particular situation that people do not deserve to the claim that there is a particular situation that they *do* deserve (or that should be brought about). Perhaps the worse off do not deserve to be worse off (just as the better off do not deserve to be better off), simply because there is no particular situation that they deserve. In other words, Rawls does not establish that inequalities are bad or regrettable (Hurley, 1993, p. 185, Parfit, 1995, p. 12). Rather, whether we think that an equal distribution would be right depends upon whether we, for independent reasons, find equality an attractive political ideal.

Of course, many people find equality an attractive ideal, especially in situations where inequalities exist that are not deserved. However, I myself think that the apparently egalitarian intuitions that many of us have are not really captured all that well by the Principle of Equality. Consider the fact that, according to this principle, we can make things better, at least in one respect, by making everyone worse off. We just have to make sure that, after the reduction, everyone is at the same (low) level (Parfit, 1995, p. 17). There will then be perfect equality and so, in one respect, the reduction will be an improvement, according to egalitarians. But how can it be an improvement in any respect, when it does not make anyone better off? Because the Principle of Equality faces this problem, I submit, we should reject it. What we are concerned about is not really equality *as such*. Rather, we are concerned about how well the worse off fare. And this concern is better captured by:

> The Priority View: Benefits to people matter more the worse off these people are (in so far as they are not worse off through any fault of their own).

According to the Priority View, a benefit of a certain size matters more if it befalls a person who is worse off (through no fault of her own) than if it befalls a person who is better off (Parfit, 1995, p. 19). Therefore, social institutions should be designed such that they (especially) favour the worse off. This will tend to create equality but, if we are motivated by the Priority View, this will (in itself) be of no interest to us. We will want to improve the lot of the worse off and, if this tends to create equality, that is an irrelevant side-effect. Furthermore, since equality is not assigned any value, there will be *nothing* good about making everyone worse off, even when this brings about an equal distribution.

Rawls himself accepts that we should not remain at an equal distribution, and he holds a version of the Priority View. However, he holds a rather peculiar version in that he claims that all the priority should go to one group, namely the group of *worst-off* people. This may seem an excessive focus on the worst off. After all, Rawls' view—the so-called Difference Principle—implies that even the smallest benefit to the worst-off group outweighs even the largest benefit to the second worst-off group—a group that may be almost as badly off as the former group (Rawls, 1971, p. 302). Partly for that reason, I suggest that we should rather accept a version of the Priority View that assigns a positive weight to everyone, but gradually assigns more weight to an individual the worse off he or she is.

To briefly summarize the conclusions reached so far: The benefits and burdens that accrue to us because of our fortune or misfortune in the genetic lottery are not deserved. Therefore, if we can provide a good moral reason to redistribute these benefits and burdens, those who have been fortunate in the lottery cannot legitimately object (on grounds of desert at least). And such a reason can be provided by appealing to the Principle of Equality or, as I suggest, the Priority View.

At this point, some people will no doubt raise the following objection. They will argue that if there were a distributor of genes or more generally of natural assets, there might be an issue of whether she ought to distribute them according to an ideal of priority or perhaps of equality but, since there is no such distributor, there is no unfairness in an unequal distribution (Gauthier, 1986, pp. 220). It is just a fact—in itself neither good nor bad—that nature favours some but not others. Therefore, there is no reason to redistribute.

However, I think that most of us believe that there is something paradigmatically unfair about some people being very badly off because of their lot in the genetic lottery. Therefore, in order for the objection to have any bite, we must be offered an alternative conception of fairness with greater appeal than those we have just considered. According to this conception, we must be entitled to the benefits we derive from our genetic makeups, even if we cannot be said to deserve these benefits.

One suggestion, then, is the following. My genetic constitution and the properties to which it gives rise are amongst my personal assets. Furthermore, I am entitled to the benefits that I can derive from these assets. Therefore, if, because of my luck in the genetic lottery, I do better than others, I am entitled to my extra benefits. It would be wrong if the state or some similar body were to redistribute them to accommodate the needs of the worse off (or, indeed, the needs of anyone else).

The crucial premise in this argument is that we are entitled to the benefits we can derive from our own personal assets. This premise can be defended in either of two ways, that I shall now briefly consider. On a Hobbesian theory of fairness, our society should be organized according to principles that we could agree on if we were to negotiate as rational egoists, that is, if each of us was trying to further his or her own interests as much as possible (Gauthier,

1986). Since rational people seeking to further their interests will bargain on the basis of their personal assets, the principles they will end up with will reflect how they differ in these assets. Suppose, for instance, that you are healthy whereas I have a disability (and we are equal in other respects). It would not be rational for you to agree to a set of principles that, if implemented, would deprive you of your relative advantage. Rather, the principles we would decide on would, in effect, morally codify your advantage over me.

However, the Hobbesian conception of fairness should be rejected. It has implications that are utterly unacceptable. For instance, it implies that since future generations cannot benefit us, we have no reason to benefit them. Therefore, we owe no moral concern to future generations, and so if we were to pollute to such an extent that it would make all life on earth impossible two-hundred years from now, we would not have done anything wrong. Furthermore, the Hobbesian theory implies that we owe no moral concern to animals or to people who are disabled to such an extent that either they cannot agree to any set of principles (because they are mentally handicapped) or cannot give us anything in return if we choose to take care of them. Since these implications are morally outrageous, we should not accept the Hobbesian theory.

But perhaps there is a better argument to the effect that we are entitled to the benefits we can derive from our personal assets. Robert Nozick has defended such entitlements on the basis of the so-called Principle of Self-Ownership; each person owns his or her own body and so only he or she has the right to decide what should happen to it. Therefore, according to Nozick, persons are entitled to the benefits they can derive from their bodies, e.g. by selling their labour, as long as they do not violate other people's property rights (including their self-ownership). So it would be wrong if the state or some similar body were to tax the income people gain from selling their labour, in order to benefit the worse off.

Although this argument may seem intuitively appealing, we need to think more closely about the Principle of Self-Ownership and the implications Nozick thinks it has. When trying to justify this principle, Nozick appeals to the fact that persons have the ability to shape their own lives according to their conceptions of what makes their lives valuable or meaningful (Nozick, 1974, p. 50). However, if it is bad that people cannot fulfil their life-plans, and this is what is supposed to justify the Principle of Self-Ownership, we may raise the following objection. People who have severe disabilities or have otherwise been unfortunate in the genetic lottery may be unable to create for themselves a meaningful life according to their life-plans. Since, according to Nozick, this is bad, why do we not have an obligation to provide for them what is necessary in order that they may live such lives? For this reason and others, I do not think that Nozick comes up with a plausible argument to the effect that people are entitled to the benefits they can derive from their personal assets. I realize that more needs to be said about the Principle of Self-Ownership, but I cannot go further into this issue here.[5]

It turns out, then, that it is difficult to justify the claim that people are entitled to the benefits they can derive from their personal assets. Therefore, we may continue to accept the view that, in cases where due to their genetic constitutions some people are worse off, they should be compensated in accordance with the Priority View. Let us now consider what this implies regarding various uses that may be made of genetic information in our society.

4. HEALTH CARE AND HEALTH INSURANCES

Evidently, people who suffer from genetically caused diseases (e.g. Lesch Nyhan's disease, Huntington's disease, or cystic fibrosis), are not responsible for these. Since such people are worse off through no fault of their own, they should be compensated. This follows from some of our deeply held moral and political values, such as those expressed in the Priority View.

While this is hardly a surprising conclusion in the light of the previous sections, it does carry important implications for the question of what kinds of health-care systems are morally and politically acceptable. For instance, it rules out a system of more or less unregulated private medical insurances, at least as long as such a system is not supplemented with a system of publicly funded health care.

Consider the fact that insurers can determine the individual risks of people seeking insurance. While, to some extent, this has always been possible, much more accurate tests will become available as the genome project progresses. When such tests become available, people at high risk will be more likely to purchase insurances or will purchase them with a higher coverage than people at a lower risk (a process called adverse selection). Therefore, insurance companies will be forced to demand that people purchasing insurances be tested, such that they can either exclude those at high risk or at least set a premium that corresponds to the risk they pose (a strategy known as medical underwriting). Thus, at least in theory, people will end up paying a premium that corresponds to their own individual risk.

Clearly, this is not compatible with the idea that people should be compensated for the burdens that accrue to them because of their misfortune in the genetic lottery. Giving priority to the worse off is quite the opposite of accepting a system that either excludes them from gaining access to health-care services or allows them access only if they pay premiums that are much higher than those paid by everyone else.

Insurers sometimes defend underwriting practices by arguing that they have a responsibility to treat people seeking insurance fairly by offering them a policy at a premium that corresponds to their individual risk. But why does fairness require that people seeking insurance should have their premiums fixed in this way? Presumably, insurers would answer that if they were to offer low risk individuals policies at higher premiums, they would be treating these individuals unfairly. But, in making this claim, they seem to assume that people are entitled to the benefits they can derive from their personal assets. They assume that if a person has a genetic constitution that makes her a low-risk individual, she is entitled to the benefits she can derive from this, including the benefit of being able to obtain a cheap health insurance. However, as we have seen, the claim that people are entitled to the benefits they can derive from their luck in the genetic lottery is dubious. Therefore, the argument in favour of underwriting practices is unconvincing (Daniels, 1994, p. 114).

It may seem that there is a rather straightforward answer to the problem of medical underwriting. A law could be passed that precludes insurers from obtaining information about individual risks, such that, for instance, they cannot require that a genetic test is presented when a person seeks insurance. But such a system would lead to a breakdown of the insurance market. Since people seeking insurance could obtain such information themselves, adverse selection would force insurers to raise the premiums, ultimately to a level where it would only be advantageous for individuals with the highest risks to buy insurance (Greely, 1992, p. 266).

A possible solution would be to require that insurance companies offer the same insurances to everyone, regardless of individual risks. Presumably, since not all risks are predictable, and since people are risk-averse, even low-risk individuals would still seek insurance. This would mean that the health-related burdens that accrue to people because of their misfortune in the genetic lottery would be shared collectively, in the sense that the health-care services they need would be subsidized by those more fortunate. Presumably, however, the state would have to subsidize some people to enable them to pay for insurance if, for instance, they were too ill to work.

Alternatively, health-care services could be publicly funded and provided by the state to everyone under due concern for fair rationing. Finally, a mixed system could be introduced. In order to decide which of these systems would be preferable in the light of a plausible moral principle such as the Priority View, I would need to go into a discussion of various empirical matters such as which system is most efficient. I cannot go into such a discussion here.

There is a further issue that needs to be discussed in this section. So far, I have focused on diseases that are exclusively (or at least primarily) caused by the presence of a specific genetic make-up in the individual who falls ill. But what of diseases that are caused by a person's genetic make-up in conjunction with specific environmental factors? Are people not responsible for them either, such that they are entitled to compensation in terms of health care?

In some cases, diseases may be triggered by a person's genetic make-up and environmental factors that he cannot avoid. In other cases, they may be triggered by a person's genetic make-up and environmental factors that he could—but has no way of knowing that he should—avoid. These cases seem sufficiently similar to cases in which a disease has only a genetic component. In neither case is the person responsible for becoming ill and so he does not deserve to be worse off. Therefore, these cases fall within the scope of the Priority View and people with such diseases should be compensated.

But there are cases in which it is more plausible to claim that we are responsible for our diseases ourselves, although they have a genetic component. Consider, for instance, so-called lifestyle choices. I may choose to smoke, drink, take drugs, or eat fattish foods, although I know that these choices may very well turn out to be fatal. So am I myself responsible if I fall ill because of such choices?

I may try to excuse myself by claiming that it is merely my bad luck that my genetic constitution is such that I develop lung cancer as a consequence of my smoking. Suppose that you and I both choose to smoke. Because of our genetic differences, I develop cancer whereas you do not. Now, the difference in what happens to us is due to our genetic make-ups, not to our choices. Does this imply that I am not responsible for developing cancer? Evidently not. It may imply, together with some plausible moral assumptions, that I do not deserve to fare any worse than you. But that is not the question. Rather, perhaps, since I chose to smoke, I am responsible for my cancer, and so would you be if you were to develop it. The idea is that, since I could have chosen not to smoke, I am responsible.

In the future, we may become more inclined to hold people responsible for the consequences of their lifestyle choices. If genetic tests become available that will inform us of our individual risks, we may be more likely to hold people who 'act foolishly' in the light of their individual risk profile responsible.

Are such attitudes appropriate? Are people responsible for the consequences of their lifestyle choices? In order to settle this issue, we would need to settle some very complicated issues in metaphysics and in ethics. If determinism is true, and if determinism rules out our being responsible for our choices, then I am not responsible if, for instance, I develop cancer because I smoke. So I should be compensated. More generally, the Priority View (and similar moral principles) will have a universal scope in the sense that, in all cases in which people are worse off, priority should be assigned them. This is because they are never responsible for their being worse off.

However, as I have already pointed out, this is not the place to try to settle these complicated metaphysical and ethical questions. So I need to consider what we should say about lifestyle choices if we accept the view that, at least sometimes, we are responsible for our choices. Are we then forced to accept that we are responsible when, for instance, our smoking causes us to develop cancer, and that therefore we are not entitled to be compensated? I think that most of us actually believe that lifestyle choices should not make a difference for the conditions on which a person gains access to health care. Perhaps this is because we believe that such choices are not products of a free will but are rather determined by other factors. However, perhaps it is for a different reason. Even if we accept that people are responsible for their choices, we may acknowledge that there are certain choices it may be very difficult for them to make. For instance, in a world of advertising and pressure to conform to certain social norms, it may be very difficult to make the most sensible lifestyle choices. So even if, in a sense, people are responsible for such choices, we may decide not to *hold them* responsible.

Alternatively, perhaps there are different degrees of responsibility. Perhaps there are certain reasons that defeat or reduce our responsibility for our choices. For instance, if there are very few options that a person has that are compatible with her staying healthy, and none of these is very desirable in terms of her own values, we may claim that if she does not choose one of these options and so falls ill, she is only partly responsible for this. Therefore, she may be entitled to be treated within one of the health-care systems that I discussed above where the burden of financing treatment is shared collectively.

5. HEALTH CARE AND THE LIMITS OF JUSTICE

I now want to turn to the question of what kinds of health-care services justice may require us to provide. In particular, I want to focus on the correction versus enhancement distinction.

As the genome project progresses, we gain knowledge about what genes cause us to have various diseases, but also about what genes cause us to have various other features, some of which are considered desirable according to our current evaluative standards. In some cases, such findings may enable us to medically enhance people. For instance, genes that code for the human-growth hormone can be used to produce pharmaceuticals that will make short people taller and, perhaps at some time in the future, gene therapy can also be used to promote this aim.

Now, while it is relatively uncontroversial that justice may require us to treat people with diseases, it is far more controversial that it may require us to medically enhance people.[6] However, just as people may be worse off than others because they are ill, they may be worse off because they think they are too short, or too fat, or just not very attractive. Furthermore, because of these features, they may be disadvantaged in their interactions with other people. In a survey of the research done on the relation between people's appearance and how they are perceived, it is concluded that:

> The social psychological effects of physical attractiveness are pervasive, strong, and generally uniform in nature. They are such that the physically attractive, whether male or female, old or young, black or white, or of high or low socioeconomic status, receive preferential treatment in virtually every social situation examined thus far. (Bersheid and Gangestad, 1982, p. 290).

This, of course, just confirms that the misfortunes that people may suffer in the genetic lottery do not stop at diseases. The point here, however, is that one way of compensating people for these non-disease related misfortunes is by medical intervention. And indeed, it may be asked, if we believe that justice requires that people who do less well because of their lot in the genetic lottery should be compensated, must we then not also believe that sometimes justice requires that they should receive enhancing medical treatment?

Presumably, some people would rather quickly dismiss this argument and claim that justice cannot possibly require us to compensate people for not being very physically attractive and the like. They would claim that this is simply not the sort of issue that justice is intended to deal with. However, this answer may be quite difficult to square with some of our deep moral and political beliefs. Many of us believe that when it comes to genetic diseases, people should be compensated. Why is this? A plausible answer is that we believe that it is unfair that they are worse off, since they are not themselves responsible for their genetic constitutions. But this answer, of course, applies just as much to looks as it does to diseases. We are no more responsible for how much beauty nature has bestowed on us than for our shares of health. And if we are to be compensated when our genetic constitutions cause us to be worse off, we are to be compensated for being unfortunate in the distribution of beauty.

Another objection is that it would be more appropriate to compensate people, not by medically enhancing them, but by offering them social assets such as money or perhaps the services of a psychologist. But, of course, we then need an explanation of why that would be more appropriate. Presumably, in some cases, it would both be a better compensation and cheaper to perform the enhancement.

Norman Daniels suggests that, on a plausible conception of justice, there will be no reason to medically enhance people. To motivate this claim, he argues that it accords with our *actual* concerns about equality. Rather, we should "... mitigate the effects of normally distributed capabilities through restrictions on other inequalities we allow" (Daniels, 1994, p. 125). So Daniels' suggestion seems to be that we should compensate people in terms of social assets. However, I am not entirely sure that he is right about what our actual concerns are. If, for instance, a child is constantly teased because his ears stick out, I think that most of us are inclined to think that he should be compensated in terms of an operation rather than in some other way. In any case, since some people may themselves prefer to be compensated in terms of medical enhancements, and since in some cases this may be cheaper, it seems that we would need to be offered a reason to accept Daniels' view. And in this respect, it is not much help to be told that we already accept it.

While it may be difficult to argue that justice gives us no reason to perform medical enhancements, it may be easier to argue that—at least in some cases—while it gives us a reason to perform them, it also gives us some reason not to perform them. First of all, there may be health risks involved in medical enhancements that would be avoided if some other sort of compensation were offered. Secondly, if medical enhancements are standardly used to compensate for misfortunes in the genetic lottery, it may become increasingly difficult to fight intolerance in society. After all, if society thinks it fit to medically remove specific physical features that used to be considered within the range of what is normal, this may seem to involve an acceptance of the view that such features are not very desirable. Finally, according to plausible theories of justice such as the Priority View, it is increasingly more important to benefit people the worse off they are and so, in general, health-care resources should be spent on treating people with severe diseases rather than on enhancing various features in people who are not equally burdened.

Justice, then, may both provide us with reasons to enhance and with reasons not to. These will have to be weighted against each other and we should not expect that we will always get the same answer. Clearly, some enhancements would give rise to benefits that are so trivial that we need not even consider them. But there are other enhancements that may give rise to more substantial benefits, and we cannot rule out in advance that when the reasons for and against are weighted against each other, we will reach the result that justice requires us to perform them. Perhaps one example would be a case in which a child is unhappy about being very short.

I should stress that in this paper I have merely considered what *justice* requires. There may be other moral concerns that are relevant when assessing medical enhancements and, if so, these will need to be considered in order to reach an overall conclusion.[7]

6. CONCLUSION

I have argued that those who fare less well because of their misfortune in the genetic lottery should be compensated in accordance with the Priority View. This implies that as we come to attribute more and more of those of our properties that are relevant for our shares of benefits and burdens to the causal influence of our genes, we should become more willing to redistribute resources to accommodate the interests of the worse off. Furthermore, it implies that a system of more or less unregulated private health insurances (that is not supplemented with a system of publicly funded health care) is unjust. And finally, it implies that, somewhat surprisingly, we sometimes have a pro tanto reason to medically enhance people in the name of justice. However, whether we ought to do so, all things considered, is a further question.[8]

NOTES

1. For the perhaps most influential discussion of the lotteries and their moral implications, see Rawls, 1971.
2. For discussion of whether determinism is compatible with having a free will, see, e.g., Honderich, 1993, and Watson, 1982.
3. Rawls' argument—the so-called intuitive argument—is discussed thoroughly in Barry, 1989, pp. 213–34.
4. But for extensive discussion of what the proper unit of egalitarian concern is, see Arneson, 1989, Cohen, 1989, Dworkin, 1981a, Dworkin, 1981b, Rawls, 1982, Sen, 1982, Sumner, 1996.
5. For a detailed criticism of the use Nozick makes of the Principle of Self-Ownership, see Cohen, 1995.
6. For instance, in recent years, it has been argued by both Norman Daniels and Allen Buchanan that justice requires no such thing; see Daniels, 1994, and Buchanan, 1995.
7. For a discussion of enhancements and gene therapy, as well as of the ethics of gene therapy in general, see Holtug, 1993, and Holtug, 1997, and for a discussion of enhancements and plastic surgery, see Holtug, 1996.
8. I would like to thank Kasper Lippert-Rasmussen for some helpful comments on a previous version of this paper.

REFERENCES

Arneson, Richard (1989), 'Equality and Equality of Opportunity for Welfare', *Philosophical Studies*, vol. 56.
Barry, B. (1989), *Theories of Justice*, Harvester-Wheatsheaf: London.
Berscheid, E. and Gangestad, S. (1982), 'The Social Psychological Implications of Facial Physical Attractiveness', *Clinics in Plastic Surgery*, vol. 9.
Bishop, J.E. and Waldholz, M. (1990), *Genome*, Simon and Schuster: New York.

Buchanan, Allen (1995), 'Equal Opportunity and Genetic Intervention', *Social Philosophy & Policy*, vol. 12.
Cohen, G.A. (1989), 'On the Currency of Egalitarian Justice', *Ethics*, vol. 99.
Cohen, G.A. (1995), *Self-Ownership, Freedom and Equality*, Cambridge University Press: Cambridge.
Daniels, Norman (1994), 'The Genome Project, Individual Differences, and Just Health Care', in Murphy, Timothy F. and Lappé, Marc A. (eds.), *Justice and the Human Genome Project*, University of California Press: Berkeley.
Dworkin, Ronald (1981a), 'What is Equality? Part 1: Equality of Welfare', *Philosophy and Public Affairs*, vol. 10.
Dworkin, Ronald (1981b), 'What is Equality? Part 2: Equality of Resources', *Philosophy and Public Affairs*, vol. 10.
Gauthier, D. (1986), *Morals by Agreement*, Clarendon Press: Oxford.
Greely, Henry T. (1992), 'Health Insurance, Employment Discrimination, and the Genetics Revolution', in Kevles, Daniel and Hood, Leroy (eds.), *The Code of Codes*, Harvard University Press: Cambridge, Mass.
Holtug, Nils (1993), 'Human Gene Therapy: Down the Slippery Slope?', *Bioethics*, vol. 7.
Holtug, Nils (1995), 'Patents on Human Genes: Is There a Moral Problem?', *Monash Bioethics Review*, vol.14.
Holtug, N. (1996), 'Are We Stuck with Our Appearance? Ethics and Cosmetic Surgery', in de Beaufort, Inez, Hilhorst, Medard and Holm, Søren (eds.), *In the Eye of the Beholder. Ethics and Medical Change of Appearance*, Scandinavian University Press: Oslo.
Holtug, Nils (1997), 'Altering Humans—The Case For and Against Human Gene Therapy', *Cambridge Quarterly of Healthcare Ethics*, vol. 6.
Honderich, T. (1993), *How Free Are You?*, Oxford University Press: Oxford.
Hurley, Susan (1993), 'Justice without Constitutive Luck', in Griffiths, A. Phillips (ed.), *Ethics*, Cambridge University Press: Cambridge.
Nozick, R. (1974), *Anarchy, State, and Utopia*, Basil Blackwell: Oxford.
Parfit, D. (1995), *Equality or Priority?*, The Lindley Lecture 1991, University of Kansas.
Rawls, J. (1971), *A Theory of Justice*, Oxford University Press: Oxford.
Rawls, J. (1982), 'Social Unity and Primary Goods', in Sen, Amartya and Williams, Bernard (eds.), *Utilitarianism and Beyond*, Cambridge University Press: Cambridge.
Sen, A. (1982), 'Equality of What?', in *Choice, Welfare and Measurement*, Basil Blackwell: Cambridge.
Sumner, W. (1996), *Welfare, Happiness and Ethics*, Clarendon Press: Oxford.
Watson, G. (1982), *Free Will*, Oxford University Press: Oxford.

28

BIOTECHNOLOGY, GENETIC INFORMATION, AND COMMUNITY

From Individual Rights to Social Duties?

Sirkku Kristiina Hellsten

Department of Practical Philosophy
University of Helsinki

1. INTRODUCTION

One of the main ethical problems in biomedical issues has been deciding who should have access to our genetic information and why. In this article I set this issue along with other ethical problems of biomedicine within a framework based on the contemporary liberal-communitarian debate, and discuss whether individual rights or social duties should have priority in the use of genetic or other health related information. However, instead of attempting to provide a clear normative stand, I want to discuss this issue from the point of view of analytical political philosophy. This means that my purpose is to clarify the complex relationship between the liberal ideal of political justice, the communitarian ethical approach and individual's moral judgement in the issues of biomedicine. My starting point is the contemporary shift of emphasis from the liberal concept of justice towards a more communitarian ethical approach. This can be seen in the Western world in the recent tendency to supplement and balance individual rights with considerations of individual responsibilities.[1] What I see should be acknowledge here is that even if the promotion of individuals' social duties is now evidently gaining emphasis in Western medical practice, the academic discussions as well as political and legal debates are still mainly based on the ethical demands of informed consent, individual autonomy and individual rights.[2] This may lead us to a curious situation in which explicitly individualistic and liberal philosophical, political and legal discourse still uses the language of rights, while in practice there is, at least implicitly, increasing social pressure for individuals to accept their social duties and give 'the common good' priority over their personal moral judgement. This is why I believe we should take seriously the communitarian social challenge, and discuss how we ought to deal with a political situation in which the individual auton-

Genetic Information, edited by Thompson and Chadwick.
Kluwer Academic / Plenum Publishing, New York, 1999.

omy and rights are still formally promoted, but in which the social contest may leave an individual with no autonomy and no choice.

Since there are many and various types and uses of genetic information, I want to clarify that in this article I shall focus on the information which concerns health issues, for instance, by revealing serious genetic diseases.[3] I also want to note that in many cases this health related genetic information can be compared to other sensitive personal health related information, which reveals factors about the individual's physical or mental condition and 'health future' and its use is therefore often equated with other such information.

2. THE SOCIAL NATURE OF GENETIC INFORMATION

Despite the fact that health related genetic information and other health related information are not always a world apart, I shall start my analysis by discussing briefly why genetic information has been seen to be so different from other kinds of information concerning our health and lives in general. What are the ethical grounds behind the legislation calling for special reticence where this particular type of private information is concerned?

Firstly, genetic information is considered to be uniquely personal. Genetic information will enable science to identify each human being as genetically unique and, simultaneously, it provides a means to predict, at least to a degree, this individual's future life, particularly in terms of physical and mental well-being. Genetic information often reveals predictors of undesirable and stigmatising conditions, such as cancers, and conditions that lead to mental illness or dementia. These may be details about one's life which one may want to conceal from others or at least which one wants to personally decide who is to told about them. Secondly, despite the fact that genetic information is extremely personal, it simultaneously concerns more than one person, and thus clearly includes social aspects. Deciphering an individual's genetic code provides the reader of that code with information not only on a particular, unique individual but about those who are genetically related to him or her. The collection and use of genetic information does not concern only those individuals from whom the samples are collected. In addition it can also affect the future of a great number of other people, such as members of a particular family (parents, children, siblings), or an even larger ethnic or racial population.[4] It should, however, be added here that this social nature of genetic information is problematic, but it is not unique. Not only genetic information, but also other kinds of health related information has this same nature. Thus, the same argumentation that is relevant to the use of genetic information is, in many ways, relevant also to other kinds of heath related information which reveals illnesses or probabilities of particular health conditions in people. There are also many other diseases besides so called genetic diseases which can be 'genetically' transmitted from one person to another. For instance, HIV/AIDS is not a genetic condition but is often transmitted from a mother to a child.

This is just to show that when we take a communitarian stand in questions concerning the use of our genetic information, the same argumentation that is used to defend the public use of any other health related information can be used here. Communitarians who promote social duties appear then to have plausible premises from which to argue that the nature of genetic or any other health information about an individual never concerns only one individual and thus cannot be regarded as personal and thus all the people whom the information concerns should be able to find out about factors related to their health.[5] In the case of genetic information, the collectivist argumentation could be taken even further. Collectivists could argue that genetic information must actually be considered as common property, because many people share the genes. Therefore any genetic information should be used for the good of the population as a whole, and not merely to the benefit of individuals living within that population.

3. THE LIBERAL DILEMMA OF INDIVIDUAL RIGHTS: TO KNOW OR NOT TO KNOW?

The use of genetic information, however, has in the traditionally individualistic Western world largely been restricted by appealing to the values of a liberal political concept of justice which promotes individual privacy and personal integrity. This means that the political and legal tradition in the West has promoted an individual's right to make decisions concerning the genetic information he or she may obtain about themselves. Even if and when there are exceptions, it has been emphasised that the use of this information, for instance for research purposes, does not inflict personal harm and injury to an individual. According to this political liberal view, individual autonomy and individual rights have priority over any social duties or any 'common good'. Liberal arguments have traditionally followed the logic which takes an individual's absolute moral autonomy as its starting premise. Since individuals are to be seen as morally autonomous agents, it can be argued that the state should not interfere with their personal values and choices. Therefore, at least in legislation, we should give individuals full control over any information concerning their genetic make-up and future health. Thus, the protection of the individual's autonomy coupled with the right to privacy is assumed to provide a precaution against the loss of freedom and against different types of harms that the revelation of highly private and sensitive genetic information can easily cause. This harm may occur if a person, for instance, is precluded from obtaining an economic or social service benefits, or employment, and/or life or health insurance.[6]

This idea that an individual should have a control over any information concerning him or her is seen to include a demand that an individual has also a 'right to know' any sensitive genetic information which affects his or her life. A person should then have a chance to find out about his or her genetic make-up and the risks involved in it if he or she so chooses. Because this 'right to know' is also based on the ideas of autonomy and self-determination, it appears to be a plausible ethical and legal demand in a liberal Western democracy. However, what makes it problematic is that the ideas of autonomy and self-determination are also often assumed to back up the 'right not to know'. In general the 'right not to know' is supported by the fact that since genetic information may reveal not only diseases that could be prevented or treated, but because it can also reveal conditions that are untreatable, this might cause anguish, harm and avoidable suffering to a person. Since the 'right to know' is based on a requirement for autonomy, it can easily be reasoned that we, as autonomous persons, must also have a 'right not to know' about the genetic factors concerning us.

The problem, however, is that not obtaining this information could have great impact on the lives of many other people. Thus, the 'right not to know' may and often does conflict with someone else's 'right to know'. Even if there is no legal duty to obtain genetic knowledge and to share it with anyone else, the choice of not wanting to know important factors concerning one's genetic condition may violate another person's right to find out something about himself or herself or even cause harm and suffering to someone else.[7] Because an approach based merely on individual rights cannot solve the conflict between the autonomy, rights and interests of different parties involved, the question arises what kind of duties (if any) doctors may have towards their patients; what duties patients have towards their family and relatives, towards group of other people with the same disease, or towards society as the provider of medical treatment and as a distributor of scarce medical resources.

If we moved away from this liberal *cul-de-sac* towards utilitarian grounds, calculation and weighting benefits and harms would become central. The classical utilitarian principle requires the maximisation of benefits, that is the most benefits for the greatest number of people (or in its negative formulation the minimisation of harm and suffering: that is the least suffering to the smallest number of people). When we talk about the use of genetic information, in most cases this would mean that an individual should not only obtain information on his or her medical and genetic condition but that he or she also has a duty to do just this and a duty to share this information with others concerned and maybe even with the whole community. After all, this would benefit many and prevent avoidable harm and suffering. The anxiety and fear caused to an individual who does not want to know about his or her health in the future would have to be seen as less important if it was set against the greater benefit that this information provides other people and society as a whole.[8] It appears then, that a utilitarian answer would require us to disregard an individual's assumed right 'not to know', because this right prevents the benefit of a great number of other people or causes them harm and suffering that could be avoided if this information was obtained and redistributed.

The promoters of liberal individualism cannot give up the priority of the individual in the name of the common good as the utilitarians have to do. Thus, promoters of a liberal view reject the idea that an individual has any social duties regarding his or her own genetic information. Instead, in order to solve the conflict between autonomy, rights and interests, they suggest that in liberal political reasoning we have to make a clear distinction between the public and private spheres of our social and moral life: that is, between political and legal procedures and individual moral judgement. Many liberal political philosophers and ethicists take here a Kantian turn and admit that we, as individual moral agents, may well have *a moral duty* to obtain relevant information about our own condition when serious diseases are suspected and that we may also have *a moral duty* to reveal this information to those who are concerned. But simultaneously they emphasise that the liberal view cannot allow that either one of these *inherently moral duties* is enforced by political or legal measures.[9] So, in general the liberals demand that in the public sphere the law is to protect individual privacy, and that the final moral choices should be made in the private sphere by autonomous individuals. Liberalism as a political rather than ethical theory, however, leaves the question of how individuals as assumed autonomous moral agents could learn their Kantian *moral duties* open. Now from the point of view of political philosophy, this motivational question is the exact starting point of the contemporary Western communitarian ethical approach.[10]

It is the aim of the contemporary communitarian approach, at least in its Western formulation, to solve the ethical conflict between the individual's integrity and the general desire to maximise the social benefits of individuals. In the case of genetic information this conflict is manifest in the dilemma between one's interest in protecting sensitive personal information and the 'general will' (of individual's as the members of particular societies) to obtain this information, for instance, in order to undertake research for the benefit of the society as a whole. It could then be stated that the contemporary communitarians are working in the grey zone of ethics between the public concept of justice and private morality. Communitarians are trying to find the motivational connection point at which political and legal norms stop and individual morality starts by trying to provide modern political theory with the missing motivational link between the moral choices made by individuals and coercive state intervention.

4. COMMUNITARIAN RESPONSE: DUTY TO KNOW, RESPONSIBILITY TO SHARE

Before the communitarian normative position can be understood, it should be acknowledged that much in the same manner as there are different formulations of liberalism or utilitarian principles, there are also many diverse communitarian views which overlap and sometimes even conflict with each other.[11] In this context I shall focus on contemporary Western, that is mainly American, communitarianism. This Western communitarianism differs from other forms of more traditional communitarianisms, because methodologically and normatively it is has its roots in individualistic rather than collectivist social practices. Within this view which, for clarity's sake, I shall be calling 'Western communitarianism', I shall make one more central distinction between two different, though again often overlapping, types of communitarian approaches, namely, the descriptive and the prescriptive.[12]

The descriptive communitarian view first presented by Alasdair MacIntyre (1984, 1988), Michael Sandel (1982, 1996), Charles Taylor (1979, 1989) and Michael Walzer (1983) started as a criticism of the universalist liberal social theory. This descriptive communitarianism works as an internal criticism of liberalism rather than as a rival normative approach. Its main argument is that the liberal approach to social justice is too atomistic and abstract, and thus cannot promote the kind of individual morality which true maintenance of liberal democratic values would require. Descriptive communitarians point out that in particular the liberal image of an individual as an autonomous moral agent fails to take into account our communal ties and values which always affect our moral choices in everyday life. Communitarian critiques of liberal political theory and practice also assume that the liberal emphasis on subjective values and formal rights in fact tends to justify and consequently increase egoism, moral indifference, political alienation and moral fragmentation in a pluralist society.

The prescriptive communitarian approach, for its part, wants to guide modern Western society back to the right track. This Communitarian social movement is at the moment particularly strong in the United States, but it has gained some support also elsewhere in Western Europe. Basically this social movement begins its normative agenda where descriptive communitarianism ends its criticism. According to the communitarian social movement, we can decrease egoism, moral and political indifference and social fragmentation, by emphasising common values and the importance of our social ties—not only in a descriptive but also in a prescriptive sense: that is, as actual ethical choices and as interdependent caring.

The communitarian social movement has often been seen to present an opposite or rival approach to liberalism. This, however, is not exactly the case with the Western communitarian movement, since Western communitarians themselves, if their own logic is followed, are the products of a liberal society and are therefore also indoctrinated into liberal values of autonomy, equality, tolerance and democracy. However, what makes the contemporary communitarianism so confusing to the promoters of liberalism is that while the communitarian account clearly emphasises social responsibility, it truly does want to greatly increase the state's involvement in moral matters. While a collectivist political theory would in general emphasise governmental responsibility and citizens' duties towards the state as a political entity, the contemporary communitarian view starts rather with local relations and individual ethics and emphasises individuals' responsibilities towards each other. Thus, the idea of social responsibility is not seen to be the task of the state as a social whole, as the welfare state ideology tends to see it.[13]

In order to better explicate this difference I want to now briefly compare the communitarian idea of social duty with that of socialistic collectivism and that of welfare liberalism, for instance, in the matters of distributing genetic information. We can note the following: first, in a collectivist system, such as socialism, every citizen has a duty, as a constructive part of the state, to provide this information and accept its public use for the benefit of the state as the political whole. Thus, the obligations are primarily to the benefit of the social whole, and only secondarily to the benefit of individuals even if they are the contractive parts of the state. The rights and benefits are not in balance because the state has most of the rights as a social collective and individuals are left with mere duties to the state. Second, in welfare liberalism the state as a whole has a responsibility to collect and make public as much of this information as possible in order to allocate the resources and to target the medical services better. Here the obligations are the obligations of the state as the political agent and they are to the benefit of individual citizens as members of the state. In addition there is an imbalance between individual rights and benefits, because individuals have most of the formal rights while all the social duties are left to the state. The modern Western communitarian view, for its part, seeks to balance the individual's rights with duties by emphasising that social duties are an individual's duties towards other members of different communities which exist within the political whole. For the communitarians the problem of collectivist socialism is in its totalitarianism, and the problem with modern welfare state ideology is its tendency to transfer too much of the social responsibility from individuals to the state. This relieves individuals from their civic duties and makes citizens passive objects of the state welfare policies (rather than active agents). What the communitarians have taken to be their main task is then the attempt to balance individual freedom with responsibility, and rights with duties in order to complete liberal theory and implement liberal values.

Communitarians assume that because of our human nature as social beings, the realisation of these liberal values have also to be backed up by our social institutions. Since communitarians do not see that individual morality should be motivated by political force or strict legal regulations, it falls to the smaller communities within a state to guide our personal ethics and moral behaviour. Thus, the main difference between the liberal political approach to justice and the communitarian ethical approach is that the communitarians focus less on the political relationship between the state and its citizens and more on the social relationship between individuals and their communities as well as on the moral relationship between individuals as members of different communities.

Therefore, it is evident that the communitarian social movement also emphasises its non-political nature. Instead of having a clear political agenda, it has its goals on the level of ethical communities. It stresses the importance of communal values, civic and family duties as well as community care. Since the state cannot be allowed to violate one's personal autonomy and integrity, the enforcement of any social duties which may violate autonomy cannot be done by political and legal measures. However, since we are all born as members of different communities which influence our values and lifestyles, these communities can be used to strengthen the desired ethical attitudes towards wider social responsibility.[14] Philosophically, the communitarian requirements of social obligations are based on the idea that while individuals might be the final moral agents, the liberal political idea of unlimited individual freedom is illusory. While liberals have taken a Kantian view of our moral duties, the communitarians follow Hegelian criticism of Kant's moral universalism and claim that there can be no individual freedom without responsibility and we can have no individual rights unless there are reciprocal social duties. Since these duties are moral in their nature, and since individuals are—despite their potential capacity for independent moral judgement—social animals, the place to learn these duties is within a social context and, if necessary, by social pressure.

Since the state cannot be the motivational force for social duties, communitarians see that the guidance for reasoned judgement, virtuous action and self-restraint has to come from the moral voices of one's community. Our responsibilities are then not merely a personal matter (we cannot always choose what responsibilities we want to accept as our own) nor are they coerced by the state. Instead they are already anchored in community. Although communitarians believe, much as the liberals do, that the ultimate foundation of morality may be commitments of individual conscience, they demand that communities help introduce and sustain these commitments. Hence it is the duty of communities to articulate the responsibilities they expect their members to accept, especially in times when understanding of these responsibilities has weakened.

The motivational moral power between the state and the individual is with a responsive community whose moral standards reflect the basic human needs of all its members. To the extent that these needs compete with one another, the community's standards reflect the relative priority accorded by members to some needs over others. The communitarian platform declares that while law does play a significant role, not only in regulating society but also in indicating which values it holds dear, our first and foremost recourse is not to the law, but rather is to affirm the moral commitments of parents, young persons, neighbours and citizens, to affirm the importance of the communities within which such commitments take place and to ensure that these commitments are transmitted from one generation to the next. Thus, teaching moral duties, social responsibilities and common values is not primarily a legal matter but rather a private matter. In fact, for communitarians, when a community reaches the point at which the social responsibilities are largely enforced by the powers of the state, the state is in deep moral crisis.[15]

5. PRACTICAL EXAMPLES

What does the communitarian emphasis on our social duties mean in practice. In their normative agenda in health care policies, for instance, some of the communitarian ethicists, mainly in the United States, argue, on somewhat utilitarian grounds here, that because a great number of patients needlessly suffer or die because of a lack of donated organs, societies ought to adopt a policy that hold organ donation as a social duty or as an 'act of obligation to community' routinely expected of each of us.[16] Communitarians also propose that people should be compelled to tell their partners if they are for instance HIV positive. It is also suggested that AIDS testing be obligatory for all pregnant women and infants. Some communitarians even go so far as opposing unhealthy personal behaviour because it increases the cost that the whole community must bear (a question arises: if we have a duty be healthy, but we fail in this, do we then have a duty to die?). Communitarians also want to protect public health by doing drug screening on people who directly affect public safety (such as pilots, train engineers, medical personnel etc.)[17] If we follow this communitarian logic in the matters of genetics, it can easily be concluded that there would be similar obligation to obtain and share the genetic information with anyone affected one way or another, and that it is our duty to choose a life style which prevents the outbreak of a particular genetic (or other) probable disease.

A practical experiment of the communitarian policies can be found in Cyprus, where there is a strong emphasis on community involvement and public awareness in matters of genetics. The difference is that in Cyprus, communitarian ethics has already been integrated with political and legal demands. One reason for this integration might be that in Cyprus there is still quite a small and homogeneous ethnic population which forms the state and thus shares its interests with the state. The communitarian examples of the Cyprus Thalassemia Program are included in the following: it is still publicly emphasised that reproductive decision-making should be voluntary and people should not be forced into any kind of reproductive practices they do not want. This individualistic rhetoric is used to avoid eugenic overtones in the genetic program. Thus, publicly it is stated that the social concern is about individuals and about families, and not about the gene pool and certainly not about race.[18] However, simultaneously the importance of community involvement in genetics is emphasised in order to bring the threatening disease under full and effective control.[19] Strategies include community involvement in the control of genetic diseases. This is significant because many aspects of social life and attitudes can be affected by this preventive program, such as marriage practices, choice of partner, attitude towards abortion, as well as other religious, ethical and legal attitudes. In order to enforce attitudes favourable to general screening, premarital testing and premarital certificates, the state has, for example, made an agreement with the Church that the Church will not interfere in these issues. The non-interference of the Church is important because the Church in general opposes termination of pregnancies on religious grounds. This way, couples rely on their own judgement and decide themselves, without any outside pressure, what is best for them. However, prior to their final decision, couples are to have proper genetic counselling and are educated by doctors and other medical personnel. The ethical question is then how much individuals' final decisions are affected by communal propaganda, if you wish, and social pressure and how much moral autonomy there really is left in the end. Thus, there is a danger that the 'responsive community' defines what is good for people and what is expected of them by educating its members to accept the desired values.[20]

All the aspects of the Cyprus Thalassemia program, such as: population screening; genetic counselling; foetal diagnosis; taxation and voluntary donations; high level of awareness and concern for community; education of the public in general and the "family units" in particular; and education in schools and education of medical, paramedical and health workers, are based on the communitarian approach to social responsibility which in the end is taken to turn out to be the best for the individual members of this society.[21] Thus, in Cyprus, the main aphorism is still presented in individualistic terms, though in rather a Aristotelian formulation: it is emphasised that informed and educated people will make 'the right choice'. This seems to justify the further social pressure upon those who, despite the information available, still tend to make 'the wrong choice'. This shows that even if communitarians promote the autonomy of one's choice, they can—in the very process of promoting it—as easily suffocate it by social pressure. There would then no longer be any freedom of choice or autonomous consent, because in practice only 'the right things' are accepted. Whether or not I want to know my genetic make up, it is my duty as a member of a particular family and particular community to take the necessary steps to learn about my hereditary traits and to share this information with others whom it might concern.

Milder forms of these same communitarian trends can also be found in the traditionally very individualistic United States. Even if the political and legal approach in the USA is still extremely concerned about protecting the individual's formal rights and thus emphasises that there can be no contractual or statutory obligation to give out one's genetic information, in practice, sharing this information at least with one's family is often advised by the medical personnel. The US Genetic Privacy Act concerning the use of genetic information shows how the individualistic legal trend and communitarian ethical approaches overlap in health care practice. When a law leaves the choice to an individual, the communitarian ethical guidance gains space. Even if it is admitted that the creation of new substantive rights or duties of citizens as family members or members of a particular group of people does not provide the solution, the communitarian ethics of social duty and communal care can still easily be endorsed in practice. Despite the lack of legal obligation to do so, an individual is often encouraged out of moral obligation to share as much of the information as it would provide the siblings, children or other relatives with the opportunity to obtain information about their own conditions or risk. However, in addition there are also exceptions in legal regulations which limit the scope of individual privacy. Individual privacy is violated when DNA samples are collected for genetic analysis, for instance, from minors, incompetent persons, pregnant women, and embryos.[22]

6. CONCLUSION

I shall conclude by noting that there is an evident tendency in the medical practice even in the Western world to shift more emphasis from the individual's rights to his or her social duties. Therefore, the pros and cons of the communitarian ethical view should be taken into critical consideration. We now have to start discussing seriously the questions of social duties and responsibility as well as the influence of social pressure on individual decision-making, even in a situation in which the political and legal rhetoric still functions within the language of individual rights and autonomy. This means that in the name of autonomous, educated choice or informed consent, the moral majority of so called responsive communities may well succeed in making individuals adapt to values and practices, which, in a different social situation might not have been the individuals' first choice (if indeed, it is their choice at all). Thus, we have to consider whether the autonomy promoted by the liberal approach is in the West now turning into a social requirement for a virtuous life and self-restraint.

When the communitarian approach is discussed in a bioethical context, particularly where issues surrounding genetic information are concerned, we should start by taking the descriptive communitarian criticism of liberalism seriously. This descriptive communitarian criticism of liberalism can actually help us to understand the influence of social pressure in individual decision-making; it even clarifies the promises and dangers of communitarian normative views by bringing out the origin of the problems involved in balancing individual rights with social duties.

The communitarian normative agenda even in its Western formulation, however, should be adopted with caution because in the process of educating virtuous individuals, it can enforce whatever policies the moral majority happen to consider desirable at any given time. Particularly, in the case of biotechnology, the communal pressure to adapt to a particular line of normative reasoning does not leave an individual much room to make morally autonomous choices, but rather, there is a pressure to accept the community values as one's social duty. While we may have duties towards each other and while we may have social responsibility, when these 'choices' are no longer made independently, the balance between 'freewill' and responsibility is once again lost. What should be kept in mind is, then, that there is neither freedom without responsibility, nor responsibility without freedom. In a situation where responsibility turns into social coercion (no matter how subtle) the use of the individualistic rhetoric in the political and legal language is misleading and may allow a state to enforce policies which otherwise would be seen as suppressing autonomy and paternalistic. What we should do then is pay more attention to how to redefine the concept of autonomy in a way which takes social context into account, without losing sight of individual morality.

NOTES

1. Chadwick 1997, 21.
2. The traditionalist communitarian approach which heavily disregards the Western demand for informed consent of an individual and instead emphasises the individual's duties towards his or her family, the community or society as a whole, is still dominant in the collectivist cultures of Asia and Africa. See the studies on biomedical attitudes of originally collectivist cultures presented for instance in *Japanese and Western Bioethics* ed. By Hoshino 1997, *Encyclopedia of Bioethics* ed. By Reich 1995, see also Tangwa 1996.
3. Other uses for genetic information which have 'the common good' as their aim rather than the benefit of just one person are, for instance criminal investigations where genetic information is used in order to track down dangerous criminals or to identify dead bodies. Our genetic information is not only stored in places which maintain our medical records, but there are other public entities which store our genetic materials. In the USA, for instance, such state organisations as the FBI as well as some other individual state programs store DNA samples from convicted sex offenders and other dangerous criminals. Also the U.S. Army has a DNA sample storage which it can use to identify criminal suspects. Other places which collect a large amount of biological material are the Red Cross and other local and the international blood banks, private sperm, ovum and embryo banks, as well as different state facilities that store blood samples for different testing purposes. See the US Genetic Privacy Act at url:http://www.ornl.gov/TechResources/...ome/resource/privacy/privacy1–5.html. Genetic information is also in general used to identify genetic relations, such as parenthood—sometimes this can be done even after the assumed genetic parent has died, as was the case with the famous French singer and actor Yves Montand.
4. On these issues see for instance Genetic Privacy Act and Commentary, HTTP://www.ornl.gov/TechResources/...ome/resource/privacy/privacy1.html, also Report on Bioethics in Europe 1992, Final Report, Ethics and Mapping of the Human Genome, The Danish Council of Ethics, 1993)
5. Introduction in Chadwick 1997, 6, Royal College of Physicians 1991, para 4.10.
6. Chadwick 1997, 14, Ethics and Mapping of the Human Genome, The Danish Council of Ethics, 1993, 14–15, Helsinki II Declaration I.5., I.6, III.4.
7. Chadwick 1997, 17–20. In general it seems that once a person knows about his or her genetic conditions and also knows about how the same condition can affect someone else, the moral pressure to tell the others concerned increases and it is more difficult not to share this information with them.
8. See for instance Chadwick 1997, 15, Harris 1993, Hare 1993, 98–146. In Harris 1993, for instance, the utilitarian principles are defined as avoidance of avoidable harm and suffering.
9. This Kantian emphasis is gaining more and more influence in modern liberal reasoning introducing the demands of individual morality to the liberal concept of political justice. See for instance the Kantian constructivism of Rawls 1971, 1993. This neo-Kantian formulation of liberalism appears to demand that where the political and legal requirements must end, the moral law and moral duties of an autonoumous agent must begin.
10. This shift from liberal individualism towards communitarian ethics in issues of genetic information appears at least in part to be connected to the social nature of genetic information as such. Since an individual's choices concerning the use of his or her genetic information may often affect the lives of many other people who are genetically related to him or her, it has become difficult to deal with this information in merely individualistic terms. The social nature of genetic information inevitably creates conflicts between rights of different individuals. In the case of genetic information, an attempt to protect one person's assumed right, for instance, to 'not to know' his or her genetic make-up can easily violate another person's 'right to know' this information if it happens also to concern him or her. Also, the harm avoided by one person in this situation may cause harm and suffering to a number of other people.

11. On the different forms of communitarianism see Avineri, S. & de-Shalit, A. 1992. There are for instance earlier forms of communitarianism of Aristotle, St. Aquinas, Rousseau, and later Burke, which emphasised the social and political nature of human beings. There are also clearly collectivist formulations of communitarianism such as traditionalism and communalism which are more typical of collectivist, non-Western cultures, such as African and Asian cultures, and there are the contemporary formulations of Western communitarianism of MacIntyre, Sandel, Taylor, and Walzer which function as critiques against overly individualistic Western policies and as the social movement of Etzione, Galston, Bellah, which want to balance rights with responsibilities. See for instance The Responsive Communitarian Platform, 1997.
12. Descriptive communitarianism is presented by MacIntyre, Sandel, Taylor and Walzer as a criticism of liberal political theory and practice. Prescriptive communitarianism or the communitarian social movement is most fully elaborated by Amitai Etzione 1993, 1995, and on the Responsive Communitarian Platform, 1997 (was drafted first by Amitai Etzione, Mary Ann Glendon and William Galston in November 1991). The application of the communitarian ideas in biomedicine can be found in Beauchamp and Childress 1984 and Emanuel 1991. See also Etzioni, A., Cassel, C., Dougherty, C., McCollister Evarts, C., Griffith, J., Nelson, J., Osterweis, M., and Wikler, D. 1997 in *A Communitarian Position Paper: Core Values in Health-Care Reform, A Communitarian Approach,*.
13. See for instance Walzer 1980, 1983, and The Responsive Communitarian Platform 1997, welfare liberal reasoning also theoretically deliberated in Rawls 1971, 1993.
14. See the Communitarian Network's homepages at http://www.gwu.edu/~ccps/cate1.html.
15. The Responsive Communitarian Platform 1997 and at http://www.edu/~ccps/RCPlatform.html.
16. See for instance James Lindemann Nelson, The Responsive Communitarian Platform, http://www.edu/~ccps/RCPlatform.html
17. See for instance Etzioni, 1997 in a Communitarian Working Paper: HIV testing for Infants and Pregnant Women: a Case study in Privacy and Public Health. For comparison, in Finland there is also an increasing tendency to test potential employees. Not only army and police, but some of the private security companies have now health and fitness testing for people working for them. Just recently security companies have agreed to start compulsory drug testing for their potential employees. Genetic testing is not far away from this.
18. Motulsky 1994, 24–25.
19. Thalassemia condition is the absence of, or a decreased rate of, synthesis of one of the globulin chains of the haemoglobin molecule. b-Thalassemia, in the homozygous state, is for Cyprus the most important of the disorders, both clinically and from the public health point of view, since it leads to chronic anaemia, bone deformities and early death. See for instance Hadjiminas 1994, 26–27.
20. In addition, by excluding the Church as one ethical authority or influence, the moral majority of the community gains more direct power.
21. Hadjiminas 1994, 30–37.
22. The Genetic Privacy Act and Commentary by George Annas, Leonard Glantz, Patricia Roche at http://www.ornl.gov/TechResources/...ome/resource/privacy/privacy1–5.html

REFERENCES

Annas, G., Glantz, L, and Roche, P. (1995) *The Genetic Privacy Act and Commentary* at url http://www.ornl.gov/TechResources/...ome/resource/privacy/privacy1–5.html.

Avineri, S. & de-Shalit, A. (1992), *Communitarianism and Individualism*. Oxford University Press, New York.

Chadwick, R. (1997), 'The Philosophy of the right to know and the right not to know' in Chadwick, Ruth, Levitt, Mairi and Shickle, Darren (eds.), *The Right to Know and the Right not to Know*, Avebury: Aldershot, Brookfield USA, Hong Kong, Singapore, Sydney.

Chadwick, R., Levitt, M., and Shickle, D.(1997), 'Introduction' (1997) in Chadwick, R., Levitt, M., and Shickle, D. *The Right to Know and the Right not to Know.* Avebury: Aldershot, Brookfield USA, Hong Kong, Singapore, Sydney.

Childress, J. & Beauchamp, T. (1979), *Principles of Biomedical Ethics* 4th ed. Oxford University Press: New York.

Danish Council of Ethics (1993), *Ethics and Mapping of the Human Genome, Protection of Sensitive Personal Information*, The Danish Council of Ethics: Copenhagen.

Declaration of Helsinki (1964), Recommendations guiding physicians in biomedical research involving human subjects, the 18th World medical Assembly, *Bulletin of Medical Ethics*, No. 128, May 1997, Bioethics Publications Ltd, Royal Society of Medicine Press: London.

Etzioni, A.(1996), *The Spirit of Community: The Reinvention of American Society*. Touchstone: New York.

Etzioni, A.(1997), *HIV Testing for Infants and Pregnant Women: A Case Study in Privacy and Public Health*, The Communitarian Network: Washington, DC.

Etzioni, A., Cassel, C., Dougherty, C., McCollister Evarts, C., Griffith, J., Nelson, J., Osterweis, M., and Wikler, D. (1997), *A Communitarian Position Paper: Core Values in Health-Care Reform, A Communitarian Approach*, The Communitarian Network: Washington, DC.

Ezekiel, E. (1991), *The Ends of Human Life: Medical Ethics in a Liberal Policy*. Harvard University Press: Cambridge.

Hadjiminas G. M. (1994), *Ethics and human genetics, proceedings 2nd Symposium of the Council of Europe on Bioethics*, Strasbourg, 30-November-2 December 1993, Publishing & Documentation Service, Council of Europe, Council of Europe Press: Strasbourg Cedex.

Hare, R.M. (1993), *Essays on Bioethics*, Clarendon Press: Oxford.

Harris, J. (1993) *Wonderwoman and Superman*. The Ethics of Human Biotechnology. Oxford University Press: Oxford.

Hoshino, K. (ed.)(1977) , *Japanese and Western Bioethics: studies in moral diversity*. Kluwer: Dordrecht.

MacIntyre, A. (1984), *After Virtue*. Duckworth: London.

MacIntyre, A. (1988), *Whose Justice? Which Rationality?* Duckworth, London.

Motulsky, A. (1993), *Ethics and human genetics, proceedings 2nd Symposium of the Council of Europe on Bioethics*, Strasbourg, 30-November-2 December 1993, Publishing & Documentation Service, Council of Europe, Council of Europe Press: Strasbourg Cedex.

Ndinya-Achola, J.O. (1995), 'Medical Ethics, History of: Africa' in Reich, W.T. (ed.), *Encyclopedia of Bioethics* (1995), revised edition vol. 2, Georgetown University, Washington. D.C. Simon & Schuster and Prentice Hall International: London, Mexico City, New Delhi, Singapore, Sidney, Toronto.

Rawls, J.(1971), *Theory of Justice*. Harvard University Press: Cambridge.

Rawls, J.(1993), *Political Liberalism*. Columbia University Press: New York.

Royal College of Physicians (1991), *Ethical Issues in Clinical Genetics*, Royal College of Physicians: London.

Sandel, M. (1982), *Liberalism and the Limits of Justice*. Cambridge University Press: Cambridge.

Tangwa, G. (1996),' Bioethics: An African Perspective', *Bioethics* Vol. 10. No 3, 183–200.

Taylor, C. (1979), *Hegel and the Modern Society*. Cambridge University Press: Cambridge.

Taylor, C. (1989), *Sources of the Self: The Making of Moral Identity*. Cambridge University Press: Cambridge.

The Responsive Communitarian Platform (1997), The Communitarian Network: Washington, DC, also at url http://www.edu/~ccps/RCPlatform.html.

Walzer, M. (1980), *Radical Principles. Reflections of an Unreconstructed Democrat*. Basic Books: New York.

Walzer, M. (1983), *Spheres of Justice*. Basic Books: New York.

Wiesing, U. (1997), 'Individual rights and genetics: the historical perspective. A comment on Ruth Chadwick's paper' in Chadwick, R., Levitt, M. and Shickle, D. (eds.) *The Right to Know and the Right not to Know*, Avebury: Aldershot, Brookfield USA, Hong Kong, Singapore, Sydney.

29

LINEAR DESTINY AND GEOMETRIC FATE

Emilio Mordini*

Psychoanalytic Institute for Social Research
21, Via Arenula, 00186 Rome, Italy

It is pictures rather than propositions, metaphors rather than statements, which determine most of our philosophical convictions
—R. Rorty, 1979, p.12

INTRODUCTION

A metaphor is a figure of speech, which employs an implicit comparison to express a concept, to describe an object, or to emphasise some of the qualities of that object. The comparison contained in a metaphor must be only suggested, avoiding using the words like or as. "In that battle Caesar fought as a lion" is not a metaphor—it is a simile; while "In that battle Caesar was a lion" is a metaphor. Any metaphor is thus made by two elements: i) a comparison; ii) an implicit suggestion. Psychiatry has investigated the suggestive power of metaphors, and some scholars, such as M.Erickson, the most influential clinician in the development of modern hypnotherapy, advocated the use of metaphors in psychotherapy.

Metaphors are used to describe better, with more nuances, what we mean, but they often serve two masters. Any metaphor conceals as much as it reveals, masks as well as unveils. When I say: "Caesar was a lion" I probably intend to hint at his courage, however I cannot avoid suggesting other qualities that our culture attributes to lions (ferocity, violence, regality, capacity to inspire fear and respect). Moreover if the listener belongs to another culture, for instance Masai culture, chances are that she perceives something rather different from what I intended to communicate. Perhaps she may appreciate the central idea of bravery, but other qualities involved in the image are likely to be misperceived. Likewise I am expected to evoke different feelings if I say "Caesar was a lion" to a child who has just seen Disney's movie " The King Lion", rather than to another who has just come from the circus. Since in general we ignore that our listener has just come from the

* Tel: + 39 - 6 - 6867495 Fax: + 39 - 6 - 68 33 552 Email: iprs@mclink.it

Genetic Information, edited by Thompson and Chadwick.
Kluwer Academic / Plenum Publishing, New York, 1999.

cinema or the circus, metaphors should always be handled with care; indeed, much more than other linguistic expressions they can carry meanings of which we are unaware. Moreover their capacity to suggest under the level of consciousness makes metaphors psychologically invasive. People tend to forget metaphorical origins of many expressions and they use metaphors as if they were referential language. As a consequence we often commit the fallacy of believing we speak of the real object while we are speaking of the compared (metaphorical) object. That is particularly important in science.

Scientific language is full of meanings that have metaphorical roots. Indeed humans are often obliged to make use of metaphors when they speak of realities far from their perceptive experience (either because they belong to different physic realms or because they are abstractions). For instance any educated person is likely to be aware that mathematical formulas describe atoms better than any three dimensional model, but when she speaks of atoms she cannot avoid that in her mind the idea of an atom is formed according to human sensorial modalities. The naive student will imagine a planetary model (the nucleus around which electrons orbit), the scholar will probably envision something similar to the restless movement of different clouds in a windy sky. Of course none of these images are closer to the truth than the others, they are all similes. Yet when people, even scholars, become unaware that it is matter of a comparison, similes turn into metaphors (viz. implicit comparison), and they eventually end up being considered as if they were objective descriptions. Moreover, each metaphor suggests a spectrum of related meanings just as a note played by an instrument necessarily involves its harmonic series. For example, people who think of the atom as a planetary system end up somehow sharing the ancient belief in a correspondence between micro and macro cosmos—even if they do not even suspect it—and their beliefs, fears, and hopes about atomic energy are probably related more to the metaphor used than to the real object.

This paper will investigate a metaphor which probably underlies many reasons of public concern towards genetics.

GENETIC DETERMINISM

There is now a growing body of empirical evidence on the contribution of genetic factors to phenotypic differences. The advent of recombinant DNA technology has resulted in a number of new discoveries. More and more we can interfere with or diagnose diseases, detect presymptomatically genes for monogenic disorders, uncover (early in life) genetic predisposition to common disorders (including cancer, and psychiatric disorders), anticipate normal phenotypic traits, and, in the near future, even foresee behavioural traits, such as novelty seeking behaviours, antisocial behaviours, and sexual orientation. Most people, scholars included, agree that the new genetics offer possibilities of great importance, even if some of the future scenarios may create anxiety and concern. Perhaps, the main reason for concern involves the predictive nature of genetics. To find out regularities and, consequently, to foresee future events, is one of the major tasks we have assigned science in Western culture, and also human biology searches for causal explanations for what we are and do. Humans have always been fascinated, although also frightened, by the idea of being able to foresee their future, and modern genetics promises to unveil some aspects of their "biologic future". The "emotional value" of genetics is therefore fully comprehensible, and the contradictory set of fears and hopes that scientific research in this field has provoked. Indeed, predictive genetics allows us to foresee (according to certain degrees of probabilities) some aspects of the development of an individual with a variety of traits.

The vast majority of geneticists think that genetic influences should be considered in probabilistic rather than in deterministic terms, and that they can only point out a susceptibility. This means that genetic influence cannot be considered the sole factor that rules biological development, but other elements, that can be subsumed under the broad heading of "environment", play an important role in determining what human beings are and do. Moreover only a few genes are directly responsible for a phenotypic trait (that could be influenced also by environmental factors), most characters are indeed polygenic, namely, they are influenced by the interaction of many genes. Biologic diversity is thus a function of several variables, and any attempt to oversimplify runs the risk of serving as a biologic alibi to cultural biases (Mordini E, 1997). A few scientists proudly claim to be determinists, namely they share the view that the genetic inheritance constrains and makes inevitable all phenotypic development. On the contrary geneticists such as R.Lewontin, S.Rose, and L.Kamin (Lewontin RC, Rose S, Kamin LJ, 1984), scientist such as J.Maddox (Maddox J, 1993), sociologists such as D.Nelkin (Nelkin D, Lindee SM, 1995), expert groups and ad hoc committees such as the Joint Working Group on Ethical, Legal, and Social Implications (ELSI) of the Human Genome Project (Joint Working Group on ELSI of the HUGO, 1995) and the Nuffield Council on Bioethics (Nuffield Council on Bioethics, 1993) have all warned against the ethical, social, and legal risks related to genetic determinism.

Nevertheless, the theologian T.Peters, in his recent book (Peters T, 1997), contests that the real target of these scholars and groups is genetic determinism. According to Peters, those who criticisize genetic determinism only tend to substitute it with a double determinism: genetics plus environment. Peters distinguishes two different myths: the myth of Puppet Determinism and that of Promethean Determinism. His argument deserves a full quotation:

> The genetic determinism of the gene myth, curiously enough, has two distinct faces. The first is the fatalistic face. I call it puppet determinism. According to puppet determinism, the DNA defines who we are, and the genes, like a puppeteer, pull the strings that makes us dance. To speak of "genetic essentialism" or to see genetics as "the ultimate explanation of human being" is to place DNA in the position of defining who we are and who we can be. To speak of genetics fatalism or to say " It's all in the genes" is to assume that genetic influences are unchangeable, that we are immutably destined to act as our DNA programs us to act. [...] The second is the future face. I call it Promethean determinism. It is a version of what I have elsewhere identified as the understanding-decision- control formula, characteristic of the modern doctrine of technological progress. Promethean determinism assigns our scientists the task of understanding just how the genes work plus that of making the decision to develop appropriate technologies based upon this understanding; and this will then give the human race control over what nature has bequeathed to us. Because the history of genes constitutes the history of human evolution, once we have gained control we will be able to guide the future evolution of the human race. We will have wrested from nature her secrets, and this will transform us from the determined into the determiners. (pp.6–7)

According to Peters those who criticise genetic determinism are actually "Promethean determinists" who carry on their battle against "puppet determinists", namely sociobiologists and other genetic fatalists. Sociobiologists advocate a biological interpretation of almost every aspect of life and they pretend to demonstrate that our genes encode quite a number of social, political, and psychological features of human beings. Their position, indeed, often leads to a sort of genetic fatalism. By and large their argument can be summarised as the following: i) if what is in our genes is unchangeable, and ii) if the majority of our biological and social traits derive from our genes, iii) then any attempt to modify and ameliorate human life by social means is just an illusion. Against genetic fatalism Joseph Alper and Jonathan Beckwith (Alper JS, Beckwith J, 1993) have recently argued that its premises are a scientific mistake and logically untenable; but genetic fatalism is not the real issue at stake. In general geneticists are not fatalists but determinists, and fatalism and determinism are quite different things[1]. Both determinism and fatalism are necessitarianist doctrines (viz., they make broad use of the concept of necessity) but the central point of fatalism is that human action is ineffectual, while determinism makes no statement about human action, just arguing that all events are caused (there is no randomness in the world, or, at least, in the area of the world observed by the particular science taken into account).

Determinism in the physical sciences does not pose major problems to the public, since people are used to considering "nature" as ruled by universal and necessary laws (of course philosophers and scientists may contest that this idea is unproblematic). Determinism creates a lot of emotional problems when applied to humans since it tends to provoke two opposite reactions in the public (Doherty P., Sutton A., 1997). On the one hand determinism stimulates fantasies of omnipotence, of absolute control (Leibnitz's scientist, who knows all the present conditions and all the scientific laws, is actually God!), and people may love to imagine that science will allow them to play God. On the other hand, determinism excites paranoid feelings (scientists can control society; we are just biologic puppets controlled by scientific puppeteers; etc.) that tend to turn into a superficial rejection of all modern biotechnologies. Determinism is a problem even for philosophers, at least for those philosophers who try to save both science (and consequently determinism) and ethics (and consequently human agency).

Schools of philosophy are used to distinguish between hard and soft determinism[2]. While hard determinism assumes that everything is caused and consequently, in principle, everything is predictable (knowing the initial conditions and the laws that are relevant); soft determinism says that to a certain degree, there is freedom, or randomness, in the universe, or, at least, we had better believe there is[3]. From a theoretical point of view soft determinism is obviously the weaker position: its strength depends on the fact that at heart we are all soft determinists. No matter which philosophical perspective we have chosen; in everyday life it is impossible either to deny the concept of cause or to accept that everything is strictly determined by causation. As Prof. J. Glover stated: "This kind of determinism [soft-determinism] is not a metaphysical dogma, but is a regulative ideal: we hope to get always closer to the determinist picture." (Glover J., 1996. p.239). In this sense even the so-called "probabilistic genetics" should be considered determinist. A determinist picture undoubtedly remains the ideal of genetics and there is no such thing as a "principle of indetermination" that can be considered constitutive of modern biology. Probabilities in genetics are just the unavoidable inconvenience that derives from the fact that Leibnitz's scientist, fully aware of the present and of all relevant scientific laws, does not exist. It does not yet make sense to appeal to stochastic processes such as those which occur in physics. Let me quote J. Glover once again: 'Sometimes people suggest a way in which chance might allow us to escape from determinism. There might be quantum effects which mean that, at least at the neurophysiological or neurochemical level, we can't make predictions about behaviour. I am doubtful about this escape for two reasons. One reason is scepticism about the claim that quantum effects actually do affect gross physical objects very much...The other worry I have about that sort of approach is that even if it could be shown that some of our behaviour was unpredictable—that indeterminism held for human decision—it doesn't seem to rescue freedom... An element of randomness does not seem to be the same as an element of freedom' (pp.244–45). The landscape of modern genetics is thus a determinist landscape (even if geneticists try to avoid any philosophical implications of it) and geneticists' main goals remain to foresee phenotypic traits from the knowledge of the genotype. In spite of any wise discourse about probabilistic and determinist biology, not only the public but even scientists think of genetics as if the DNA were the book of destiny. The discoverer of the DNA structure and the Nobel Prize winner, J.Watson, states: "We used to think our fate was in our stars. Now we know, in large measure, our fate is in our genes" (quoted from Peter T, 1997, p.6). Similarly, another distinguished scholar, the French geneticist and European MP, J.F. Mattei, expressed the same concept (Mattei JF, 1991):

> I would simply point out that we have reached a new level, that of predictive medicine. It is a fascinating and frightening field because, with the techniques of molecular biology, we will one day be able to detect strengths and weakness of individuals even though they are buried deeply within the nucleus of their cells. We shall soon, (and this is already being done) foretell what fate holds in store for us before life even starts, as if we are the fortune-tellers of modern times. Southern blots now replace the crystal ball. (p.91)

Thus the metaphor that we find in genetics is the metaphor of the DNA as a place where our destiny has been written. The two crucial elements of this metaphor are: i) the idea of destiny, and ii) the fact that the destiny has been written. Let's analyse separately these two elements.

LINEAR DESTINY

Of course destiny is just a concept (or, if you are nominalist, just a word). I mean that there is no such thing as a destiny that our eyes can see, our tongue can taste, or our nose can smell. Destiny is a concept that implies that the word "future" is not only applicable to a specific verbal tense, but also to a state of affairs that "already exists in the present, namely when it should not exist by definition". If the future is something concrete it means that in the present there are seminal principles that somchow give reality to the future. For instance the genes of a newborn are supposed to cause the phenotypic traits of the adult: they are somehow the future in the present[4]. Briefly, what I want to show is that when speaking of destiny, the first issue we meet concerns the structure of causation. Which kind of causation are we speaking of in genetics?

Causation is the relation between two events, when there is one event which occurs, and produces, determines, or necessitates the other. The idea of causation seems to be reasonable and intuitive. Yet philosophers know that there is no way to demonstrate that in principle causes exist or, to put it differently, that each effect requires a cause[5]. Even the word "destiny" means nothing but a particular relation of causation. It comes from the Latin destinatio, which comes in turn from the verb destino, to fasten[6]. To destine thus literally means to fasten one thing to another. The idea underlying the word "destiny" is that events are fastened to each other as links of a chain. Genetics often makes use of this linear model of causation. According to a linear model of causation each event is a member of a series, being both the effect of a previous event and the cause of the next one. Causation can be consequently analysed as a linear sequence of discrete events. Of course the model can be complicated considering various concomitant sequences, that interact with each other and sometimes cross. In genetics, the simplest "complex model" provides for two concomitant sequences: genes and environment; other, and more complex, models consider neurodevelopment, societal influences, cultural heritage, and so on. The idea that causation is a matter of discrete events joined by links is clearly constitutive of Mendelian (unigenics) genetics[7], but it remains as the landscape of multifactorial and polygenic genetics. Both the linear model of causation, and the image of the chain of destiny, come from Stoic philosophy (Von Harnim H, 1964). Stoics believed that destiny (η ειμαρμενη) was the rationale of the world, that it was the reason (λογος) according to which the past was, the present is, and the future will be. Stoics thought that destiny was a chain of causes, namely an order and a connection that can never be forced or transgressed. No thing is without cause, and randomness does not exist in the universe.

Stoics have developed a complex doctrine about causes. They called "antecedent causes" (αιτια ου ουχ ανευ), those causes that link events to each other in the causal chain[8]. Crysippos derives the existence of antecedent causes from logic. If something existed without causes—he argued—each proposition cannot be either true or false, since what lacks causes will be neither true nor false. But each proposition is true or false, thus there is nothing without causes. The problem of foreseeing was also central in Stoic philosophy. In fact any linear model of causation implies that one should be able to infer, or to foresee, from the existence of one event of the series, the existence of the others. Here, time is the crucial issue. Is a temporal relationship between the events required? From a trivial point of view the answer should be yes. Yet in a fully deterministic universe time's arrow is two-directional[9]. Any statement is simply true or false, no matter if it pertains to the past, the present, or the future. Nevertheless this position implied the existence of necessary futures, a thing that was denied by Stoicism. The issue of necessary futures is particularly intriguing. Stoics were definitely determinist, but they rejected the idea of necessary futures. Actually, the idea of necessary futures belongs to a naïve determinism, very far from the sophisticated Stoic philosophy. According to Stoics, the future is not logically necessary since it always derives from conditional syllogisms, thus it is the result of various possibilities. That position created some interesting problems, in particular considering divination, a practice that was accepted by Stoic philosophers. Crysippos, one of the most influential Stoic philosophers, suggested that any kind of "true" divination should have the structure of a negative conditional syllogism (*tollendo tollens*), that means that any anticipation can be considered necessary only a posteriori, that is after its realisation[10]. This elegant solution is particularly intriguing since it corresponds, in the last analysis, to the solution given to modern geneticists in response to the challenge posed by genetic fatalism (Mordini E., 1997b).

The Stoic conception of destiny penetrated Roman culture and remained a part of the Latin culture of any educated persons during the Middle Ages and the Renaissance. The linear model of causation was adopted by modern scientists after the crisis of the Aristotelian-Thomistic model of causation and, through Galileo, Newton, Leibnitz, and Kant has survived till the present. At any rate, discussing the history of this model lies outside of the purpose of this article. What I would like to emphasise is that the idea of linear causation, which underlies several aspects of modern genetics, is not at all new, but it comes from a specific philosophic perspective (Stoicism). This is my first argument.

FATA SCRIBUNDA

The other element of the metaphor according to which genetics is the book of destiny is that destiny is written in the DNA.

In speaking of linear destiny we have hardly touched upon the essence of the issue. Speaking of its scriptural aspects we are at its core. There is no need for many examples to demonstrate that the scriptural metaphor is constitutive of genetics. Watson and Crick first interpreted and translated the genetic code. The DNA encodes the messages; there is a text that should be read; the enzymes that promote the synthesis of RNA are called transcriptasis, the whole process is considered a process of translation, and mutations are mistranscriptions or mistranslations. In short, almost all chemical processes connected to DNA are expressed—in the scientific language—in terms of writing or reading. Where does this metaphor come from? What are the implicit meanings, and suggestions, carried by it?

The idea of a written destiny is ancient in Western culture. Undoubtedly one of its roots is in Jewish culture, the culture of the book. The mystical and esoteric tradition of interpretation of the books of the Old Testament—the "kabbalah"—is part of the Western tradition. However the most important tradition, which connects writing to destiny, comes from Classical Greece.

In the Greek tradition, there was a close link between writing and the hereafter[11]. There are at least two elements on which scholars (Piccaluga G, 1988) have focused[12]:

1. All gods of the nether world (differently from Olympic gods) can write and read. Hades, the King of the Infernal Regions, was called by Aeschylus "he, who writes on the tablet[13]" (δελτογραφος) (Aesch. Eum.275) . Hades and Thanatos (Death) "catalogue everything in their kingdom", wrote Hesopus (Aesop. fab. 133). Persephones, the Queen of the nether world, marked on the door of her realm the names of those who are about to die, moreover she had a written list of those who were to be punished in the hereafter. Radamanthos, the Judge, first read the faults written on the body of each deceased, then he wrote, always on the body of the dead, the punishment decreed (Luc. Cat. 24, Aesh.frg. 530,21 M, 24, Eur. 506 M., Plat. Gorg. 526 B);
2. The relationship between the deceased and gods of the nether world is often mediated by writings, as it is also demonstrated by the defixionum tabellae, laminates sent to the gods, containing blessings or, more frequently, curses[14], written by people acquainted with the deceased. All the dead took in their mouth a small coin—an obolus—for Charon, the ferryman[15] of the Infernal Regions. At last those deceased who belonged to Orphic sects took with them complex instruction written on golden tablets, to escape from their destiny in the hereafter (Janko R, 1984). But the fact that the existence in the nether world is characterised by writing ended up assimilating all human existence to a sort of writing. Life "writes" its story on human bodies (by means of wrinkles, scars, wounds) and one of the processes that must happen in the hereafter before the re-birth is to erase these writings, as well as the writing done on the mind, that is memory (Plato Axioch. 366 C).

Amongst the gods of the nether world we also find the gods of destiny. One of the ancient representations of destiny was that of the Moirai. In the oldest vascular pictures they were usually represented as three women that, provided with spindle and distaff, spin each human life, cutting the thread of life at due time[16]. The Moirai accomplish the task of defining for each individual, from the very beginning and without appeal, his or her path; a path that, beyond the specificity of each fate, must at any rate end in unavoidable death. The destiny defined by the Moirai is not a metaphysical entity, but it is biological reality. It has to do with embodied persons: biologic, concrete existence.

The more alphabetisation spread over Classic culture, the more the gods of destiny became literate. From the Hellenistic period on, the Moirai were always represented with a seal, or with tablets. They were represented sometimes reading the book of destiny to a man about to die, or taking note and registering those who had just arrived in the hereafter. The Moirai's writing should be indelible: their tablets were made of bronze, iron, and even diamonds. Day by day, they slowly turned into the bureaucratic secretaries of the Inferno, and, even if they were still sometimes represented with their spindle and distaff, the writing tools became essential in their vascular representations (Cumont F, 1942).

Another personification of written destiny, even if less characterised than the Moirai, was Potmos. Potmos was represented by Homer as a friend of Thanatos. But already in the fifth century B.C., he had "learned" to write, and Pindarus spoke of him as "he, who writes destiny and death" (Pind. Nem. VI).

Romans added to their group of divinities of destiny, analogous to the Moirai (the *Parcae*), an indistinct group of gods that they called Fata Scribunda: "those, who write the fates". They were always present during the celebration of a new birth, to register on a book the destiny of the newborn (Piccaluga G, 1988).

During the Middle Ages, the Renaissance, and the Modern Age this metaphor spread and the scholar, who is interested, could follow its path till our days, through Dante, L.Ariosto, W.Shakespeare, Voltaire, E.A.Poe, W.Dilthey, H.G.Gadamer, J.L.Borges, I.B.Singer, and U.Eco, to cite a few.

The metaphor of a written destiny thus has ruled Western culture for at least 2500 years, and has expressed itself through a wealth of different nuances and images. It is therefore understandable that this metaphor has been highly pervasive, and that we have traced its crucial presence also in modern biology.

Which are the chief meanings that it carried? Let us try to infer them from the metaphor's ancient roots. First, (written) destiny is unchangeable. "What I wrote, I wrote" (quod scripsi, scripsi) claimed Pilato (John, XIX, 22); writing cystalyzes a will, and renders it unchangeable. Second, destiny is nothing but the specific way by which each individual reaches her death. The divinities of destiny are all infernal divinities. Third, destiny is written on our bodies. Destiny regards human bodies and human bodies are slowly written by destiny till the last sentence. Forth, each destiny has been assigned before birth. One comes to the world with one's own biological destiny already written. Fifth, destiny can be read, and consequently it can be known. Sixth, if destiny is written, it has been written in a language; thus there is a necessity for a professional to be able to interpret this language.

All these statements are extraordinarily similar to those attributed to genetic determinism, even in its soft forms. This is my second argument.

THE MOIRA AND GEOMETRIC FATE

The destiny was called by Stoics the Heimarmene (ηειμαρμενη). The origin of this word is in the root meir (μειρ) that one finds also in the word Moirai. The origin of the root "meir" is in the verb μειρομαι, to share. As we have just shown, the metaphor of a written destiny and the conception of linear causation are mutually congruent and actually the image of the thread, spun and cut by the Moirai, unavoidably brings to mind the idea of the chain of destiny. Besides, both Stoicism and the representation of the destiny as a written text developed in the same period: in the Hellenistic-Roman age and in late classicism. But what was there prior to it, namely prior to the Moirai? In the ancient time, before the flourishing of Greek culture, there was just the Moira, namely one, sole, mysterious, and almighty goddess of fate.

The Moira was the allotment, viz. the specific part of life given to each individual. In origin the word "moira" just meant the part of the common land that the community assigned every citizen at his birth. Already in Homer the term is used as a metaphor for fate. Let us try to make explicit the comparison held in this metaphor. One can express it as the following: "Fate is like a piece of land (temporarily) assigned to each individual". Actually, the word "moira" was used to define any area delimited by means of geometric measures (namely an area defined according to a logos). For instance the Greek word for the degrees of an angle was always "moira", and, indeed, the land assigned each citizen was defined by geometric methods. I thus propose to call the metaphor of the Moira the geometric metaphor for fate[18]. Human life is not like a chain of events, but rather it is like land that God has entrusted us[19]. Each piece of land has its own borders and each individual lives within his or her land. The borders and the shape of each piece of land are predetermined: they are unchangeable and no one—god nor human—can do anything to modify them. However no human[20] can really know the borders nor the whole shape of the land that has been assigned. Some borders and aspects of the land are indeed similar for all humans, but many others are specific for each individual. Three forces rule human life in the moira: i) human will; ii) gods' will; iii) randomness.

Human Will is an important factor in determining individual fates. Within her moira, it is up to each individual where to set herself. The classic instance of that is Achilles' myth. Achilles knows that if he kills Hector, his life is also destined to end soon. After the suggestion of his mother, the goddess Theti, he thus accepts that he must avoid any direct fight against the Trojan hero. One day, however, Hector kills Achilles' beloved friend Patroclus. Now Achilles can choose both to avenge Patroclus and then wait to be killed in turn by Paris' arrow, or he can renounce vengeance and thus continue to live. Both events are possible, namely they are in Achilles' moira: his destiny depends on his decision. As we all know, Achilles chose to avenge Patroclus, and his Moira was realised since he died, murdered in turn by Paris. Could Achilles have made a different choice? Yes, at least according to the myth. Of course one can argue that, because even one's character is part of the Moira, the choice was not really free. The objection is partially true: indeed it would have been very strange if Achilles had chosen not to avenge his friend. Yet, at least in principle, he was not compelled by any causal chain of events to take one decision rather than another. In short, within the Moira, individuals are free enough, the only thing that they should avoid doing is trying to trespass the borders of the moira. This sin, called hubris, is the worst—perhaps the only real sin—in Greek religion, because it is the sum total of arrogance, impiety, and pretension.

Gods' Will is the second important force that determines the way in which each person "stays" within her moira. Humans should try to understand gods' messages, to anticipate their wishes, and to mitigate their stormy fury. Sometimes it may be difficult, but "the World is full of gods"—as Thales of Miletus stated—and the eternal game between different, and often, opposite divine wills results in extra freedom for humans.

The third force that plays an important role in determining one's moira is fortune, or randomness. The Greek word for fortune was τυχη, which comes from the root τυχ, the same as the verb τυνχανω, which means to strike, to hit. Fortune is thus a stroke (of luck or of apoplexy!).

To conclude, let's try to summarise the chief features of the geometric metaphor of fate: i) fate is a space rather than a point, and there is a specific delimited space for each one of us; however it is chiefly up to us what place in this space we are occupying in each moment of our life; ii) the model of causation used is heuristic rather than algorithmic, and it means that it does not imply any previous and implicit acceptance of specific metaphisical assumptions; iii) there is no constitutive contradiction between a quasi-determinist interpretation of biologic events and the hypothesis of human free agency.

My third argument is thus that the geometric metaphor of fate is more consistent with current biology than the metaphor of linear destiny. In particular the model of causation used in the geometric metaphor is more sophisticated and more adept at giving reason to the complex interactions between genotype, phenotype, individual, and societies (Gindro S, 1985). Indeed, geometric fate is compatible with probabilistic genetics; it overcomes and takes away the meaning of the old and silly debate between genetic determinists and environmental determinists; eventually, it suggests that the human attempt to understand genes and to foresee their effects is meaningful. Actually, genetic fatalism is definitely excluded by the geometric metaphor of fate. Anyone has the right and the moral duty to exploit her moira[21], and from an ethical point of view, genetic research is totally justified[22].

CONCLUSIONS

The tasks of this papers were i) to unveil an important metaphor underlying modern genetics, the metaphor of written destiny, and to illustrate its chief features; ii) to show that this metaphor has influenced and influences the current philosophical and ethical debate on genetics; and iii) to suggest that another metaphor, geometric fate, could be in principle more consistent with modern science. Obviously I do not think that there is any possibility of substituting the metaphor of linear destiny with the metaphor of geometric fate, and, at any rate, it is not my intention at all to propose such an enterprise. Metaphors have their own life and history, and we had better let them live in peace. As I am not a professional philosopher, I am a practising psychoanalyst, and—perhaps for this very reason—I am used to thinking that we need to explore fantasies of the past to understand events of the present. As Rorty (1979) claimed: "Just as the patient needs to relive his past to answer his questions, so philosophy needs to relive its past in order to answer its questions" (p.33). My goal was to show that the better we know how to gather the wealth of nuances and gradations of the metaphors that underlie modern science, the more it will be possible to find a path through today's bioethical issues. Still quoting Rorty (1979):

> The only point on which I would insist is that philosophers' moral concern should be with continuing the conversation with the West, rather than with insisting upon a place for the traditional problems of modern philosophy within that conversation (p.394).

I could not find better words to express what I think to be our task as bioethicists.

NOTES

1. To confuse fatalism with determinism is often a rhetoric expedient to criticise the latter. A classic example of this artifice were Cicero and Alexander of Aphrodisias, who, respectively in *De Fato* , and in Περὶ Ειμαρμενη, used against Stoic determinism the so-called "Lazy Argument" (*ignava ratio*, αργὸς λόγος): "If you are sick, it is just the first link of a causal chain according to which you will die or you will recover. In any event your destiny is already decided and it is useless asking for a doctor". The logic mistake is obvi-

ously that asking for a doctor could be the following link of the causal chain (even if, to be honest, it is difficult to say whether to determine the recovery or the death).

2. A third position has been proposed by the Italian philosopher and psychoanalyst, S.Gindro. Gindro advocates the use of the Kantian *als ob* (as if). According to Gindro any scientist should conduct her research *as if* humans were ruled by a rigid determinism, and, meanwhile, *as if* they were free. Only considering at the same time these two opposite perspectives it would be possible to manage the paradox of determinism in human sciences and biology. Eventually what Gindro seems to suggest is that this problem cannot be solved by linear and rational processes, but that it needs a non-Aristotelian approach (Gindro S, 1993).
3. J.Glover states: "If we were to follow hard determinism, we would adopt in all our relationships the detached clinical attitude of a psychiatrist towards a patient, or a social worker towards a client. Life would be a great deal less satisfactory if we had only these rather detached relationships. Also, it is not at all clear that it would be possible to disengage from our deeply rooted normal human attitudes." (pp.241–242)
4. Obviously, this reasoning presents numerous fallacies, but it is part of the "popular current wisdom" about the future. The issue of genetics, future, and childbirth is largely discussed in Mc Gee G. (1997).
5. Kant discussed this point in the third antinomy of the Pure Reason. The principle of sufficient reason, *viz.* each effect requires a cause, is the typical example of those synthetic propositions *a priori* that Kant showed to be "regulative", namely about our way to think of the world, rather than "constitutive", namely about the reality of the world. Undoubtedly we think of the world as if any effect had a cause, but there is no manner to demonstrate that the world really works in such a way.
6. For instance *ancoris destinare* meant: to anchor. Anyway *destinare* meant already in Latin to destine, to allot, to appoint, to decide, in the figurative sense.
7. The logic structure of Mendelian genetics is: gene ➨ enzyme ➨ chemical reactions ➨ phenotype.
8. Actually Stoics also considered other kinds of cause: "immanent causes" (αιτιαδι'ó), that are the perfect causes, and that depend on us. The dispute on determinism in the Hellenistic period is particularly relevant to the current debate on biologic determinism and it would deserve to be reread. The literature on the Stoic concept of causality is obviously huge. The interested reader could fruitfully start from the classic A.C.Pearson *The fragments of Zeno and Cleanthes* London, Cambridge, 1891.
9. It is worth noting that the experiment of cloning from somatic cells seems to demonstrate that, in biology, this is at least partially true.
10. As we have said, Stoics believed that any statement about the future had the structure of a conditional syllogism: *if A is B, it is C. A is B, therefore it is C (ponendo ponens)* or *if A is B, it is not C. A is B, it is not C (tollendo tollens).* In the negative kind (tollendo tollens) the consequence (*viz.* the foreseen) does not derive positively from the antecedent, namely the antecedent does not transmit any necessity to the consequent (since C does not exist, it is just an unrealised possibility). The fortune-teller thus infers the future just from the absence of a sign, namely she does not foresee the future (on the contrary the future should be necessary) but she foresees what the future cannot be (and it does not exclude that the future may be a lot of other things). This is the reason why to foresee the future is and is not possible according to Crysippos (Cicero *De Fato* VI-VIII).
11. The destiny of any mortal is to die; the link between writing and the hereafter thus already shows the inner sense in Greek culture of the comparison between writing and destiny. Besides, Plato in *Phoed.* called writing the grave of the speech.
12. I would like to draw the reader's attention to the fact that when these myths formed (probably well before the VII century B. C) Greek society was largely illiterate, and even in the classical period educated people, able to write and read correctly, were a minority. It makes the relation between writing and the hereafter still more intriguing.
13. Actually Δελτος did not mean only "tablet", but any place where one could write (slate, book, papyrus). The word was used in particular for legal writings (laws, contracts, wills).
14. The *defixionum tabellae* were thin tablets that were fixed into the ground next to the buried corpses, usually to suggest to gods of the nether world some terrible punishments to be inflicted on the deceased. They are one of the chief sources of our knowledge about popular religion in the Greek-Roman period.
15. Actually any coin is a piece of written metal that can be interpreted only by those who can read. Moreover the fact that the coin was put in the dead mouth is meaningful: the written word substituted the spoken word. The poet Luciano, in the later classicism, ironically noted that it would have been sufficient not to pay this obolus to avoid passing away.
16. The image is already present in Homer (*Il.*XX, 127; XXIV 209; *Od.*VII 197). In the most ancient vascular pictures they were represented also chanting, emphasising the creative and irrevocable power of music.
17. The Moirai spun also the gods' thread, even if their fate was not death. Gods' fate was their character (as in Heraclitus' sentence).

18. D.Neri proposed to call this metaphor the "topographic metaphor of the destiny". In substance I agree with him, but I still prefer the word "geometric" as it is closer to the Greek spirit
19. But see also *Genesis 2,19*: humans have been created to take care of God's land
20. Romanticism tried to turn *hubris* into Prometheism, namely a positive value, but *hubris* had never had a positive sense for the Greeks. If a hero committed *hubris*, it happened although and never because he was a hero.
21. One could even remember the parable of the talents in the Gospels (*Matt.25, 13:30*).
22. There remains the issue of *hubris*. Are there certain limits that modern research should not trespass?

REFERENCES

Alper J.S., Beckwith J. (1993), Genetic Fatalism and Social Policy . The Implications of Behaviour Genetics Research. *Yale Journal of Biology and Medicine*, 66, 511–24

Andrews B.L., Nelkin D. (1996), The Bell Curve: A Statement, *Science*, 271, 13–14

Cumont F. (1942), *Recherches sur le symbolisme funéraire des Romains* PUF, Paris

Doherty P., Sutton A., eds, (1997), *Man-made Man. Ethical and Legal Issues in Genetics,*Fourth Cort Press, Dublin

Gindro S. (1985), La bilancia d'oro, *Psicoanalisi Contro* Vol. II, 11: 1–3

Gindro S. (1993), La diagnosi e la persona, *Psicoanalisi Contro* Vol. X, 1: 9–54

Glover J. (1996), The implication for responsibility of possible genetics factors in the explanations of violence, in Bock G.R., Goode J.A. (eds), *Genetics of Criminal and Antisocial Behaviour*, John Wiley & Sons Ltd, Chichester: 237–255

Janko R. (1984) Forgetfulness in the Golden Tablets of Memory, *Classical Quarterly*, 34: 89–100

Joint Working Group on ELSI of the HUGO (1995), Genome Research Risks Abuse, Panel Warns, *Nature*, 378, 529

Lewontin R.C., Rose S., Kamin L.J.(1984), *Not in our Genes: Biology, Ideology, and Human Nature*. Pantheon, New York

Maddox J. (1993), Has Nature Overwhelmed Nurture? *Nature*, 366, 107

Mattei J.F. (1991), The use of Prenatal Diagnosis for Psychiatric Diseases. In Srám RJ, Bulyzhenkov V, Prilipko L, Christen Y (eds): *Ethical Issues of Molecular Genetics in Psychiatry*. Springer-Verlag, Berlin: 87–94

Mc Gee G. (1997), *The Perfect Baby. A pragmatic approach to genetics*, Rowman & Littlefield Publish., New York

Mordini E. (1997a), Brain Research and Human Diversity, in Mordini E., Pasquini P. (eds), *Ethics and Neuroscience*, Annali dell'Istituto Superiore di Sanità 33, 4:469–473

Mordini E (1997b): *Commentary on "The Stoic Concept of Mental Disorders"* Phil Psychiat Psych 4, 4:297–301

Nelkin D., Lindee S.M.(1995), *The DNA Mystique: The Gene as a Cultural Icon*, WH Freeman & Company, New York

Nuffield Council on Bioethics (1993), *Genetic Screening Ethical Issues*, Nuffield Council on Bioethics, London

Peters T. (1997), *Playing God? Genetic Determinism and Human Freedom*, Routledge, New York

Piccaluga G. (1988), Potmos impara a scrivere, *Scrittura e Civiltà* 12: 5–46

Rorty R. (1979) *Philosophy and the Mirror of Nature*, Princeton University Press, Princeton

Von Harnim H. (1964), *Stoicorum veterum fragmenta* (IV vol.), Teubner, Stuttgart

30

A NEW FRAMEWORK FOR THE USE OF GENETIC INFORMATION

Jacquelyn Ann K. Kegley

California State University, Bakersfield

In this paper I will seek to do two basic things: (1) Critique the current biological and ethical frameworks which seem to inform most attempts to deal with genetic information and its use; and (2) Present a sketch of a new framework which I believe will be more adequate in dealing with questions of genetic information. What, then, do I believe needs to be changed in the usual approachs to genetic issues? First, the concepts of "gene," "disease," and "patient," need to be de-individualized. An often unexamined and unanalyzed genetic essentialism and determinism, which also imply an individualistic bias, must be exposed and critically analyzed. A more adequate framework needs to be developed which recognizes the complexity of the human genome and especially the inter-dependent and interactive nature of genetic mechanisms. Further, there needs to be a better understanding of the notion of the "genetic causes of disease" that sees disease and illness as a result of the subtle interplay of individual physical characteristics (a variety of biological and psychological factors, not just genetic); environmental factors; and social factors including the cultural.

Further, a new ethical/legal framework needs to be forged which better understands the interaction and inter-relationships of individual values, rights, and needs with communal values, rights and needs. Above all, the individual decision-maker needs to be seen as a relational being who usually operates within a network of intimates and other persons who are often profoundly impacted by the decisions made and whose concerns sometimes constitute the goals, interests and values of the that decision-maker.

This framework would see genuine, autonomous individuals and genuine supportive communities as necessary complements and as crucially interrelated.

This new ethical framework would emphasize a relational context for ethical analysis and would argue for dialogue, interpretation and negotiation of values, rights, goods and goals so those "significantly impacted by the genetic decisions" might receive due consideration and respect. Indeed, dialogue and negotiation of the uses of genetic information and its meanings seems especially important in the genetic area where biological kinship is a key concept, where impact of decisions on others is rarely avoidable and where decisions are being made under conditions of great uncertainty. Facts and information

Genetic Information, edited by Thompson and Chadwick.
Kluwer Academic / Plenum Publishing, New York, 1999.

need to be shared, interpretation of perspectives is essential and values need to be dialogued and understood in terms of short-term and long-term goals and the various genetic possibilities in relationship to and within specific communities.

1. THE OPERATING BIOLOGICAL FRAMEWORK

The operating paradigm of the Human Genome Project is that of "genetic essentialism," which sees the human self as a molecular, genetic entity whose physical, behavorial, social characteristics are ultimately determined and thus explainable in genetic terms. In the words of Richard Dawkins, in his 1976 book, *The Selfish Gene*, "human beings are survival machines- robot machines that are blindly programmed to preserve the selfish molecules know as 'genes" (Dawkins, 1976, p. 24).

"Genetic essentialism" is a "genetic determinism," and, in the case of the Human Genome project seems to involve three central assumptions: (1) genotype determines phenotype; (2) genes determine capacity in each person and the boundary conditions established are permanent and unchanging, e.g. intelligence is fixed and non-malleable; and (3) genes determine tendencies, both behavorial and physiological or disease tendencies (Strohman, 1996). Three further assumptions manifest themselves in the growing field of "genetic medicine," namely, that (1) genes determine diseases; (2) genes govern the aging process; and (3) genetic analysis provides diagnosis and therapy of disease and aging.

What, then, is problematic about this framework? First, underlying this genetic essentialism and determinism is a simplistic and radical individualistic framework which sees the "gene" as an independent, isolated, individual monadic unit. Such a view ignores the many complexities of genetic processes as well as the vast range of ignorance about these processes that still exists. Nobel Laureate Barbara McClintock wrote succinctly about the mysterious flexibility of genetic processes as follows: "We know about the components of the genome. We know nothing about how the cell senses danger and initiates responses that are often truly remarkable" (McClintock, 1984). Noble Laureate I. I. Rabi makes another point, namely, "Our knowledge is a little island in a great ocean of nonknowledge" (Rabbi, 1978, p. 30).

To believe that genotype determines phenotype is a large oversimplification. Thus, for example, population biology "sees complex traits, including disease, as highly interactive and impossible to reduce to genetic elements alone" (Strohman, 1996). Other scientists argue that efforts to measure the relative effects of heredity and environment on behavior misconstrues these two factors as independent rather than as interactive forces, and thus, there is an underestimation of the environmental factors on gene expression (Strohman, 1996). The idea that genotype determines phenotype is not even true for twins, where there are subtle differences between phenotypes and change over time. One need only understand the history of the manifestation of schizophrenia among twins to know that the genotype/phenotype relationship is far more complex.

Second, the belief that genes determine tendencies also fails in light of careful genetic analysis. A number of critics have faulted the dubious "risk and statistical analysis methodology" used in the Human Genome project. They argue that a statistical construct may have meaning for populations but no simple meaning for individuals (McGuffin and Katz, 1990). Biologist, Stephen Gould, has pinpointed the nature/nurture distinction as a false dichotomy which confuses correlation with causation. He writes: "Genes influence many aspects of behavior but we cannot say that such behavior is caused by genes in any direct way. We cannot even claim that a given behavior is say, 40% genetic and 60% environmental...Genes and environment interact in nonadditive ways" (Gould, 1992, p.48).

Further, the facts seem also to argue against any notion that genetic information remains functionally identical across human populations and through time. For example, studies of certain hereditary diseases show multiple mutations for each gene. This has been a clear problem for BRCA1 Breast Cancer Testing (Wadman, 1996). Even the paradigm case of a single gene disease, e.g. Cystic Fibrosis, seems to be encountering this phenomenon. (Burroughs, 1989). Strohman and other biologists add another kind of argument against the "genetic determinism" assumptions of the Human Genome Project, namely, that our most "feared" diseases such as cancer and heart disease are primarily "diseases of civilization" (Strohman, 1996).

Given these problems of the currently operating genetic framework, what should be our genetic message? First, we need to better understand the notion of a genetic map. Maps, as we know, are selective, i.e. they focus on and link certain features of the world. It is misleading then to focus on the neutrality of the Genome map to suggest that once a gene is located, its interpretation will be objective and independent of context. Christopher Willis, in his book, *Exons, Introns, and Talking Genes: The Science Behind the Human Genome Project*, observes that "simply determining the sequence of all this DNA will not mean we have learned everything there is to know about human beings, any more than looking up sequences of notes in a Beethoven sonata gives us the capacity to play it" (Willis, 1992, p. 10). Genes, if indeed, they are like words, are dependent on context and are open to more than one interpretation. Thus, we should not be mislead to think that a mapped gene is merely a straight forward detail.

The message, then, about genetics and about the DNA and genes is one of complexity and uncertainty and mutual interaction with environmental factors. Yet, the Human Genome Project and the rationale for "genetic testing" and genetic medicine is based on an oversimplified, individualistic, deterministic framework which is clearly misguided and dangerous. The message that scientists, health care practitioners and philosophers should be carrying to the public and public policymakers is one of caution, of complexity and uncertainty, of shared vulnerabilities and responsibilities and of shared risks involved in the complex interactions of genes and environmental factors. The message should be one of basing decisions on "good science," and that means that the bottom line in any ethical/legal agenda for the use of genetic information should be a new understanding of the "genetic causes" of disease, namely, that the cause of a disease is a complex issue and environment and other factors cannot be ignored (Walstein, 1990). This understanding provides important guidance to the decisions about use, control and access of genetic information. Certainly it argues that genetic testing never occur without counseling and consent, and without careful discussion among the relevant parties about the "meaning" of the tests in their individual, familial, social and communal context. Indeed, individuals and families should play a primary role in clarification of such meaning.

2. THE PROBLEMATIC INDIVIDUALISTIC ETHICAL /LEGAL FRAMEWORK FOR GENETIC DECISION-MAKING

Parallel to the problematic, individualistic, genetic deterministic framework of the Human Genome Project is an equally troublesome, operative, individualistic ethical/legal framework for genetic decision-making. This paradigm operates most strongly in the United States, but is also present in Western Bioethics and Law in general. We need to identify the key assumptions of this paradigm. The first is that each person has the right to be an autonomous, rational, self-interested individual, a single decision-maker who acts as authoritative judge and jury of his or her own choices. "Autonomy" is defined as "making one's own choices," generally without interference. Closely associated with "autonomy" is the notion of "personal liberty," primarily interpreted as "negative liberty," i.e., in terms of noncoercion and non-interference. "Non-interference" means that others, whether individuals or society, should remain "neutral" in influencing the final decision by the autonomous individual.

A second central central belief of the paradigm is that of the individual as a "bearer of rights." Each persons has certain basic rights, e.g. life, property, privacy, free speech, and free religious practice, which must be protected by law and government. Further, the rights are seen as entailing an obligation on the part of others to honor and fulfill these rights to which a person is entitled. Infringement of rights demands legal recourse and compensation. The context of rights is often adversarial, rights in themselves can conflict and individual claims cannot always be accommodated in the context at issue.

A third key assumption of the individualistic paradigm is that groups or communities are viewed as collections of self-interested individuals and the interests of any community (family, group, society) are considered to be the sum of the interests of the individuals who compose it.

All of these assumptions, along with those of the genetic essentialist framework inform present analyses of the use of genetic information and do so, I believe, in ways harmful to both individuals and communities. How, then, do these assumptions mislead us in dealing with genetic information?

3. IMPLICATIONS OF THESE FRAMEWORKS FOR ANALYSIS OF GENETIC ISSUES

The individualistic frameworks we have outlined do impact on genetic issues in some very problematic ways. First, the notions of "patient" and "disease" are automatically "individualized." Thus, a "patient" is an individual with a disease who is receiving treatment. But "genetic disease" does not necessarily fit this pattern because genetic problems may be shared with family members and future offspring and the genetic information and proposed treatment(s) may have significant benefits or liabilities for others. Thus parents contemplating proceeding with a pregnancy which involves a baby with a genetic defect must think of the effect on future or present siblings in terms of their health and prospects and may well need to involve family members. A good test case was that of a U.S. couple named the Hamptons[1] who discovered that their first pregnancy involved twins, both of whom had Cystic Fibrosis. They understood that they had to face the prospect that they would "outlive" their children. They also had to face perceptions among those in the community in which they lived that "they were crazy to carry the pregnancy through to fruition. Further, since the presence of the Cystic Fibrosis gene in the "family," was unknown, this led to sharing information with the broader family and to the establishment of the maternal grandfather as a carrier. This led, in turn, to the need to overcome "guilt" and other negative feelings. And, when another pregnancy occurred, again with a fetus affected with Cystic Fibrosis, the decision had to involve "family" prospects and resources. In this case the parents decided that another Cystic Fibrosis child would be an impossible burden on family resources. A third pregnancy, with a normal fetus, also involved decision-making in a complex familial and societal context and, in this case they went ahead with the pregnancy. Clearly the "individualized notion of disease" and of "informed consent" is very problematic in providing good guidance to this kind of complex decision-making situation. What is needed is a framework which emphasizes "shared vulnerability" and decision-making in an interpersonal process in which workable solutions are sought via dialogue among all those concerned with and impacted by the genetic situation. Similar "familial" and "cross-generational" issues occur in making decisions about genetic testing and treatments for those who may have a BRCA 1 and BRCA 2 inherited mutation.

A second problematic aspect of the operative individualistic paradigm for ethical/legal analysis results from the strong emphasis on non-interference with individual decision-making. Such an emphasis leads most naturally to the prevalent notion of "non-directive genetic counseling." However, such a concept falters against the fact that much genetic information is poorly understood both by health care practitioners and by patients and their significant others. "Non-directive counseling" could, given these circumstances, lead to bad advice and resulting misconceptions that can have serious implications for others affected by the decision-maker's actions. For example, parents of children with sickle cell trait often confuse the trait with the disease which, in turn, leads to the false perception of the child as "seriously ill" or "different." Such children are at high risk for a serious disorder of psychosocial development,"the vulnerable child syndrome" (Pearson and O'Brien, 1972). The paradigm of an individual decision-maker as authoritative judge of all choices may not be an effective or moral strategy in this situation or others like it. Individuals "alone" facing difficult decisions involving complex information may act too hesitantly or too impulsively and without a fully informed base for judgment.

The complexity of the BRCA1 and BRCA2 breast cancer mutations is another case of providing genetic information in a situation of great uncertainty, misconception and misunderstanding. There is disagreement about the risks associated with the mutations and about prophylactic treatments that might be appropriate. A good case of impulsive and misguided action in such a situation is that reported in the *Jerusalem Post* of a thirteen year old girl who got pregnant, believing that this would protect her against the cancers predicted for those with the BRCA mutations (Shiloh, 1997). Given the complexity and uncertainty of genetic information, as outlined above, health care providers cannot be seen as neutral dispensers of genetic information and procedures. There needs to be careful conversation about the meaning of genetic information, about the accuracy and predictability of such information and it needs to be explained and understood in the "lived context" of those who must make the decisions and then "live with them." I agree fully with the judgment of Glen McGee when he writes: "The so-called 'non-directive' genetic counseling which requires the physician and genetic counselor to dispense neutral information about diseases without making evaluative commentary is just as dangerous as paternalistic medicine" (McGee, 1997, p. 93).

The "individualistic notions" of patient and disease are also inadequate for dealing with the "asymptomatic patient," who has a genetic defect, but who does not now or may not ever have the full-blown disease manifestation. In this case social context and actions impact heavily and these persons surely are not "protected and autonomous individuals." Thus, the "asymptomatic" patient's social status as "sick" leads to self-image problems, loss of employment and insurance benefits, and, as in case of sickle cell trait, exclusion from military service and other job opportunities (Rosenstein, 1983, p. 139). A further interesting aspect of this arises with the question of prophylactic treatment and insurance coverage. In the United States, insurance covers things considered "medically necessary" and this, in turn, means to treat "an actual existing disease." Further insurance often excludes "pre-existing conditions." In light of this, how should "genetic disease" be defined?

Three possible definitions have been suggested. One suggestion is to define "genetic predisposition" as the "disease", but this leads to the insurance coverage of various kinds of prophylactic treatment and perhaps the swamping of the insurance business. A second possibility is to define the "genetic risk" as the "disease", but then one needs to know what level of risk constitutes a disease and what kind of risk. For example, there is "life risk" which is the percentage chance that an individual will develop the disease in a lifetime based on an average life span of 88. This means one "always has the disease" and all persons with the "risk", fetus, infant, old man, would be covered by insurance. Another kind of risk is "current risk" which is the risk at a given age over the course of an individual life. Thus one could have a 1% risk at birth, but a 50% risk at age 42, given the family history. At what level of risk, then, would one be considered to have a "genetic disease?" These problems well illustrate the complexities of genetics and the inadequacies of an individualistic analysis of genetic disease, genetic information and its use. Indeed, one strong conclusion made by those who have addressed these issues is that societal interests must be taken into account. One must weigh both individual and societal future expense of not pursing effective and timely treatment, while also weighing the scarcity of resources and spending money on unnecessary treatments (Glazer, 1997).

Analyzing "genetic disease" in terms of these kinds of possibilities also, I believe, lends credence to a notion of a person as "in process," who has possibilities of overcoming the present, and whose situation can change with the context. Given the uncertainities and probabilities involved in genetic information this is an important concept .

4. SUGGESTIONS FOR A NEW FRAMEWORK

Clearly the individualistic framework will not serve us well in dealing with the many issues involved in the use of genetic information. What is needed is a new philosophical view which contains the following elements. First, "persons," whether patients or other decision-makers must be seen as complex relational networks, composed of a variety of public physical aspects—material, neural, genetic, behavorial, social, cultural, political, economic—and a mix of private, inner aspects—sensual, emotional, mental, intentional. Further, all these aspects must be seen as interacting with and influencing each other in complex and multiple ways. Individual persons are holistic and should not be reduced to any one of their aspects, whether genetic or endowed rights (Kegley, 1994). Further, as hinted above, "persons" must be seen as in process, creatures of time, who are capable of stagnation, development, growth and able to take advantage of opportunities of the future. Persons can and should be able to think in terms of both short term and long term goals and events. This is especially important for genetics where one deals with probabilities, new horizons of information and a future of new technologies and treatments. All of us know the uncertainties of prognosis for many genetic-based disorders, e.g. Down's Syndrome, spinal bifida, etc. One should not give up too easily on the prospects for the flourishing of various afflicted individuals, especially if there is strong social, familial and other kinds of support. Indeed, diversity should be valued since it is often the key to survivability in the genetic scheme of things.

Secondly, illness or disease must *not* be seen as located solely in a physically, malfunctioning body, but rather be viewed as impacting persons at a multiplicity of levels (Kegley, 1997, pp. 182–187). In addition to deficit, perhaps multiple deficit in various systems of the bodily organism, illness impacts psychological functioning. It often is a fundamental shattering of everyday personal assumptions and a blow to one's total personhood and life plans as an individual. Illness is also clearly social in nature; it involves loss in terms of social functioning and an inability to meet individual and societal expectations. Illness often involves failure of obligations and disruptions of relationships (Kegley, 1997). In dealing with genetic disorders and genetic information all of these aspects of disease and illness need to be taken into account. What kind of dysfunction does the genetic information indicate? What kind of abnormality is it? Is it biological, psychological, sociological, cultural, or even a matter of individual perception? Or, is it a combination of all of these factors? Further, given this genetic information, what type of "flourishing" still might be possible for the affected individual, given that individual's social structure and support systems?

Thirdly, in the new framework, "persons" must be seen as relational in a fundamental sense, namely, as a being-with-others. Individuals are usually engaged in life in a mutual journey with others with whom they share values, goals, goods, liabilities and risks. These relationships can be biological and/or social and "kinship," of either kind or both is a powerful aspect of human life. Thus, anthropologist, Robin Fox, writes: "The relationship to ancestors and kin have been the key relationships in social structure; they have been the pivots on which most interaction and most claims and obligations, most loyalties and sentiments turned" (Fox, 1967, p. 86). A closely connected relational aspect of human individuals is their nature as "narrative beings," i.e. those who construct their own stories, interpreting and interconnecting past, present and future (Nelson, 1992). These narratives usually, if not always, include others, especially those with whom the individual has some genetic connection. Our narrative connections, which include the genetic ones, give cohesiveness and quality to our individual lives and make us feel situated and recognized as individuals.

Our kinships, both social and biological, are a basic dimension of our human identity and of our flourishing. As such, they must be taken into account in using genetic information and in decision-making which involves genetic information.

Much ethical analysis of genetic and other decisions, as indicated, emphasizes the self-chooser, usually in isolation from others, and certainly as a bearer of rights who must be protected from the interference of others in actions and decision-making. These emphases are certainly important and not to be disregarded, but they should be placed in context. Part of that context is the recognition that ascriptions of rights are, in a most fundamental way, ascriptions of others and dependent upon others. The recognition of rights is communal and the accompanying moral duties and obligations are grounded in complex social practices such as parenting and the practice of medicine. It may well be useful to view rights as fallbacks, i.e., as efforts to obtain something that could not be better assured by ties of affection and loyalty or by moral duties deeply ingrained in our complex social practices such as parenting and health caring (Murray, 1997). As for self-choice, the notion of an isolated self-chooser is belied by the recognition of human individuals as essentially "beings-with-others." And, as Charles Taylor has argued in *The Ethics of Authenticity*, if choice is its own rationale, then morality is trivialized (Taylor, 1993).

The new framework should see decision-making, whether in the genetic realm or another, as necessarily involving more than an individual. It would be seen in terms of individuals in dialogue, a dialogue that could include the primary individual, significant others, some (such as potential offspring) represented through others, health care professionals, counselors, and others with significant interest in the outcome. The relevant individuals could indeed be identified via dialogue. Such dialogue should be based upon at least three premises. First, there should be efforts to broaden the dialogue and decision-making context with at least three goals in mind: (1) to include those others likely to be significantly impacted by the decision in question or whose values, interests and goals are crucial aspects of the individual decision-makers own life story; (2) to gather more relevant information for the decision to be made including the facts particular to the case and context; and (3) to stress a democratic model of moral problem solving which recognizes that moral problems in clinical practice cannot and should not be solved by expert judgment alone (Fins, Bacccheta, and Miller, 1997). This third goal is very much in concert with a view know as "Clinical Pragmatism" and the efforts of the task force, appointed jointly by the Society for Health and Human Values and The Society for Bioethics Consultation to develop a set of standards for Bioethics Consultation. In the Task Force's recent discussion draft they argue that the *"qualified facilitation model"* is more appropriate for health care ethics consultation in U.S. society. They write: "This is because a *qualified* facilitation role taken by the bioethics consultant is consistent with the fact of pluralism and the political rights of individuals to live by their own moral values" (Society for Human Health and Human Values, 1997).

Further, another emphasis in the dialogue about genetic information presupposes respect for each voice and individual, while seeking mutually agreed upon choices in the situation at hand. There would be a moral commitment to mutual interpretation and understanding so that each voice would be "genuinely heard." To facilitate this kind of interpretation and understanding is, of course, a special skill, and theTask Force on Standards for Bioethics Consultation advocates "process and interpersonal skills" as a second core skill for anyone who undertakes the task of "facilitating decision-making and "helping patients, families, surrogates, health providers, or other involved parties address uncertainty or conflict regarding value-laden issues that emerge in health care" (Society for Health and Human Values, 1997).

Finally, the dialogue model of decision-making would be governed by the principle of "loyalty to loyalty," the principle of ever-broadening community in the sense of fostering conditions for the increased autonomy and flourishing of individuals and groups (Kegley, 1997). This means there is a need to take account of the "full" genetic, environmental, social, organizational and individual context of the situation and of the individuals most critically affected by the decisions. This is, of course, an ideal that gives us goals to work toward in dealing with the complexity of using genetic information in a "fitting" manner.

The new framework sketched here clearly needs to be developed in much more detail. However, I have argued that the nature of genetic information and genetic disease creates a serious problem for any individualistically based framework. Indeed, such a framework, especially as manifested in "genetic essentialism," poses a very real threat to individuals who suffer from "genetic disability," and given that such traits are carried by us all, it threatens us all. This means that together we must forge a new paradigm to guide the judicious use of genetic information. It should be one that fosters human autonomy in the context of community and that respects individuals both as unique and yet dependent upon and formed by relational context. Only then might we be able to adequately deal with genetic disease, genetic health and human persons as genetic, but also so much more.

NOTES

1. This case was presented as one of the sequences in "A Question of Genes: Inherited Risks," a PBS special. The sequence was entitled Case Study # 1 "New Choices, New Dilemmas." The special was underwritten

by SmithKline Beecham and the U.S. Department of Energy and copyrighted 1997 by Oregon Public Broadcasting.

REFERENCES

Burroughs, B.R. (1989), "Cystic Fibrosis miscalculation", *American Journal Maternal Child Nursing,* 14(6), 416.

Dawkins, Richard. (1976.), *The Selfish Gene* , Oxford University Press: London

Joseph J. Fins, Matthew D. Bacchetta, and Franklin G. Miller.(1997), "Clincal Pragmatism: A Method of Moral Problem Solving", *Kennedy Institute of Ethics Journal,* Vol. 7, No. 2, 129–145.

Fox, Robin. (1967), *Kinship and Marriage: An Anthropological Perspective,* Penguin Books: Baltimore, Maryland

Glazer, Alexandra K. (1997), "Genetic Predispositions, Prophylactic Treatments and Private Health Insurance: Nothing is Better Than a Good Pair of Genes", *American Journal of Law & Medicine,* XXIII. 1. 45–68

Gould, Stephen Jay.(1992), "The Confusion Over Evolution", *New York Review of Books,* 19 November, 48

Kegley, Jacquelyn Ann K. (1994), "Peirce and Royce on Person", in Parret, Hermann (ed.), *Peirce and Value Theory,* John Benjamin: Amsterdam

Kegley, Jacquelyn Ann K. (1996), "Dealing With the Fate of Genetic Information", *Medical Law.* 15,377: 715 f.

Kegley, Jacquelyn Ann K. (1997), *Genuine Individuals and Genuine Communities,* Vanderbilt University Press: Nashville, Tennessee

McGee, Glenn. (1997), *The Perfect Baby: A Pragmatic Approach to Genetics,* Rowman, Littlefield Publishers: Boulder, Colorado

McClintock, Barbara. (1984), "The significance of responses of the genome to challenge", *Science.* 226:792–801

McGuffin, Peter and Randy Katz. (1990), "Who Believes in Estimating Heritability as End in Itself?", in Walsten, Douglas (ed.), "Insensitivity of the Analysis of Variance to Herdity-Environment Ineraction", *Behavorial and Brain Science, 13. 109–161*

Thomas Murray. (1997), "New Reproductive Technologies and the Family", in Cynthia B. Cohen (ed.), *New Ways of Making Babies,* Indiana University Press: Bloomington, Indiana

Pearson H.A. and R.T. O'Brien. (1972), "Sickle Cell Testing Programs," *Journal of Pediatrics,* 81:1201–1204

Rosentein, Mark A. (1983), "Employee Selection Based on Susceptibility to Occupational Illness", *Michigan Law Review,* 139

Shiloh, Dina. (1997). "The DNA Dilemma", *The Jerusalem Post.*, 8 August, 8

Society for Health and Human Values. (1997), "Discussion Draft of SHHV-SBC Task Force on Standards for Bioethics Consultation"

Strohman, Richard C. (1996), "The Nature of Polygenic Disease: Toward a Holistic Theory of Biology", in Kegley, Jacquelyn (ed.), *Genetic Knowledge, Human Values and Human Responsibility,* In Press

Taylor, Charles. (1993), *The Ethics of Authenticity,* Harvard University Press: Cambridge, Massachusetts and London

Wadman, Meredith. (1996), "Panel Softens Cancer Gene Test Warning", *Nature,* 380

Walsten, Douglas. (1990), "Insensitivity of the Analysis of Variance to Heredity-Environment Interaction", *Behavior and Brain Sciences,* 13: 109–161

Willis, Christopher. (1992), *Exons, Introns, and Talking Genes: The Science Behind The Human Genome Project,* Basic Books: New York

INDEX

Ablist, 246–247
Abortion, 4, 18–19, 23–24, 26–27, 73, 106, 180, 208, 210, 226, 236, 246, 248, 265, 304
Accountability, 86–87, 189
Adolescents, 175–177, 179–181
Adverse selection, 58–60, 62–63, 66–68, 110, 290
Adversity bag, 280
African, 169, 245, 275–281, 307
African-american, 123, 248
Alzheimer disease, 151
Angwa', 280
Ante-natal clinic, 246, 250
At risk, 13, 18, 44, 58–59, 116, 176, 207–208, 211, 215, 231, 236–237, 248, 261, 263–265, 280
Attitudes, 6–7, 13, 25, 28, 48, 145, 168, 175–178, 197–198, 210, 248, 276, 283, 291, 302, 304, 306, 319
Autonomy, 6, 80, 150, 153–154, 156–157, 170, 215, 235–236, 239, 249, 255–257, 259, 297–302, 304–306, 324, 328

Bangladeshi, 245
Bearer of rights, 324, 327
Bias, 3, 5, 7, 13, 42, 63, 86, 204, 247, 321
Bio-colonialism, 277
Bio-piracy, 278
Bio-prospecting, 277–278
Biodiversity, 144–145, 147, 167, 169–170, 173, 260, 278, 280
Bioethics, 38, 163–164, 168–170, 173, 175, 211, 306, 311, 323, 328
Bioinformatics, 85–89, 91–92
Biopower, 214
Biotechnology, 45–46, 91, 111, 117, 129–132, 135–136, 139, 141, 146, 149–150, 152, 156, 163–173, 175, 249, 305
Body surfaces, 250
Brazil, 2, 146, 164, 166–168, 170–173
BRCA 1, 325
BRCA 2, 325
Breast cancer, 32, 38, 41–42, 58, 60, 62, 105–106, 121–128, 155, 203–204, 215, 221, 236, 323, 325

Cameroon, 280–281
Capitalism, 246
Caribbean, 166–167, 245
Carrier, 33, 44, 47, 61, 110, 149, 175–181, 207, 236, 243–244, 249–250, 264–265, 324
Causal chain, 317–318
Chinese, 14, 245
Chorionic villus sampling, 208, 246, 267, 271
Cloning, 46, 164, 168, 199, 278, 281, 319
Colon cancer, 57, 106, 221
Combinatorial chemistry, 85, 91–92
Commercialisation, 169, 172, 278
Commodification, 130, 133, 136–137, 152
Commodity, 36, 169
Common disorders, 191, 193, 195–196, 310
Communal, 35, 37–38, 145, 167, 277, 301–302, 304–305, 321, 323, 327
Communication, 46, 89, 173, 184, 196, 215, 224, 231, 249, 251
Communitarianism, 277, 301, 307
Community, 3, 14, 19–20, 25, 36–37, 89, 92, 128, 131, 141–142, 146–147, 151–152, 164, 167, 172, 185, 195–198, 243, 245–250, 257, 264, 276, 300, 302–307, 316, 324, 328
Community awareness, 245
Complexity, 8, 129, 158, 321, 323, 325, 328
Consent, 42, 48–49, 61–62, 116, 154, 157, 169–172, 175–177, 184, 187, 189, 263–264, 297, 304–306, 323, 325
Consequentialism, 81, 210
Correspondence problem, 244
Crouch, 19
Cultural factors, 247
Culture, 34–35, 38, 89, 114, 124, 127, 150, 153–157, 165, 168–169, 247, 250, 263, 275–276, 279–281, 306–307, 309–310, 314–316, 319
Cyprus, 250, 304, 307

Cystic fibrosis, 105, 175–180, 209, 226, 262, 264, 268, 289, 323–324

Death, 8, 18, 99, 122, 125, 141, 213, 215, 219–233, 245, 250, 280, 307, 315–316, 318–319
Decision-analytic model, 208, 210–211
Decision-making, 186, 210, 217, 237–239, 241, 257–258, 269, 304–305, 323–325, 327–328
Deregulation, 77
Desert, 35, 80–81, 134, 288
Destiny, 98–99, 313–319
Diagnosis, 18, 23, 25–28, 72–73, 75, 105–106, 108, 153, 168, 173, 179–181, 207–208, 212, 218, 220, 223–224, 226–227, 236–238, 240, 245–246, 249–250, 261, 263, 271, 304, 322
Dialogue, 198, 271, 321, 325, 327–328
Directive on the Legal Protection of Biotechnological Inventions, 141, 143
Disability, 9–10, 14, 17–29, 42, 59–60, 110, 129, 153, 208–209, 216–217, 224, 228–230, 243, 246–247, 250, 263, 265, 268, 281, 283, 286–287, 289, 328
Disabled individuals, 18
Discourse, 3, 7, 72, 82, 211, 245–246, 248–251, 255, 257–259, 262, 277, 279, 297, 313
Discrimination, 17, 19, 22–26, 28, 31–32, 41–43, 45–46, 57, 59, 68–69, 71, 80–81, 111, 175, 180, 247
Discriminatory, 18, 21, 23–24, 26, 28, 32, 37, 44, 80–81, 83–84, 100, 212, 246
Dispositional properties, 279
Distributive justice, 13–15, 34, 37, 134
DNA, 43, 76–77, 86–90, 93–94, 107–108, 116, 118, 130, 132, 135–136, 142–144, 163–165, 167, 173, 202, 212, 216, 305–306, 310–311, 313–315, 323
Dominance, 35, 127, 145, 156
Down's syndrome, 183–187, 189, 208, 267
Drugs, 81, 91–92, 115, 122, 131, 135–136, 140, 166, 168, 172, 247, 270, 291
Duchene muscular dystrophy, 18
Dysfunction, 327

Egalitarianism, 277, 279–281
Electronic computing, 87, 94
Employees, 7, 102, 110–111, 113–114, 194, 307
Employers, 10, 32, 44–45, 60, 101–102, 110–111, 113, 117, 119, 227–228, 240, 307
Environment, 3, 20, 59, 89, 98–100, 121, 124–125, 127–128, 143, 150, 152–154, 157, 165, 167, 170–171, 180, 213–214, 220, 244, 248, 272, 284–286, 311, 313, 322–323
Epidemiology, 109, 220, 222, 228
Equality, 21, 34–37, 156, 257, 279, 286–288, 293, 301
Ethics, 9, 13, 82–83, 102, 140–141, 145–146, 151–152, 155, 163, 170–171, 173, 218, 235, 256–257, 263, 268–270, 272, 291, 294, 300, 302, 304–306, 312, 327–328
Ethnicity, 68, 246, 250
Etzioni, 307
Eugenics–15, 18, 73, 83, 98, 153, 156–157, 165
European Convention on Human Rights, 143
Extraordinary intelligence, 279

Familial polyposis, 221
Family cancer history, 48–49, 52, 54–55, 66
Family traits, 223
Fatalism, 217, 311–312, 314, 318
Fate, 47, 114, 127, 217, 222, 232, 237, 244, 286, 313, 315–319
Fertility, 4, 8, 12, 248, 250
Fetal disability, 18
Finkelstein, 20–22, 24
Folk construct, 245
Framework, 22, 28, 37, 72, 75, 143, 153, 165, 167, 172–173, 175, 199, 255, 277, 297, 321–328
Free exchange, 35–37
Freedom, 6, 9, 13, 66, 74, 144, 153–154, 168, 176, 222, 235–237, 239–241, 256, 299, 302–304, 306, 312, 317
Future–2, 5–6, 8–10, 13–15, 17, 19, 27, 37, 46, 55, 67, 72–73, 75–77, 80, 99–101, 109–111, 113, 115–117, 135–136, 154, 170, 172–173, 175–176, 178–180, 186, 190, 195–196, 198–199, 202, 210–211, 216–217, 220–222, 224, 226–233, 235–240, 256, 258, 260–262, 270, 272, 279, 284, 287, 289, 291–292, 298–300, 310–311, 313–314, 319, 324, 326–327

Gene patenting, 129–136
Gene scan, 220, 222
Gene sequence, 90, 129–133, 135–137, 163
Gene test, 72
Genetic counselling, 106, 157, 196, 239, 244, 246, 249, 304
Genetic determinism, 7, 13, 98, 153, 210, 311, 316, 322–323
Genetic deviation, 71
Genetic disease, 13, 17, 33, 44, 46, 106, 112, 124–125, 175–176, 179–180, 196, 218, 222, 225–226, 228, 230, 262, 275, 293, 298, 304, 324–326, 328
Genetic disorder, 45, 73, 176, 208, 215, 217, 228–229, 236, 239–240, 250, 256, 261–263, 267–272, 327
Genetic engineering, 9, 29, 46, 71–73, 141, 153, 163, 168, 171, 173, 198, 203, 227
Genetic essentialism, 6, 98–100, 311, 321–322, 328
Genetic information, 31–33, 42, 44–45, 57–60, 69, 77, 79–81, 83–84, 97–103, 108–109, 113–118, 133, 135, 139, 141–143, 145–146, 169, 175, 207–208, 210, 214–215, 217–218, 221, 228–229, 238–240, 243, 245–247, 249–251, 272, 275–276, 278–281, 284, 289, 297–300, 302–303, 305–306, 321, 323–328
Genetic knowledge, 106, 110, 114, 235–239, 241, 244, 250–251, 267, 299

Genetic lottery, 284–286, 288–290, 292–294
Genetic policy, 5, 251
Genetic risk, 36, 49, 73, 109–110, 128, 175–176, 179–180, 227, 237, 240, 247, 326
Genetic screening, 9, 13, 18, 79–80, 83, 175, 186, 210, 212, 214–215, 268
Genetic testing, 15, 18, 25–26, 28, 31–33, 38, 41–51, 54–55, 58–62, 68–69, 75–77, 97, 106–114, 116–118, 125–126, 156, 158, 175–177, 180–181, 191, 193, 195–196, 198, 208–209, 229, 235–241, 256, 262–264, 307, 323, 325
Geneticization, 109, 121, 123–124, 127
Genome, 9, 15, 36, 57, 72, 74–77, 87, 90, 105, 109, 111, 118, 123, 130, 133–134, 149, 157, 164–165, 168–170, 175, 207, 213–215, 218, 220, 222, 249, 256, 263, 265, 268, 284, 286–287, 290, 292, 306, 311, 321–323
Genomics, 89, 91–92, 155, 165
Germ-line gene therapy, 29, 255–258, 260–264
Goods, 34–38, 137, 140, 191, 321, 327
Greek religion, 317
Gujarati, 245

Habermas, 255, 259, 264
Haemoglobin disorders, 243–245, 250, 264
Handicapped, 18, 20, 23, 179, 289
Health care, 3, 11, 13–14, 31–33, 36–38, 44, 69, 71, 110–112, 114–115, 117–118, 127, 152, 154–158, 180, 191, 193, 196, 237, 248, 256, 263–264, 270, 283–284, 289–290, 292, 294, 303, 305, 325, 327–328
Health education, 247–248
Health insurance, 37–38, 42, 45, 57–61, 69, 71–77, 106, 109–110, 227, 240, 281, 284, 290, 294, 299
Health related information, 74, 97, 99, 102, 297–298
Healthcare, 191, 193, 196–198, 226
High-throughput screening, 85, 91–92
History–2, 5–6, 10, 32, 35, 38, 42–43, 46–55, 59–64, 66–68, 76, 87, 93, 106, 109, 111, 113, 115, 118, 145, 153, 157, 164, 196, 208, 222–224, 230, 237, 311, 314, 318, 322, 326
Homosexuality, 215–216, 276
Human genome, 15, 36, 57, 87, 105, 109, 111, 118, 123, 130, 133–134, 157, 164–165, 169–170, 175, 220, 222, 249, 256, 263, 265, 268, 284, 286, 306, 311, 321–323
Human health, 111, 163, 165, 168, 170, 328
Human rights, 32, 37, 74, 77, 143, 255, 264
Human tissue trade, 169
Huntington disease, 57, 68
Huntington's chorea, 211, 213

Identity, 7, 29, 43, 132, 157, 207, 210–211, 213, 250, 327
Illness, 71, 79, 91, 98, 100, 121–123, 126–128, 179, 217, 220, 224, 229, 271, 280, 298, 321, 326–327
IMF, 281

Individual right, 6, 10, 157, 249, 297, 299, 302–303, 305
Individualism, 279, 300, 306
Individualistic, 297, 299, 301, 304–307, 321–326
Informed decision, 177, 180, 183
Insurance, 31–34, 36–38, 41–55, 57–69, 71–77, 79–81, 83–84, 102, 106, 109–114, 118, 125, 227, 230, 240, 278–281, 284, 290–291, 294, 299, 325–326
Intellectual property, 131, 139–141, 143–146, 151, 163, 167–170, 172, 278
Interview, 48–49, 61–63, 67–68, 102, 190, 204–205, 245, 248
Inventions, 129–133, 135, 139–143, 145, 276

Journalism, 202–203
Journals, 13, 166, 202–204
Justice, 4, 13–15, 17, 22, 27–28, 31, 34–37, 111, 130, 134, 168, 227–228, 269–270, 284, 292–294, 297, 299–302, 306

Kant, 303, 314, 319
Kantian, 132, 300, 302, 306, 318
Kibam ke ngeh, 280
Kinship, 249, 280, 321, 327
Knowledge as power, 276

Law, 7, 10, 14, 17, 21–23, 32, 37, 41–43, 59–60, 77, 83, 85–86, 101–102, 117, 129–132, 136, 139–146, 152, 154, 168–173, 177, 246, 255, 258, 264, 269, 276, 281, 290, 300, 303, 305–306, 312, 319, 323–324
Legislation, 4–5, 33, 37, 42, 59–60, 72–73, 97, 101–102, 111, 143, 153, 168, 177, 196, 256, 259, 262, 264, 277, 298–299
Length of life, 209, 232
Lesch-Nyhan syndrome, 18, 21
Levelling tendency, 277, 279
Liberalism, 300–302, 305–306
Life choices, 231
Life insurance, 32, 38, 41–44, 46–55, 58–69, 74, 106, 125, 227, 230, 280
Life risk, 326
Life-hopes, 236, 238–239
Lifespan, 219, 224, 226–227, 231–232
Loyalty to loyalty, 328

Macromolecules, 91–92, 94
Malaria, 167–168, 222
Map, 89, 91, 105, 108, 133, 323
Marketplace, 34–36, 133–134, 139, 143–144
Media, 149, 155, 185, 196, 203–205
Medical care, 3, 11, 35–36, 114, 118
Medical records, 69, 109, 113–118, 306
Metaphor, 114, 281, 309–310, 313–319
Minority ethnic groups, 245
Moira, 316–317
Molecular biology, 5, 85, 88, 92–95, 313
Moral imperatives, 276–277

Morality, 5, 28, 123, 142–143, 146, 158, 169, 268, 276, 279, 300–303, 306, 327
Mortgage, 38, 63, 230
Moslem, 246–247, 250
Multiple choice, 188, 243–245, 250

Nanotechnology, 85
Nature/nurture, 322
Neonatal screening, 246, 249
New genetics, 17–18, 24, 27–28, 193, 196–198, 246, 249, 264, 310
Normal, 19, 24, 26, 71, 176, 209, 215, 218, 226, 256, 263, 279, 293, 310, 319, 324
Nunez, 19

Old eugenics, 18
Oligonucleotide probe arrays, 107–108
Oncomouse, 142
Ovarian cancer, 42, 47–52, 54–55, 57–58, 61–63, 66, 68

Pakistani, 245, 247
Parfit, 7, 26, 287–288
Patenting, 129–136, 142, 151, 164, 167–169, 172–173, 276–277
Patents, 129–137, 139, 141–143, 146, 163–164, 167–168, 170
Patriarchy, 246
Pharmaceutical industry, 72, 91
Pharmaceuticals, 92, 126, 140–141, 144–145, 151, 155, 170–171, 292
Phenotype, 213, 318, 322
Planning, 26, 72, 91, 106, 176, 180, 188–189, 192, 196, 208, 230–231, 236–237, 246, 273
Possible people, 27, 214
Postmodernism, 243, 251
Poverty, 171, 184, 247–248
Power, 13, 34–37, 55, 90, 93, 95, 101–102, 127, 133, 145, 193, 212, 215, 218, 248–249, 259, 264, 276–277, 279, 303, 307, 309, 319
Predictability, 224, 226, 228, 279–280, 325
Prediction, 52, 63, 66, 76, 81, 99–100, 111, 136, 210–211, 219–220, 224–225, 228, 279, 312
Predictive information, 109
Preferences, 209–211, 213, 237
Pregnancy, 25–26, 73, 175–176, 178–181, 184–186, 188, 208, 220, 236, 240, 244, 248, 261–263, 267, 270–272, 324–325
Prenatal diagnosis, 18, 23, 25–28, 153, 179–180, 246, 249–250, 263
Prenatal testing, 18, 23–25, 27, 106, 157, 176, 263, 267–272
Prevention, 5, 18–19, 23–24, 27, 29, 73, 75, 125–127, 143, 168, 230, 275
Privacy, 42, 44, 46, 55, 58–59, 74, 83, 113–118, 129, 215, 299–300, 305–307, 324
Private sector, 34, 75, 77, 149–150, 152, 156, 167–168, 195
Probabilistic genetics, 312, 318
Profit motive, 276
Prognostication, 220–224, 226, 228–232
Property rights, 129, 131, 139–141, 144, 170, 289
Prophylactic mastectomy, 125
Protein folding problem, 85
Protein sequencing, 85, 89
Proverbs, 279
Public awareness, 202, 243–244, 250–251, 264, 304
Public consultation, 191, 199
Public education, 35, 179
Public opinion, 41, 43–47, 50–51, 191, 196–198
Public participation, 191
Punjabi, 245

Qualified facilitation model, 328
Quality of life, 25, 79, 122, 209–212, 217
Questionnaires, 184, 194, 243, 250

Racial discrimination, 247
Racism, 3, 5–7, 13, 246–248
Rational drug design, 85, 92
Rationality, 76, 208, 218, 235, 238, 259
Rawls, 13, 286–288, 294, 306–307
Reductionism, 263, 275
Regulation, 34, 75, 97, 100–103, 111, 116, 136, 139, 146, 150–151, 153–154, 156–157, 163–164, 168–170, 172, 196, 214–215, 255, 261–262, 264, 302, 305
Relational, 21, 29, 321, 326–328
Right not to know, 299
Right to know, 52, 249, 299, 306
Rights, 6, 10, 13, 17–18, 20–21, 23–24, 28, 32, 35, 37, 74, 77, 117, 129–132, 135, 137, 139–141, 143–146, 153–154, 157, 163, 167–168, 170–172, 243, 246–247, 249–250, 255, 257, 259, 264, 268, 278, 289, 297–303, 305–307, 321, 324, 326–328
Risk, 13–14, 18, 29, 31–33, 36–37, 43–44, 47, 49, 58–64, 66, 68, 71–77, 80, 83, 86, 99–100, 106, 108–110, 113, 116, 122–128, 143, 151, 167–168, 171, 175–181, 187, 195–196, 207–211, 213, 215, 218–219, 221–222, 226–228, 231, 236–237, 240, 244–245, 247–249, 258, 260–265, 267, 271, 279–280, 290–291, 293, 299, 305, 311, 322–323, 325–327
Role of government, 168, 171
Romantic author/creator paradigm, 144

Sandel, 301, 307
Scientific community, 167, 172
Screened populations, 243
Screening, 9, 13, 15, 18, 23–26, 28, 75, 79–80, 83, 85, 91–92, 121, 126, 146, 153, 175–176, 179–181, 183, 185–190, 196, 207–210, 212, 214–215, 217, 226, 246, 249, 264, 268, 303–304
Security and welfare, 35, 37–38
Self-knowledge, 217, 231, 235, 237–238
Sëm, 277

Sickle cell anaemia, 243, 248
Size of life, 221, 226, 232
Social conditions, 18–19, 117–118, 263
Social construct, 18–20, 22, 29
Social construction of knowledge, 248
Social duty, 302–303, 305
Social responsibility, 301–302, 304–305
Sociologies, 243, 251
Sociology, 157, 195
Somatic cell gene therapy, 256, 260–263
Spheres of justice, 34–37
Spina bifida, 18, 20, 23, 25, 209, 268
Spirit of omnivorous discovery, 276
Stigma, 28, 112, 177, 245, 247
Stigmatisation, 71, 73, 240
Stoics, 314, 316, 319
Subjectivism, 243, 251
Super-human power, 279
Survey, 7, 43–45, 48–49, 58, 61–62, 68–69, 94, 154, 157, 176, 185, 245, 248, 292

Taboos, 276, 278
Tay Sachs, 18
Tay-Sachs disease, 21, 221, 226
Taylor, 301, 307, 327
Technological frames, 249
Thalassaemia, 243, 245, 268
Thalassemia, 176, 304, 307
The western world, 157, 276, 278–280, 297, 305
Theories of personal identity, 211
Theory of justice, 13, 22, 31, 34
Tontins, 280
Treatment, 19–20, 29, 38, 73, 77, 80–81, 105–106, 108, 110, 113, 115, 122, 146, 154, 165, 176, 207–208, 220, 222, 226, 230, 236–237, 249, 256, 260–261, 264, 275, 277, 285, 287, 292, 300, 325–326
Trouble fund, 280

Universal health care, 31, 33, 36
Utilitarianism, 268
Utility, 81, 130, 140, 167, 208–209, 218

Virtue ethics, 268–270
Virtuous patient, 271
Vonnegut, 21–22

Wales, 193–194, 264
Walzer, 31, 34–38, 301, 307
Welfare, 4, 11, 13, 31, 33–38, 73, 142, 171–172, 228, 287, 302, 307
Wells, 19
Western culture, 276, 279, 281, 310, 315–316
Western outlook, 276
Witchcraft, 277, 279
Workplace, 180, 220, 228
World bank, 281

Zamora, 278

www.ingramcontent.com/pod-product-compliance
Ingram Content Group UK Ltd.
Pitfield, Milton Keynes, MK11 3LW, UK
UKHW020135160926
475607UK00003B/47
9780306460524